Biocatalysis in Polymer Science

ACS SYMPOSIUM SERIES **840**

Biocatalysis in Polymer Science

Richard A. Gross, Editor
Polytechnic University

H. N. Cheng, Editor
Hercules Incorporated Research Center

American Chemical Society, Washington, DC

Library of Congress Cataloging-in-Publication Data

Biocatalysis in polymer science / Richard A. Gross, editor, H. N. Cheng, editor.

p. cm.—(ACS symposium series ; 840)

Includes bibliographical references and indexes.

ISBN 0–8412–3781–6

1. Polymers—Biotechnology—Congresses. 2. Polymers—Synthesis—Congresses.

I. Gross, Richard A., 1957- II. Cheng, H. N. III. American Chemical Society. Division of Polymer Chemistry, Inc. IV. American Chemical Society. . Meeting. (220th : 2000 : Washington, D.C.). V. Series.

TP248.65.P62 B54 2002
668.9—dc21 2002026223

The paper used in this publication meets the minimum requirements of American National Standard for Information Sciences—Permanence of Paper for Printed Library Materials, ANSI Z39.48–1984.

Distributed by Oxford University Press

PRINTED IN THE UNITED STATES OF AMERICA

Foreword

The ACS Symposium Series was first published in 1974 to provide a mechanism for publishing symposia quickly in book form. The purpose of the series is to publish timely, comprehensive books developed from ACS sponsored symposia based on current scientific research. Occasionally, books are developed from symposia sponsored by other organizations when the topic is of keen interest to the chemistry audience.

Before agreeing to publish a book, the proposed table of contents is reviewed for appropriate and comprehensive coverage and for interest to the audience. Some papers may be excluded to better focus the book; others may be added to provide comprehensiveness. When appropriate, overview or introductory chapters are added. Drafts of chapters are peer-reviewed prior to final acceptance or rejection, and manuscripts are prepared in camera-ready format.

As a rule, only original research papers and original review papers are included in the volumes. Verbatim reproductions of previously published papers are not accepted.

ACS Books Department

Contents

Preface....................xi

1. **Biocatalysis in Polymer Science: An Overview**....................1
H. N. Cheng and Richard A. Gross

Development of Novel Methodologies

2. **Generation of Environmentally Compatible Polymer Libraries via Combinatorial Biocatalysis**....................34
Dae-Yun Kim, Xiaoqiu Wu, and Jonathan S. Dordick

3. **Evolving and Screening Enzymes for New Activities on Polymer Substrates**....................50
Douglas C. Youvan, William J. Coleman, and Edward J. Bylina

4. **Polymer-Modulated, Diffusion-Controlled Enzyme Kinetics on Monolayers**....................57
Keiji Tanaka and Hyuk Yu

5. **Recent Developments in the Field of In Vitro Biosynthesis of Plant β-Glucans**....................65
Joséphine Lai Kee Him, Henri Chanzy, Ludovic Pelosi, Jean-Luc Putaux, and Vincent Bulone

6. **Cyclohexanol Biodegradation Genes: A Pathway of Opportunities**....................80
Hiroaki Iwaki, Yoshie Hasegawa, Masahiro Teraoka, Tai Tokuyama, Laetitia Bernard, and Peter C. K. Lau

Synthesis of Monomers and Macromers

7. **Chemoenzymatic Synthesis of Nucleoside-Branched Poly(vinyl alcohol)....................93**
Y. Tokiwa, H. Fan, T. Raku, and M. Kitagawa

8. **Effect of DMSO on Regioselective Synthesis of Vinyl Sugars Catalyzed by Protease....................99**
M. Kitagawa, T. Raku, H. Fan, and Y. Tokiwa

9. **Enzyme-Catalyzed Synthesis of Well-Defined Macromers Built around a Sugar Core....................107**
Rajesh Kumar and Richard A. Gross

Polyesters and Polycarbonates

10. **In Vivo and In Vitro Metabolic Engineering of Polyhydroxyalkanoates Biosynthesis Pathways....................120**
Alexander Steinbüchel

11. **Controlling the Polymer Microstructure of Biodegradable Polyhydroxyalkanoates....................124**
Aaron S. Kelley and Friedrich Srienc

12. **Enantioselective and Regioselective Polymerization with Lipase Catalysis to Polyesters....................128**
Shiro Kobayashi and Hiroshi Uyama

13. **Enzyme-Catalyzed Direct Polyesterification....................141**
Kurt F. Brandstadt, John C. Saam, and Ajit Sharma

14. **Biocatalytic Synthesis of Novel Functional Polycarbonates....................156**
Kirpal S. Bisht and Talal F. Al-Azemi

15. **Lipase-Catalyzed Polytransesterification Reactions....................172**
Ajay Kumar, Bhanu Kalra, and Richard A. Gross

Polysaccharides

16. **Synthesis and Modification of Carbohydrates through Biotechnology..........188**
Peter R. Andreana, Wenhua Xie, and Peng George Wang

17. **Enzyme-Catalyzed Reactions of Polysaccharides..........203**
H. N. Cheng and Qu-Ming Gu

18. **Modification of Cellulose Solids by Enzyme-Catalyzed Transesterification with Vinyl Esters in Anhydrous Organic Solvents..........217**
Jiangbing Xie and You-Lo Hsieh

19. **Grafting Renewable Chemicals to Functionalize Chitosan..........231**
C. Justin Govar, Tianhong Chen, Nai-Chi Liu, Michael T. Harris, and Gregory F. Payne

20. **Lipase-Catalyzed Grafting Reactions on Polysaccharides..........243**
Qu-Ming Gu

21. **Lipase-Mediated Selective TEMPO Oxidation of Hydroxyethylcellulose..........253**
Shanghui Hu, Wei Gao, Rajesh Kumar, Richard A. Gross, Qu-Ming Gu, and H. N. Cheng

22. **Reaction-Diffusion of Enzyme Molecules in Biopolymer Matrices..........265**
Yu Cheng and Robert K. Prud'homme

Other Examples of Polymer Biocatalysis

23. **Enzymatic Degradation of Polyester Urethanes Studied by Multi-Angle Laser Light Scattering..........286**
D. Himel, D. P. Norwood, and W. F. Reed

24. **Peroxidase-Mediated Free Radical Polymerization of Vinyl Monomers..........297**
B. Kalra and R. A. Gross

Indexes

Author Index..311

Subject Index..313

Preface

The rapid pace at which biotechnology continues to develop is extraordinary and promises to impact virtually all aspects of life. New knowledge in biotechnology has already delivered on many promises, offering exciting new medicines, rapid tests to give information on the status of our health, and large improvements in the productivity of agricultural systems. Biotechnology also offers biological systems that can be used as catalysts (biocatalysts) for the synthesis and modification of various chemicals. Although options for biocatalytic routes to therapeutics such as proteins and DNA are actively under study, many opportunities to integrate biocatalysts into specialty and commodity chemical processes have received relatively little attention. This book documents the growing awareness and research activity that is exploring how biocatalysts can be used for the synthesis, modification, and degradation of polymers. Scientists who are participating in these activities find themselves working at the interface between biology and polymer science. Ideally, these activities will be conducted in environments that foster the seamless convergence of concepts, ideas, and research tools at the interface between these fields. Teams of researchers with core training in microbiology, genetic engineering, biophysics, agricultural science, plant physiology, food science, and other biological disciplines bring new tools, molecules, and fresh ideas to the fields of polymers and materials science. In return, polymer chemists and material scientists bring expertise in chemistry, materials design, structural analysis, processing, polymer solution properties, solid-state polymer physics, and an array of different characterization tools.

The synergy gained by working between these fields is marvelous and exciting. Work at this interface is growing at a rapid pace and is expected to make major contributions to science during the next ten years.

This book was developed from an international symposium held at the 220th National Meeting of the American Chemical Society (ACS) in Washington, D.C. on August 20-24, 2000. The purpose of the symposium was to create a forum for dialogue and exchange of information between scientists who are seeking to create new knowledge and technology by applying biological catalysts to polymer chemistry. Leaders in this field gathered to present their latest findings. Participants represented many countries and included scientists from academics, industry, and government laboratories. The symposium was very successful with 46 oral and poster papers and approximately 100 people in attendance. The success of the symposium encouraged us to embark upon this editorial task.

A total of 24 chapters are included in this book. Chapter 1 (by Cheng and Gross) provides an overview of this field, covering the new developments as well as the major reaction types. The remainder of the chapters is grouped into five categories.

- development of novel methodologies
- synthesis of monomers and macromers
- polyesters and polycarbonates
- polysaccharides
- other examples of polymer biocatalysis

These chapters provide the reader with representative examples of research activities that are currently ongoing in this emerging new field. It is hoped that this symposium volume will be a useful reference book for scientists who are actively engaged in related research. In addition, this book should serve as a useful entry point for those contemplating the use of biocatalysis in polymer science.

We thank the authors for their contributions and their patience while the manuscripts were being reviewed and revised. Thanks are also due to the ACS Division of Polymer Chemistry, Inc. for sponsoring the symposium and to the ACS Books Department for agreeing to publish this symposium volume.

Richard A Gross
NSF I/UCRC for Biocatalysis and Bioprocessing
of Macromolecules
Polytechnic University
6 Metrotech Center
Brooklyn, NY 11201
http://chem.poly.edu/gross
rgross@poly.edu

H. N. Cheng
Hercules Incorporated Research Center
500 Hercules Road
Wilmington, DE 19808–1599

Biocatalysis in Polymer Science

Chapter 1

Biocatalysis in Polymer Science: An Overview

H. N. Cheng[1] and Richard A. Gross[2]

[1]Hercules Incorporated Research Center, 500 Hercules Road, Wilmington, DE 19808–1599
[2]NSF I/UCRC for Biocatalysis and Bioprocessing of Macromolecules, Polytechnic University, 6 Metrotech Center, Brooklyn, NY 11201

Biocatalysts are increasingly being used in polymer science, resulting in many new reactions, new processes, and new commercial possibilities. Both enzymatic and whole-cell processes have received much attention, and new or improved methodologies appear frequently. Ongoing biocatalytic work appears to be concentrated along several themes: polymerization, polymer modification reactions, polymer degradation, and synthesis of monomers and reactive oligomers. An overview is provided herein of this exciting area of research. The papers included in this symposium volume are particularly highlighted.

I. Introduction

Biocatalysis entails the use of whole organisms and the enzymes derived from them to carry out chemical reactions and processes. In nature these biological reactions are ubiquitous, and in fact without them there would be no living things. The potential of these reactions for organic synthesis has been recognized for quite some time, as attested by the large number of books available in print *(1)*. Thus far, most of the applications of these reactions (often called biotransformations) are in the pharmaceuticals, fine chemicals, and agrochemical areas.

Although the use of biocatalysis in polymer science lags behind the other areas, there is no shortage of opportunities either for fundamental research or for product development. In fact, the materials found in nature, such as proteins, polysaccharides, and polynucleotides, are mostly polymers, and (potentially) countless enzymatic and microbial reactions are available to carry out biotransformations. In the past 10 years there has been a fair amount of research activity to develop new methodologies, new reactions, and new processes in order to exploit this exciting technology in the polymer area *(2)*. These developments can be roughly grouped into two categories: enzymatic (and chemo-enzymatic) approaches and whole-cell methods. In either case, the advances in molecular biology and industrial microbiology have had a strong and favorable impact.

This article aims to provide an overview of this subject. It starts with a review of most of the new and improved methodologies, and continues with a survey of the major reaction types being used today, viz., polymerization, polymer modification reactions, degradation, and synthesis of monomers and reactive oligomers.

II. Development of Novel Methodologies

Because of the multidisciplinary nature of this field, the research that is being done is diverse and takes on several directions. It helps to organize these developments in five sections: a) enzyme-related developments, b) whole-cell approaches, c) new reactions or products, d) new bio-processes, and e) other biocatalytic developments.

II.A. Enzyme-Related Developments

Although enzymes can be used for the polymer industry, sometimes the reaction is slow and the activity is low in non-aqueous media. The enzyme may also be expensive when compared to equivalent chemical approaches. The stability and the purity of an enzyme are also issues to be dealt with.

II.A.1. Improved Enzymes

One of the reasons for the current excitement about enzyme technology is the emergence of the Directed Evolution technique, developed by Arnold and her coworkers *(3)*, and gene shuffling techniques, developed by Stemmer et al at Maxygen *(4)*. In Directed Evolution, a gene sequence is replicated through error-prone PCR, thereby introducing a low level of point mutations in a library of mutants. In gene shuffling, a number of related genes are cut and reassembled. The resulting gene libraries are converted to enzymes through a suitable host and screened for a specific characteristic. The best mutants are chosen and become the starting points for a second generation of mutants. After a few generations, it is often possible to obtain a mutant enzyme where the characteristic is much improved.

A key contributor to the success of these techniques is the screening of the mutants *(3-5)*. The characteristic to be optimized may be enzyme activity, enzyme specificity, high temperature stability, solvent resistance, or any property where an assay can be devised. An assay should be easy to implement in the high throughput mode and yet correspond to the characteristic in question. An example of a new solid-state screening methodology (coupled to image analysis) is given in the chapter by Youvan et al *(6)*.

A related approach is the search for new microorganisms and new enzymes in nature. High on the list are the extremophile organisms *(7)*. The enzymes from these organisms can often withstand extreme conditions of temperature, pressure, pH, or salt concentrations and may be used for chemical reactions under these conditions. These enzymes may also be suitable starting points for further Directed Evolution or gene shuffling efforts.

The importance of these developments cannot be over-emphasized. If these techniques are commercially successful, they can potentially address most of the difficulties of enzyme technology in polymer science. Thus, if an enzyme activity could be enhanced, say by tenfold, the cost per weight of the enzyme would decrease by one-tenth. It has been shown that enzymes can be evolved to improve their stability in organic media and/or at high temperatures *(3)*. Finally, since only a specific gene is chosen, the enzyme impurity issue is minimized. It may be noted as a caveat that even after successful evolution or gene shuffling, the improved gene sequence still needs to be expressed in a suitable host, and fermentation and process optimization are needed to produce the enzyme.

II.A.2. Immobilization Techniques

Immobilized biocatalysts can be defined as biocatalysts physically confined or localized in a certain defined region of space with retention of their catalytic activities *(8)*. The advantages are the simplification of work-up procedure (as the biocatalyst can be removed by filtration), changes or improvements in catalytic activity, enhanced stability, and the possibility of repeated or continuous use. The biocatalysts can be enzymes, microbial cells, plant or animal cells. Several immobilization methods are commonly used: a) Binding to a carrier, e.g., through covalent bonding, physical adsorption, electrostatic interaction, or biospecific binding. b) Crosslinking with bifunctional or multifunctional reagents, or derivatization with a reactive group and subsequent polymerization. c) Entrapment in a gel, microcapsules, liposomes, hollow fibers, or ultrafiltration membranes. d) Combination of the above methods.

The carrier can be one of a number of materials, e.g., ion-exchange resins, polyacrylic resins, Celite® diatomaceous earth, cellulose, porous glass particles, alumina, silica, poly(vinyl alcohol) cryogels, and hollow fiber membranes. For example, Novozymes A/S immobilizes lipase B from *C. antarctica* on a poly(methyl methacrylate) resin. This product, marketed as Novozym-435, has shown extraordinary versatility for a broad range of esterification reactions. Another carrier that has been found to be useful is poly(ethylene glycol) (PEG) *(9)*. Attachment to PEG can increase the thermal stability of an enzyme and improve its solubility and activity in organic solvents.

Biocatalytic plastics have been made by acryloylation of a suitable enzyme, followed by free radical polymerization with vinyl monomers, in the presence of a surfactant *(10)*. Up to 50% enzymes can be incorporated into a plastic in this way. Enzymes have also been incorporated into polyurethane foams *(11)*.

Whole cells are often immobilized in alginate, κ-carrageenan, or polyurethane gels or membranes. These techniques are often used for microbial synthesis of organic compounds. An example is immobilized baker's yeast for reduction reactions *(1a)*.

II.A.3. Solubilization Techniques

Since most chemical reactions occur in the organic phase, it would be useful to solubilize the enzyme in an organic medium. A number of methods have been reported.

a. A surfactant may be used to solubilize the enzyme in an organic solvent and to increase its activity *(12)*. An example is the "surfactant coating" work by Goto et al *(13)* using several lipases and the non-ionic surfactant glutamate dioleyl ester ribitol amide, which has shown enhanced reaction rate for esterification reactions.

b. Dordick et al *(14)* added a surfactant to the protease subtilisin Carlsberg from *Bacillus licheniformis* and extracted the enzyme-surfactant ion pair from an aqueous solution to an organic solvent, like isooctane. This catalyst system remains active in isooctane and has been used for acylation reactions. In their chapter, Xie and Hsieh used this method to solubilize their enzyme for cellulose modification reactions *(15)*.

c. Another strategy is to solubilize the enzyme and substrate in a polar aprotic solvent. Although many enzymes exhibit decreased activities in polar aprotic solvents, this approach is useful for some substrates.

d. Another method to enhance the solubility of enzymes is to conjugate it with PEG, as described in the previous section.

II.A.4. Optimization of Enzyme Reactions and Processes

For most reactions, it is possible to optimize the reaction conditions in order to improve the productivity of the enzymes, in terms of units of polymer obtained per unit of enzyme used. Each reaction, however, needs to be studied individually. For example, an extensive amount of work has been done on lipase-catalyzed polyester synthesis *(2a)*. The following reaction parameters have been studied: solvent, monomer concentration, enzyme source, enzyme concentration, reaction water content, and temperature. These parameters have been shown to affect reaction rate, polymerization yield, and molecular weight.

Another example of enzyme kinetic studies is given in the chapter by Tanaka and Yu *(16)*. They studied the lipase-catalyzed hydrolysis reaction of umbelliferone stearate on monolayers and demonstrated that the reaction was entirely diffusion-controlled.

II.B. Whole-Cell Approaches

The whole-cell approaches typically use metabolic pathways of microorganisms (or plant or animal cells) to produce desirable products. Fermentation processes are usually involved in the scale-up and the manufacturing of these products.

II.B.1. Enrichment Culture and Screening

Enrichment culture techniques are commonly used in microbiology. They permit the selective cultivation of one or more specific type of microorganisms from a mixed culture or inoculum taken from the wild. Frequently, one particular organic compound is engaged as the sole carbon source. In this way, specific bacteria can be isolated that exhibit a certain metabolic activity related to the organic compound. In addition, the media or the conditions used in the enrichment culture may influence the choice of the microorganism. Mutation may be induced (e.g., through irradiation or chemical agents) to enhance diversity. The microorganism obtained can be used for a biotransformation, or the enzymes associated with the metabolic activity may be isolated.

2. Genetic Engineering

Genetic engineering techniques have had a large impact on microbial reactions. An area that has attracted a lot of attention is the conversion of biomass into alcohol *(17)*. For example, *Klebiella oxytoca* has been engineered to include genes corresponding to both cellulose and hemicellulose. The microorganism can degrade pre-treated waste paper and convert it to ethanol in 83% yield *(18)*. Several companies and organizations have wood- or waste-to ethanol technologies in the commercial or near commercial stages *(19)*.

A powerful use of molecular biology is metabolic engineering. A given cell can be altered to produce a new enzyme which is metabolically active in the cell, thereby extending a metabolic pathway by 1 or 2 desirable steps. Another way to use this technology is to delete the gene sequences corresponding to undesirable enzymes. Thus, the desired product can be made with higher yield and with less byproducts. As an example, the DuPont Company and Genencor International are collaborating on a process to convert hydrolyzed cornstarch to 1,3-propanediol using a recombinant microorganism *(20)*. The 1,3-propanediol is a raw material for the polymerization of the polyester, poly(1,3-propylene glycol terephthalate).

Several papers have appeared wherein multiple genes have been inserted into microorganisms. In their chapter, Lau et al cloned the genes for five enzymes that could convert cyclohexanol to adipic acid through a five-step enzymatic pathway in bacteria *(21)*. Translational coupling of the requisite genes ensured a balanced production of the biocatalysts and maintained proximity of the required cofactors for efficient biotransformation. Steinbüchel also cloned four genes corresponding to the enzymes in the synthesis of poly(hydroxyalkanoic acids) into *E. coli* and showed that homopolyesters and copolyesters could be made in vivo *(22)*. Wang et al developed the equivalent methodology for carbohydrate synthesis *(23)*. As many as five gene sequences were cloned and inserted into *E. coli*. In this way, a carbohydrate can be synthesized that is the product of up to five enzymatic reactions.

II.B.3. Plant Cells and Cell Extracts

Plant cells and cell extracts can also carry out polymer reactions. Bulone et al used detergent extractions to remove membrane-bound glucan synthases from

Arabidopsis thaliana and *Rubus fruticosus (24)*. A fraction enriched in (1-3)-β-glucan synthase (and a porin) was obtained from *A. thaliana* that produced callose. They also described a procedure that allowed the in vitro synthesis of cellulose from cell-free extracts of *R. Fruticosus* cells.

II.C. New Reactions or Products

Given the range and the scope of the reactions involved in polymer science, it would be highly desirable to discover and to devise new biocatalytic reactions and to generate new products from these reactions. Many chapters in this book give excellent examples of new polymeric reactions or products.

It may be noted that nature exhibits wide microbial enzyme diversity *(1f)*. There are a large number of biochemical reactions yet to be discovered. It is known that at least 94 organic functional groups are found in natural products, but only a little more than half has been studied with respect to their biosynthesis or biodegradation *(1f)*. The current interest in functional genomics and bioinformatics may also be helpful to uncover new reaction pathways *(25)*.

As regards new method development, a promising approach is the combinatorial biocatalysis work reported in Dordick's chapter *(26)*. He and his coworkers polymerized combinations of 4 activated diesters and 12 diols, and the resulting polymer library was screened for molecular weight and potentially for other physicochemical properties as well. In the literature, a combinatorial experiment was reported *(27)* giving the effects of different combinations of enzymes on the degradation of guar, a natural polysaccharide. The molecular weight and optical clarity were measured in that case.

II.D. New Bioprocesses

Biocatalysis provides opportunities but may also incur a different set of problems. One problem relates to the solvent medium: the biocatalysts are usually water soluble, but chemical reactions are frequently done in organic media. Recently, advances in non-aqueous enzyme reactions have extended the process conditions for biotransformation. The extended reaction environments include organic solvents *(28)*, biphasic organic solvent-aqueous mixtures *(29)*, reverse micelle systems *(30)*, and supercritical fluids *(31)*. These different reaction environments have facilitated downstream product separation and enzyme re-use.

In order to optimize enzymatic processes, new tools and methodologies are needed. For example, continuous processes using ultrafiltration membranes have been developed *(32)*. The enzymes are large and are retained in the ultrafiltration membranes, whereas small-molecule substrates diffuse in and out of the membranes. Another example is the "window of operation" graphical tool, developed by Woodley, whereby reaction and catalyst constraints are identified and bioprocess operating regions are defined *(33)*.

Many oxidoreductases require cofactors (e.g., NADH and NADPH) which are expensive. In order to decrease cost, it is necessary to recycle the cofactors. A number of ways have been devised to address this problem *(34,35)*:
1. Enzyme regeneration approach, using formate, is often done in a membrane reactor, and the cofactor is attached to poly(ethylene glycol) to prevent its leakage from the reactor.

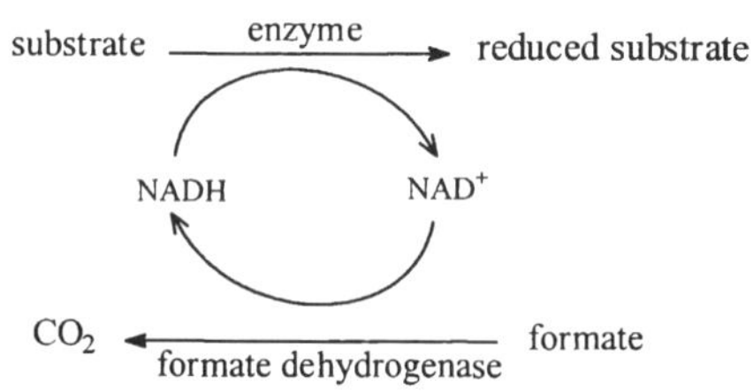

2. Substrate coupled approach, using a second substrate like isopropanol and an enzyme like alcohol dehydrogenase.
3. Electrochemical regeneration.

Instead of these approaches, an alternative is the whole-cell approach, using the cellular machinery to regenerate the cofactors.

II.E. Other Biocatalytic Developments

In addition to the methodologies described above, there are many other exciting developments in biocatalysis. A selected review of some of them is given here; not all of these developments are applicable to polymers.

Catalytic antibodies have been around for quite some time *(36)*. An antibody is a protein formed in an organism in response to a chemical compound. It is envisioned that an antibody elicited against a stable transition-state analog of a reaction should catalyze that reaction. This concept has been realized in practice, especially for ester hydrolysis and pericyclic reactions. Thus far, this technology has not been applied to synthetic polymers.

A company called Altus Biologics, Inc. purifies and crystallizes enzymes, which are then crosslinked with glutaraldehyde *(37)*. The resulting crosslinked enzyme crystals (called CLECS®) show noted improvement in heat, pH, and solvent stability. However, because the channels leading to the active sites are usually confined, the substrates tend to be small molecules. An alternative approach has been devised by Alnis Biosciences *(38)*. It has developed several techniques to wrap polymers around enzymes, thereby improving their stability.

A large body of work has been devoted to enzyme mimetics *(39)*. A recent example is the synthesis of an inorganic complex with similar activity as galactose oxidase, a well-known oxidoreductase *(39e-g)*. Another example is the effort to reduce the size of an active enzyme (called microzyme). This interesting research area has been recently reviewed *(40)*.

A different enzyme mimetic is the "synzyme", developed by Wandrey et al *(41)*. The idea is to attach a homogeneous inorganic catalyst to a soluble polymer. Whereas the turnover frequencies of this catalyst are typically much lower than those of enzymes, multiple catalytic sites can be incorporated in one polymer molecule, thereby giving comparable space-time yields in catalytic reactions.

Recently a team of chemists has made conjugates of hydrocarbon polymers with enzymes that organize into densely packed films on surfaces *(42)*. These conjugates may find use as sensors or catalyst domains in nano-scale devices.

III. Polyesters and Polycarbonates

These polymers can be readily synthesized via hydrolase enzymes, particularly lipases and esterases. A number of reactions have been successfully exploited, e.g., self-condensation polymerization, polytransesterification, ring-opening polymerization, combined condensation and ring-opening polymerization, and transesterification. A lot of work has also been done using microbial approaches, and these are discussed separately in Section III.E.

III.A. Enzymatic Condensation Polymerization

III.A.1. Self-Condensation

A monomer may contain two functional groups A and B, where A can react with B with the evolution of a small-molecule leaving group. When it happens with the formation of a polymer, this reaction is referred to as A-B type condensation polymerization. The simplest examples are A-B type polymerization of β-, δ-, and ε-hydroxyesters, where A = alcohol, B = carboxylate ester, x ~ 2-5, and the reaction is carried out in organic media *(43).*

$$HO-(CH_2)_x-COO-CH_3 \xrightarrow{\text{lipase}} H-[O-(CH_2)_x-CO]_nO-CH_3$$

When the reactant is unsubstituted, polymers are formed. Substitution on the monomer tends to favor the formation of lactones (via intramolecular cyclization) (*vide infra*, Section VIII.A).

In their chapter, Brandstadt et al studied the self-condensation of 12-hydroxydodecanoic acid under different reaction conditions using porcine pancreatic lipase *(44)*. The M_n varied from 400 to 12,500. In contrast, 12-hydroxystearic acid that has a secondary alcohol did not polymerize.

III.A.2. Condensation from different monomers

Enzymes can be used to catalyze the condensation reaction of diacids (BB) and diols (AA) to form oligoesters and polyesters. This type of polymerization is sometimes called polytransesterification. The reaction is facilitated if a diester is used instead of a diacid, and is further enhanced if the diester is activated (e.g., in the case of enol or vinyl esters):

$$HO-(CH_2)_x-OH + HOCO-(CH_2)_y-COOH \xrightarrow{\text{lipase}} H-[O-(CH_2)_x-O-CO-(CH_2)_y-CO]_n-OH$$

$$HO-(CH_2)_x-OH + CH_3OCO-(CH_2)_y-COOCH_3 \xrightarrow{\text{lipase}} H-[O-(CH_2)_x-O-CO-(CH_2)_y-CO]_n-OH + CH_3OH$$

$$HO-(CH_2)_x-OH + CH_2{=}CH-OCO-(CH_2)_y-COO-CH{=}CH_2 \xrightarrow{\text{lipase}} H-[O-(CH_2)_x-O-CO-(CH_2)_y-CO]_n-OH + CH_3CHO$$

$$HO-(CH_2)_x-OH + CF_3CH_2-OCO-(CH_2)_y-COO-CH_2CF_3 \xrightarrow{\text{lipase}} H-[O-(CH_2)_x-O-CO-(CH_2)_y-CO]_n-OH + CF_3CH_2OH$$

Again, a lipase can be used as the biocatalyst, e.g., porcine pancreatic lipase (PPL), *Candida antartica* lipase B (Novozym®-435 from Novozymes A/S), and *Mucor meihei* lipase (lipozyme).

In their chapters, Dordick and Kobayashi both employed this type of condensation to make new polymers. Dordick used combinatorial synthesis and screened many diesters and diols *(26)*. Kobayashi focussed on regioselectivity of monomers with multiple OH groups (e.g., glycerol and sorbitol) *(45).*

III.B. Enzymatic Ring-Opening Polymerization

III.B.1. Polymerization of Lactones

This type of polymerization does not generate a small-molecule leaving group. Enzymes have been used extensively for the polymerization of lactones to polyesters *(47-49)*.

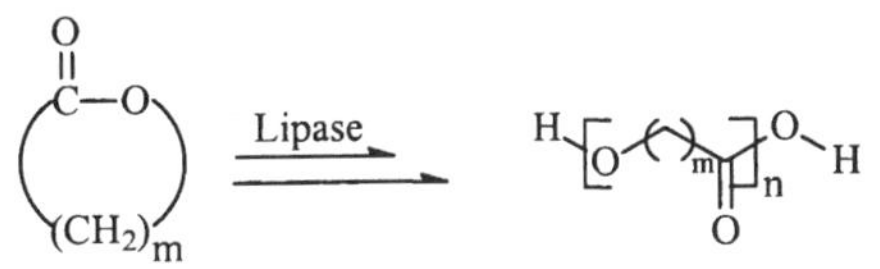

It may be noted that conventional chemical methods and enzymatic methods can both be used for the 4-, 6- and 7-membered lactone polymerization (m=2,4,5) *(46)*. However, for macrolactones (m>9), the chemical methods give slow reaction rates and low-molecular-weight polymers *(47)*. In contrast, the enzyme-catalyzed polymerization of these macrolactones gives faster rates and high polymers. Kobayashi et al *(48)* and Gross et al *(49)* have examined these reactions and optimized the reaction conditions.

Copolyesters can be made in a similar way via enzyme catalysis by using two or more lactones (m= 7 or 14, and n = 4, 5, 10, or 11) *(50)*.

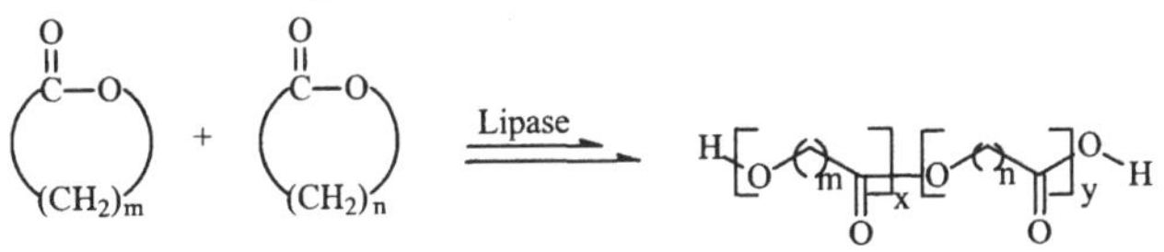

In his chapter, Kobayashi reported the synthesis of copolyesters of β-butyrolactone/δ-caprolactone with ε-caprolactone/12-docecanolide and observed their enantioselectivity as a function of monomer structure, solvent medium, and reaction time *(45)*.

III.B.2. Polycarbonate Synthesis

Ring-opening polymerization of cyclic carbonates can be achieved through enzyme catalysis to form polycarbonates *(51)*. The most common monomer is trimethylene carbonate (TMC, also known as 1,3-dioxan-2-one).

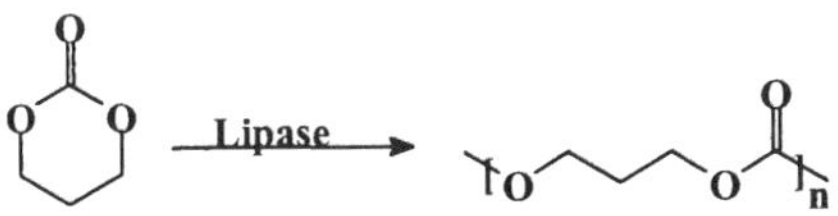

In their chapter, Bisht and Al-Azemi *(52)* copolymerized TMC with 5-methyl-5-benzyloxy-carbonyl-1,3-dioxan-2-one to produce random copolymers with different compositions. Lipase was used as the catalyst, and the polymer yields were high in all cases.

Poly(ester-co-carbonate) can be made in a similar manner using a cyclic carbonate monomer and a lactone *(53)*. An example is the copolymerization of pentadecylactone and TMC via Novozym®-435 (from Novozymes A/S). A random copolymer was obtained.

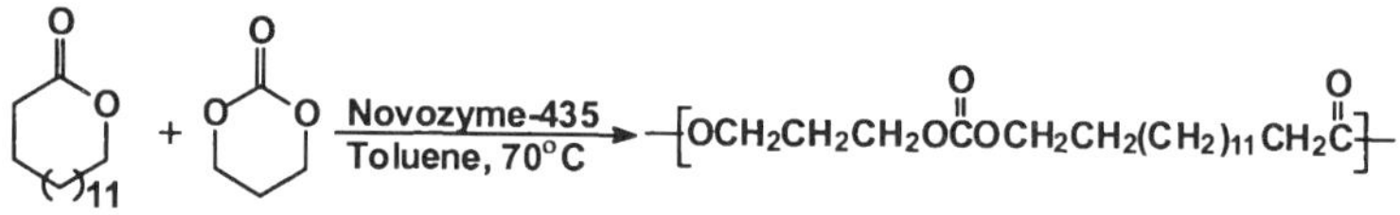

III.C. Combined Condensation and Ring-Opening Polymerizations

The two types of lipase reactions can be combined to concurrently catalyze condensation and ring-opening polymerizations *(54)*. An example is given below of the concurrent copolymerization of macrolactone, divinyl ester, and glycol with the help of *P. cepacia* lipase (lipase PC) *(54b)*.

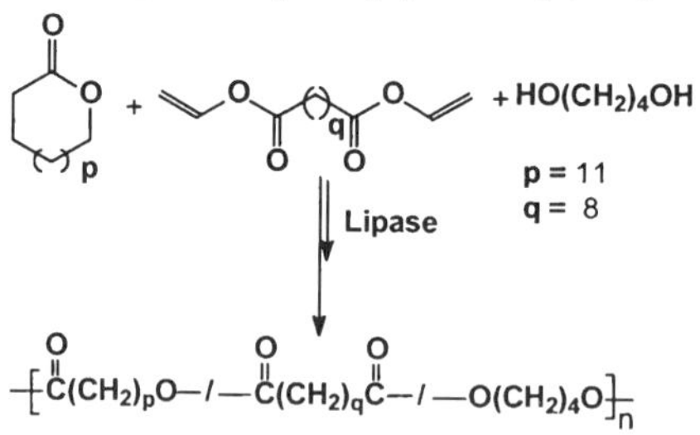

III.D. Enzymatic Transesterification

Recently Gross et al *(55a,b)* reported the ability of Novozym®-435 to catalyze transesterification reaction between preformed polyesters. Thus, when polypentadecalactone and polycaprolactone were heated in bulk at 70-75C with Novozym®-435, transacylation reaction took place. Depending on the molecular weight of the starting polymers and the reaction time, the resulting copolyester might have multiblock distribution (for higher molecular weight polymers and shorter reaction times) or random distribution (for longer reaction times). In their chapter, Kumar et al *(55c)* have extended this reaction to trimethylene carbonate, and also looked into the reaction mechanism.

III.E. Microbial Approaches

Polyesters can be made by microbial approaches. Poly(γ-glutamic acid) is being produced in *B. subtilis* and *B. licheniformis (56)*. At one time Zeneca Inc. made a random copolymer of 3-hydroxybutyrate and 3-hydroxyvalerate using *Alcaligenes entrophus (57a,b)*. A fair amount of effort has been expended to lower the cost of this process *(57c-e)*. Microbial polythioesters have also been reported *(58)*.

In his chapter, Steinbüchel reviewed his work involving metabolic engineering of poly(hydroxyalkanoic acid) (PHA) biosynthetic pathways *(22)*. He first demonstrated that a biosynthetic pathway employing purified enzymes is operational in vitro. He then expressed the new pathway in vivo in recombinant strains of *E. coli*. The genes were cloned into a plasmid, and the recombinant cells were cultivated and shown to produce homopolyesters and copolyesters when fed with the corresponding hydroxy fatty acids.

Kelley and Srienc developed a system in *R. eutropha* to produce PHA with different compositions that would phase separate *(59)*. By controlling when and how much hydroxyvalerate is added, they could control the microstructure of the copolymers of hydroxybutyrate and hydroxyvalerate. They obtained core/shell polymer granules in some cases.

IV. Phenolic and Vinyl Polymers

Peroxidases belong to the class of oxidoreductases and can catalyze the oxidation of a donor using H_2O_2. Laccases can function in the same way, using oxygen. These enzymatic systems can be employed to polymerize phenolic compounds or vinyl monomers.

IV.A. Polymerization of Phenol and Aniline

The use of peroxidase and laccase enzymes for the polymerization of phenol and pheno-containing structures is well known *(60)*. The resulting structures are similar to that of the phenol-formaldehyde resins. Thus, these reactions have been considered to be an alternative to Novalac resins, but without formaldehyde.

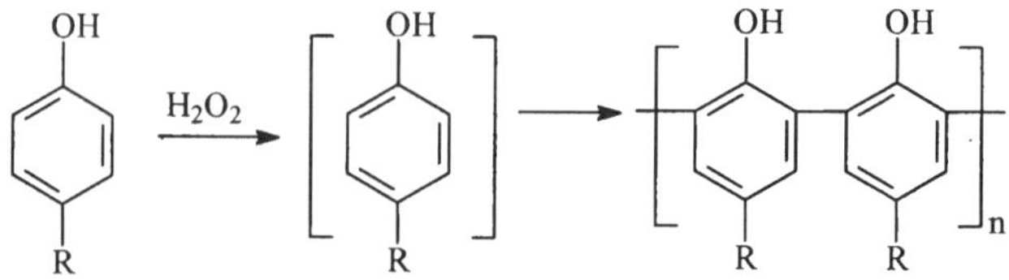

Recently, Kobayashi et al have shown the use of laccase and oxygen to polymerize 2,6-dimethylphenol *(61a)*. This is an enzymatic alternative to the synthesis of poly(phenylene oxide). The same polymer can also be made through the reaction of laccase/oxygen or peroxidase/H_2O_2 on syringic acid *(61b)*.

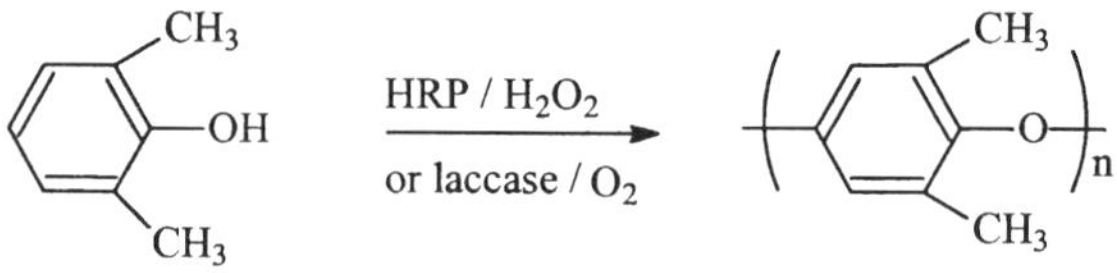

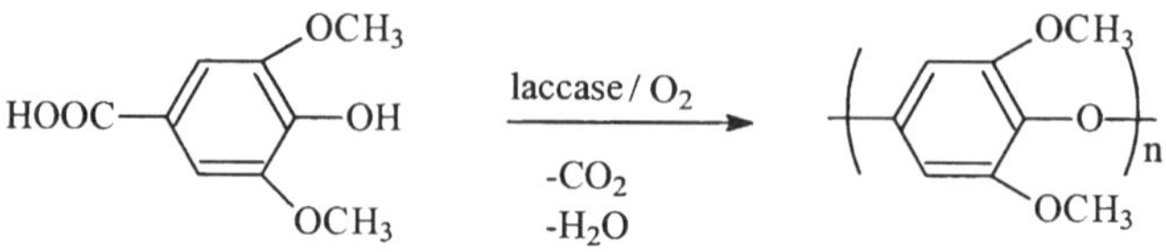

Tripathy et al have published a series of papers on the synthesis of novel water-soluble polyaniline through peroxide-catalyzed oxidative free radical coupling of appropriate monomers *(62)*.

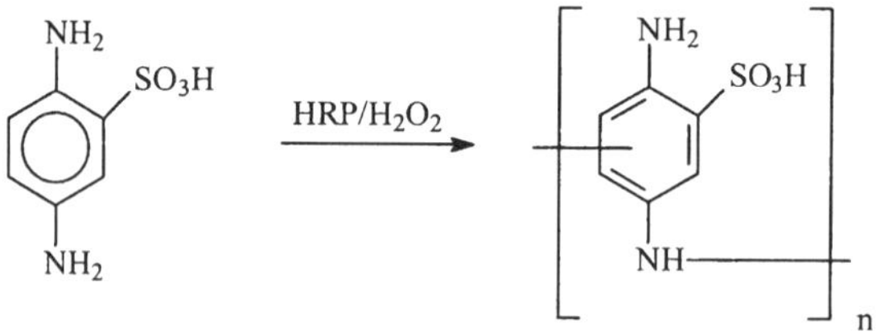

IV.B. Polymerization of Vinyl Polymers

Both peroxidases and laccases have been used to catalyze free-radical polymerization of vinyl monomers *(63)*. Most of the studies reported to date deal with acrylic systems. The chapter by Kalra and Gross (64) gave a good review of this reaction. The polymerization can be carried out in solution or as emulsions. Polyacrylamide and polyacrylates are atactic, but poly(methyl methacrylate) is syndiotactic.

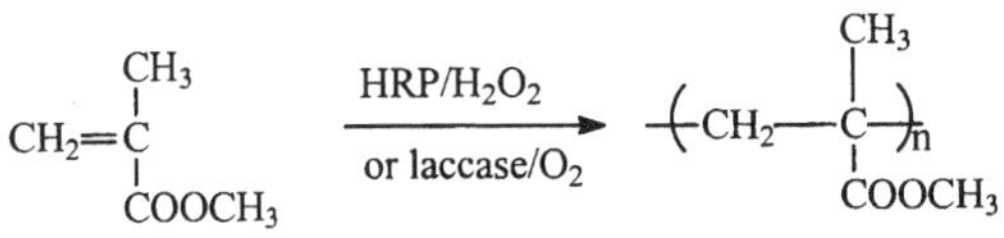

V. Synthesis of Oligosaccharides and Polysaccharides

Polysaccharides and oligosaccharides are a difficult class of compounds for chemical synthesis because of the presence of different hydroxy groups and stereochemistries. Typically such reactions require protecting and de-protecting steps which are time-consuming and expensive, except for pharmaceutical and other high-end applications *(65)*. Enzymes have been found to be a potentially cheaper alternative. Wong, Whitesides, and others have used these techniques extensively for the synthesis of carbohydrates and oligosaccharides *(66)*.

In the polysaccharide area, Kobayashi et al *(2c,67)* reported the use of a hydrolase to polymerize disaccharide fluorides into both natural and non-natural polysaccharides. An example is cellulose:

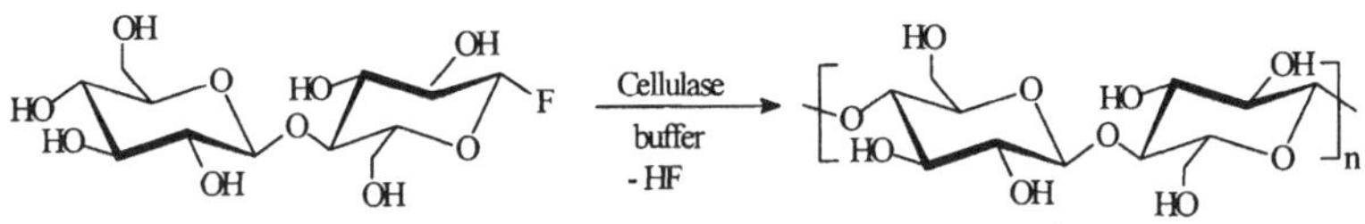

ß-D-cellobiosyl fluoride **LMW cellulose**

The molecular weight is low because once the degree of polymerization (DP) reaches about 8-10, the cellulose precipitates from solution.

Higher molecular weight polymers were obtained from the polymerization of β-xylobiosyl fluoride (to form xylan, DP~23), and the oxazoline derivative of N,N'-diacetylchitobiose (to form high-molecular-weight chitin) *(2c,67)*.

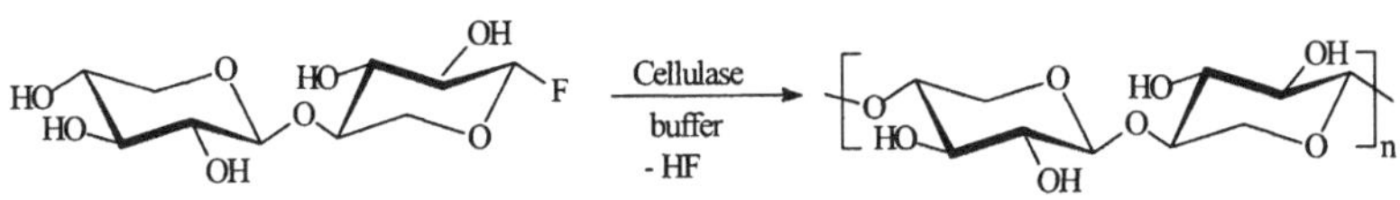

ß-xylobioxyl fluoride **LMW xylan**

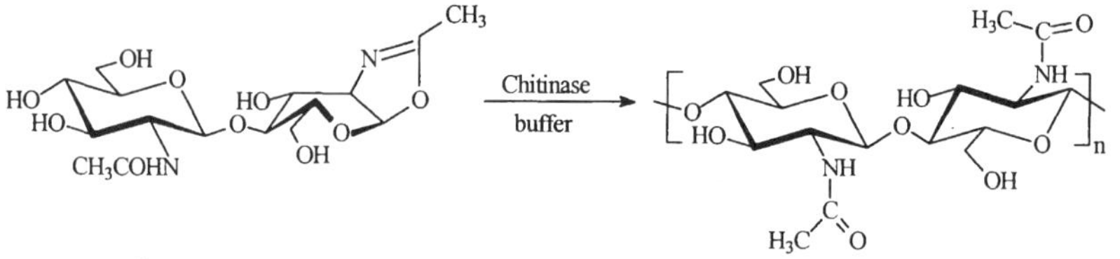

Oxazoline derivative of N,N'-diacetyl chitobiose **HMW Chitin**

Microbial methods have been employed to make polysaccharides. A celebrated case is bacterial cellulose, known since 1876, but enjoying a resurgence of interest *(68)*. Recently, different mutant strains of *Acetobacter xylinium* have been isolated, and much more is known of the molecular biology and the control of the reaction *(69)*. Other microbial polysaccharides that are produced commercially include dextran, gellan, pullulan, and xanthan *(70)*.

Through the use of molecular biology, Neose Technologies has devised microbial methods to make oligosaccharides and polysaccharides. Examples are the synthesis of fructooligosaccharide and heparan *(71)*. Wang et al have reported the synthesis of oligosaccharides using *E. coli* whereby the genes corresponding to several enzymes are cloned onto the same plasmid *(23)*.

Plant cell extracts have also been used for polysaccharides. For example, Bulone et al in their chapter reported the synthesis of callose using detergent extracts from *Arabidopsis thaliana* and cellulose from cell-free extracts of *Rubus fruticosus* cells *(24)*.

VI. Enzyme-Catalyzed Polymer Modifications

Since enzymes are capable of reacting with small organic compounds, the same reactions can often be used on polymers. The main motivation is to improve the properties of the polymer in order to add value for specific applications. Thus far, most of the publications in this area deal with water-soluble polymers, particularly polysaccharides. Some of the modification reactions are summarized in Table 1. Also given are several properties of interest and some typical applications.

In this symposium volume, the chapters by Xie and Wang *(72)*, and by Cheng and Gu *(73)* give many examples of polysaccharide modifications. These include the addition of charge, addition (or removal) of polar group, hydrophobic modification, and redox reactions. In addition, Xie and Hsieh made a number of cellulose esters using ion-paired protease-surfactant complexes in pyridine *(74)*. Payne et al used tyrosinase to graft quinone and peptides onto chitosan *(75)*. Gu grafted hydrophobic esters onto cellulosic derivatives using lipase enzymes *(76)*. Hu et al oxidized hydroxyethylcellulose using a lipase-mediated TEMPO reaction *(77)*.

Table 1. Typical Water-Soluble Polymers and Modification Reactions

structure	modification	property	application
starch	oxidation	solubility	thickener
cellulose	reduction	rheology	gelling
cellulosic derivatives	ester formation	water retention	emulsifier
chitosan	amidation	gel formation	(super)absorbant
polyethers	acylation	tensile	flocculant
polyelectrolytes	glycosylation	texture	(micro)encapsulant
	quinone addition	interfacial effects	stabilizer
	peptidoglycan	surface properties	biocomposite
	polymerization	barrier properties	biomaterial

In addition, there is a large body of literature on enzymatic reactions of natural polysaccharides, particularly food gums *(78)*. Examples of food gums are pectin, alginate, agar, carrageenan, furcellaran, ghatti, gum arabic, karaya, tragacanth, konjac, guar, locust bean gum, and (above all) starch. The microbial gums may be included as well, e.g., dextran, gellan, pullulan, and xanthan. The enzymes that operate on these gums are usually found in nature and are specific to given substrates. A detailed discussion would easily fill up a separate book and is beyond the scope of this article. Interested readers may consult the references *(78)*.

VII. Polymer Degradation

Many enzymes in nature are designed to break down materials, and therefore hydrolysis and degradation are frequently encountered enzymatic reactions. For this purpose, hydrolases are usually involved. Thus, it is well known that lipases and esterases can degrade some polyesters, proteases can degrade proteins, amylase can degrade starch, cellulase can degrade cellulose and cellulosic derivatives, and many other enzymes are specific for other natural polymeric substrates. Whereas these polymers can be degraded via chemical means (e.g., acid or base hydrolysis), the enzymatic approach entails milder reaction conditions, lower temperatures, and generates less byproducts. In addition, because many enzymes are specific in their mode of action or site of hydrolysis, enzymatic degradation products sometimes have structures different from chemical degradation products. An example from the literature is the hydrolysis of carboxymethylcellulose (CMC) through cellulase *(79)*.

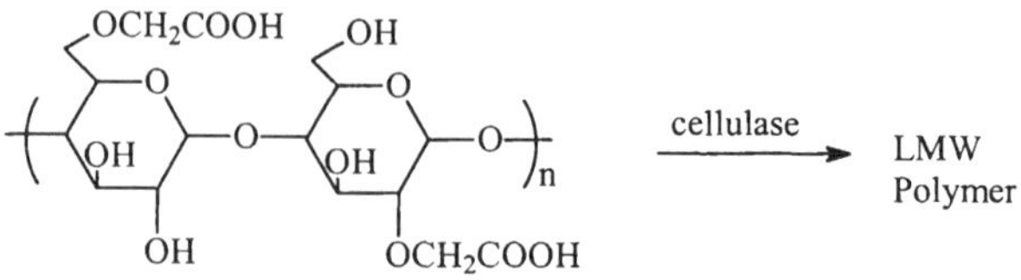

In this case, the enzyme preferentially hydrolyzes the glycosidic bond between two unsubstituted glucose residues. The least reactive glycosidic bond is the one where adjacent glucose residues are substituted. Thus, by varying the reaction conditions, one can get low-molecular-weight CMC with specific chain end units.

Lyases are also very useful in degrading polymers. For example, pectate lyase *(80)* and alginate lyase *(81)* are both very efficient in lowering the molecular weight of pectin and alginates, respectively. Some oxidoreductases can also be employed for degradative reactions. For example, laccase, lignin peroxidase, and manganese peroxidase are all known to oxidatively degrade lignin and can be used to bleach wood pulp *(82)*.

Microbial degradations are also well known and commercially used *(1f)*. In fact, bioremediation is a huge topic that by itself can fill multiple book volumes *(83)*. The techniques of biotechnology have brought new possibilities to this area.

In this symposium volume, the degradation of polysaccharides is represented by Cheng and Prudhomme's paper *(84)*, where the kinetics and the enzymatic degradation mechanism of three guar derivatives are studied. Norwood et al reported the degradation of a polyester urethane using a polyurethanase isolated from *Comomonas acidovirans (85)*. In a paper by Hu et al, the oxidative cleavage of a polysaccharide and a polyether was studied using lipase-mediated TEMPO reactions *(77)*. Additional examples of enzymatic degradation of polymers are given in the chapter by Cheng and Gu *(73)*.

VIII. Synthesis of Monomers and Reactive Oligomers

Enzymatic and microbial reactions have been used to synthesize specific monomers and reactive oligomers. In order to be economically viable, these approaches have to show advantages over equivalent chemical approaches. Several examples (grouped in four categories) are given in this section.

VIII.A. Chiral or Enantio-Enriched Monomers

Many reactions take advantage of the enantiospecificity of enzymes to produce monomers suitable for polymerization. An example of enantio-enriched monomers is the lactone. Lipases in organic media have been successful in lactonization of γ-hydroxy esters, δ-hydroxy esters and ε-hydroxy esters (n=2,3,4) to produce the corresponding lactones with enantio-enrichment *(86)*. A competing reaction is ring-opening polymerization which is favored when R = H (below).

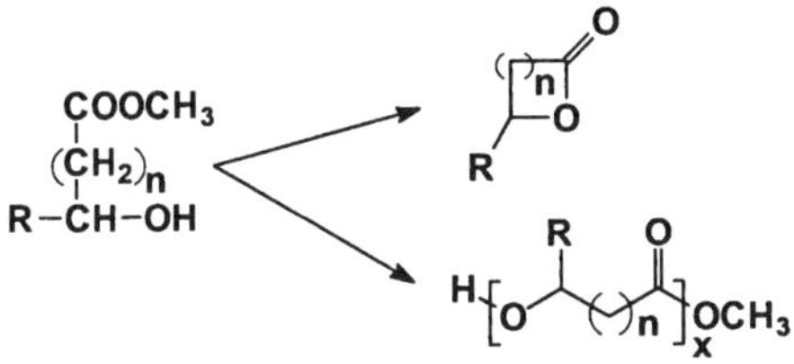

The patent literature gives many more examples of monomer synthesis via biocatalysis. For example, General Electric used chiral resolution with enzymes to produce optically active binaphthol through asymmetric hydrolysis of racemic diesters *(87)*.

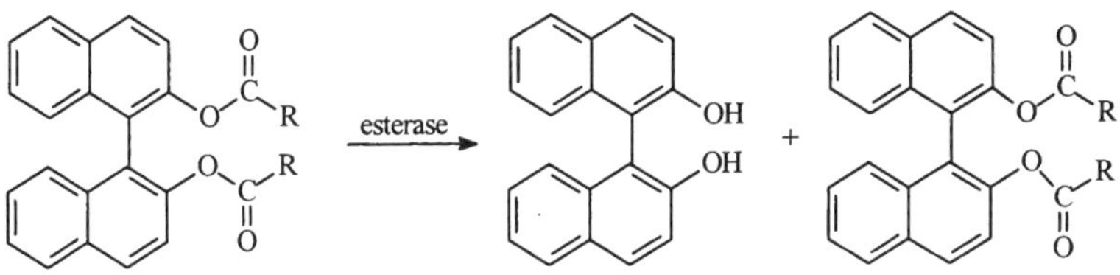

Another application of enzymes is the ester interchange reaction involving diphenyl carbonate and an alcohol or a diol *(88)*. An example is shown below of an optically active alcohol. The products obtained in both reactions are useful monomers for engineering plastics.

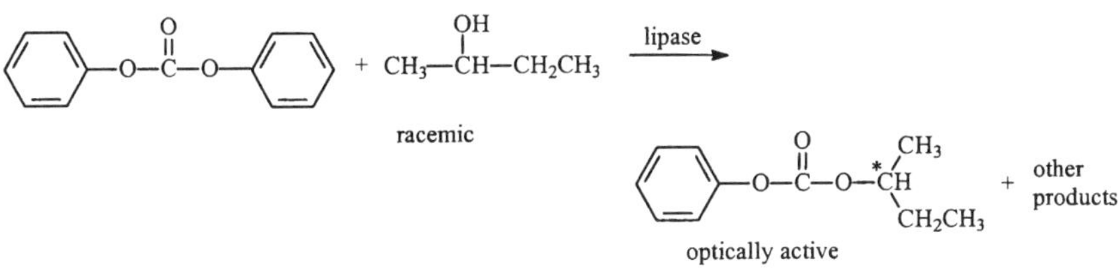

VIII.B. Monomers from Specific Biocatalytic Reactions

Some reactions utilize the chemo- and regio-specificity of enzymes. An example is the reaction of lactose and α-glycosidase on water-soluble alcohols *(73)*, e.g.,

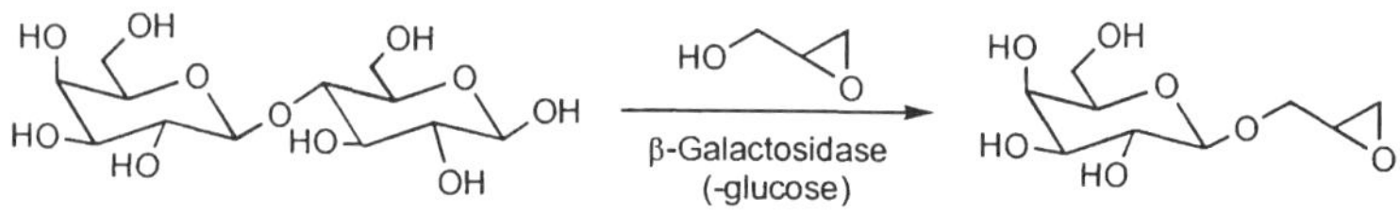

The product can be polymerized as a polyether or used as a derivatizing agent.

Another example of an industrially important enzyme is haloalkane dehydrogenase, which is used by the Dow Chemical Company to degrade an undesirable byproduct, 1,2,3-trichloropropane, in the manufacturing of epichlorohydrin *(89)*. The products of this reaction can be easily treated with base to produce epichlorohydrin, thereby recycling a waste product and improving overall yield of epichlorohydrin.

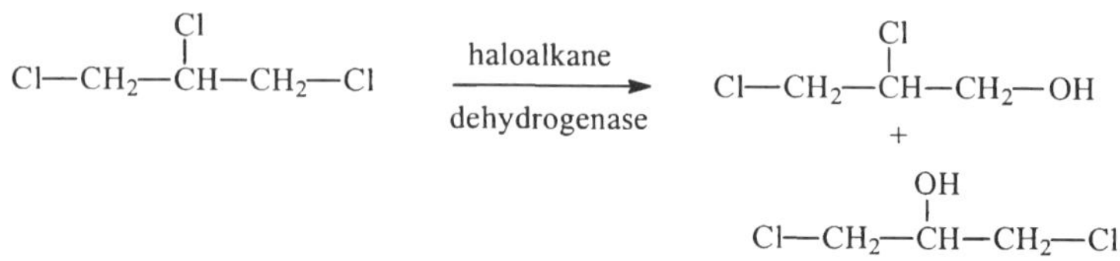

Microbial approaches often permit monomer syntheses that are difficult (or impractical) to accomplish with conventional chemistry. For example, Lau et al have cloned in bacteria the genes corresponding to five enzymes that can convert cyclohexanol to adipic acid *(21)*. DuPont-Genencor's approach to convert cornstarch to 1,3-propanediol using a recombinant microorganism has been discussed earlier (Section II.B.2).

In the literature, microbial methods and fermentation have been used to convert glucose to 4-hydroxybenzoic acid through genetically engineered *E. coli*. The products are isolated via anion exchange resins *(90)*. This compound is a monomer for liquid crystalline polymers, Other microbial reactions for monomer synthesis have also been reported *(91)*.

VIII.C. Vinyl Monomers

Enzyme catalysis is a good way to synthesize some vinyl monomers. One class of compounds is the vinyl sugar esters, obtained through the acylation reaction of a sugar, an activated ester, and a protease in polar aprotic solvents *(92)*. An example is shown below.

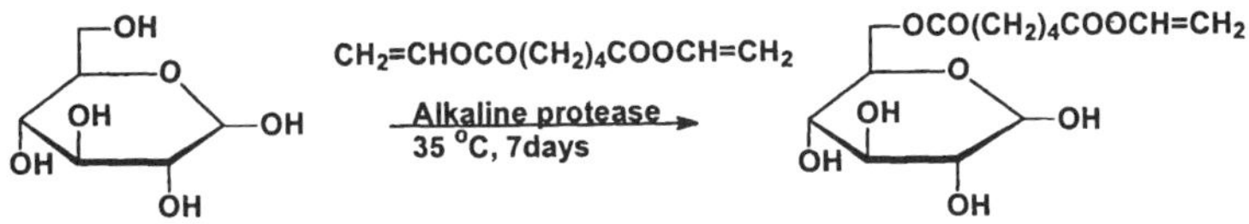

In their chapter, Kitagawa et al showed that the regioselectivity of this reaction depended on the solvent used. In 4:1 DMF/DMSO, the reaction preferentially added to the primary alcohol in glucose and in mannose, but to the secondary alcohol in galactose and allose *(93)*. In their chapter, Tokiwa et al used the same acylation reaction and grafted the vinyl functionality onto nucleosides *(94)*. The monomers were then polymerized.

Another example is the use of vinyl acrylate on sucrose in pyridine. The resulting acrylate gives a superabsorbing polymer upon polymerization *(95)*.

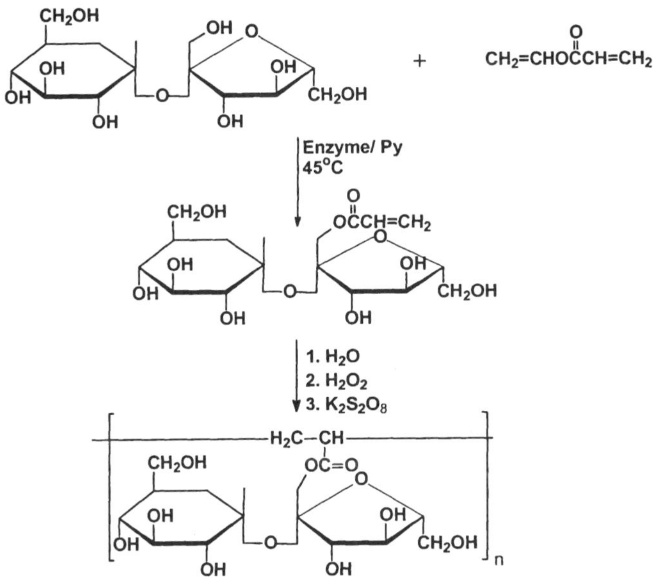

A good example of industrial applications is the silicone acrylate, made by Degussa AG, as a paint additive *(96)*. The equivalent chemical reaction frequently requires temperatures above 100°C and needs stabilization of the reaction mixture with a free-radical scavenger to suppress unwanted polymerization. The chemical catalyst must be removed for many applications, thereby adding cost. The reaction mixture is also deeply colored. The enzymatic process (using lipase, esterase, or protease) can bypass all these problems.

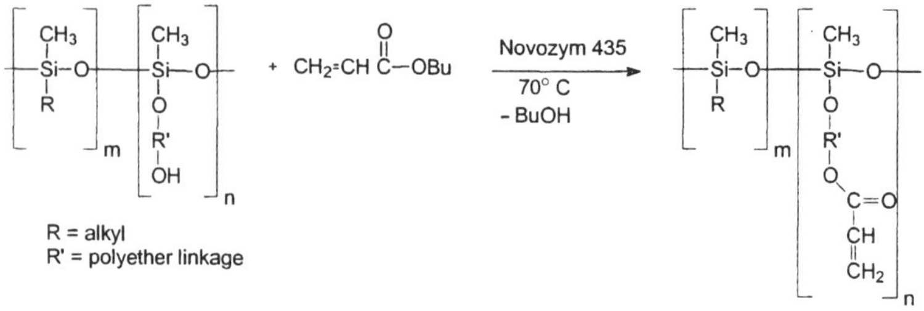

VIII.D. Reactive Oligomers

These are also known as macromers and are low-molecular-weight polymers that can be polymerized, or otherwise reacted with a polymer. A good platform for macromer is the ring-opening polymerization of lactones. Thus, Kobayashi et al *(97)* polymerized dedecanolactone in the presence of hydroxyethyl methacrylate or vinyl methacrylate. The end result in either case is a methacrylate-terminated polyester.

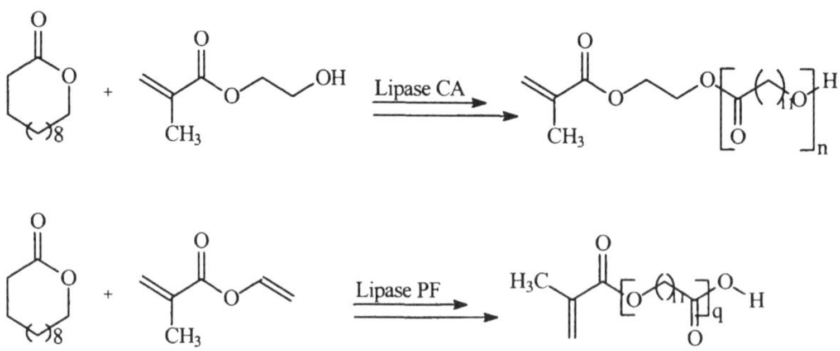

In a similar way, Cordova et al *(98)* have polymerized caprolactone in the presence of potential chain initiators, including vinyl and phenol functionalities. Again, lipase was used as the biocatalyst.

A number of reactive oligomers were made through lipase-catalyzed polymerization of compounds containing hydroxy and carboxy functionalities, in the presence of a methacrylate *(99)*. An example is shown in the following scheme. Free radical polymerization of the methacrylate produces a comb polymer.

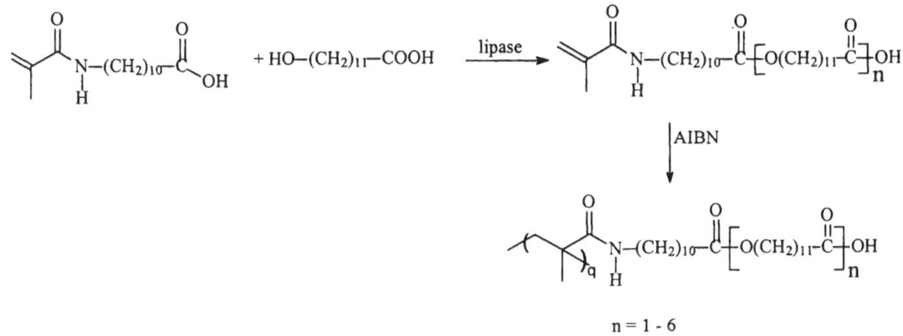

In the chapter by Kumar and Gross *(100)*, they used lipase-catalyzed transformation to construct well-defined macromers around a sugar core. The method is flexible and can be used to generate a wide range of macromers and heteroarm stars.

IX. Other Polymeric Systems

Whereas protease enzymes hydrolyze peptide bonds under physiological conditions, they can also be made to catalyze peptide synthesis under different reaction conditions *(101)*. Enzymatic peptide synthesis has the advantage over chemical methods on account of its specificity, the mild reaction conditions, and minimum need for side-chain protection. It has been pointed out that the stepwise chemical approach on polymeric supports (used in peptide synthesizers) is limited practically to peptides with < 50 amino acid residues. The synthesis of larger proteins may require block-wise enzymatic coupling of synthetic fragments *(102)*. Thus, despite the widespread use of peptide synthesizers, the enzymatic (or chemo-enzymatic) methods may still be useful in peptide synthesis.

The DNA synthesizer is available commercially. Nevertheless, a number of approaches to make polynucleotide mimetics are being developed. For example, a chemo-enzymatic approach towards a polynucleotide mimetic was given in Tokiwa's chapter *(94)*. They grafted a vinyl group onto nucleosides (Section VIII.C), and the vinyl group was then polymerized to generate a poly(vinyl alcohol) backbone with nucleoside branches. Another enzymatic approach towards a polynucleotide mimetic was reported in the literature by Dordick et al *(103)*, using two enzymes to generate nucleoside-based polyphenol.

There has been recent interest in silicon bioscience. A protein from marine sponge has been identified that can convert alkoxysilanes at neutral pH to silicates and silsesquioxanes. Genetic engineering and biotechnology offer the prospects of identifying the structural determinants of the silicone synthesis-directing activities *(104)* Recently, Dow Corning and Genencor signed a collaborative agreement to create a new area of materials science *(105)*.

X. Conclusions

This article provides an overview of this large and vibrant area of research. Hopefully the reader is impressed with the diversity of this topic, from both biological and scientific points of view.

One of the welcome developments in this field is the attention it has received from industry. Indeed, several companies are now seriously looking at this technology *(106)*. It is important for the industrial scientists to recognize where biocatalysis provides significant advantages relative to conventional chemical approaches. Cost versus benefit is the key to commercial viability. Some promising areas for biotransformation include reactions or products that require chemo-, regio-, or enantio-selectivities, or products that cannot be made via chemical means. If the current chemical processes encounter problems such as waste generation, color formation, and high temperature operation, biotransformation perhaps can make a difference. Another opportunity is the synthesis of monomers and reactive oligomers. In view of the rapid progress achieved in the past several years, we expect continued vitality of this field in the future and increasing use of this technology in industrial product and process development.

References

1. For example, (a) *Biocatalysis for Fine Chemicals Synthesis*; Roberts, S. M., Ed.; Wiley, Chichester, UK, 1999. (b) *Enzyme Catalysis in Organic Synthesis*; Drauz, K.; Waldemann, H., Eds.; VCH, Weinheim, Germany, 1995. (c) *Biotransformations in Organic Chemistry*, 3rd Ed.; Faber, K.; Springer, Berlin, Germany, 1997. (d) *Applied Biocatalysis*; Cabral, J. M. S.; Best, D.; Boross, L.; Tramper, J., Eds.; Harwood, Char, Switzerland, 1994. (e) *Enzymes in Synthetic Organic Chemistry*; Wong, C. H.; Whiteside, G. M.; Elsevier, Oxford, UK, 1994. (f) *Biocatalysis and Biodegradation*; Wackett, L. P.; Hershberger, C. D.; ASM Press, Washington, DC, 2001.

2. For example, (a) Gross, R. A.; Kumar, A.; Kalra, B. *Chem. Rev.* **2001**, *101*, 2097. (b) *Enzymes in Polymer Synthesis*; Gross, R. A.; Kaplan, D. L.; Swift, G., Eds.; Amer. Chem. Soc., Washington, DC, 1998. (c) Kobayashi, S.; Uyama, H.; Kimura, S. *Chem. Rev.* **2001**, 101, 3793.
3. (a) Arnold, F. H.; Wintrode, P.; Miyazaki, K. Gershenson, A. *Trends Biochem. Sciences (TIBS)* **2001**, 26, 100, and references therein. (b) May, O.; Nguyen, P. T.; Arnold, F. H. *Nature Biotechnol.* **2000**, 18, 317. (c) Moore, J.; Arnold, F. H. *Nature Biotechnol.* **1996**, 14, 458.
4. (a) Stemmer, W. P. C. *Proc. Natl. Acad. Sci. USA.* **1994** ,91, 10747. (b) Crameri, A.; S-A. Raillard; E. Bermudez; W. P. C. Stemmer. *Nature.* **1998,** 391, 288. (c) Penning, T. M.; Jez, J. M. *Chem. Rev.* **2001**, *101*, 3027.
5. For example, (a) *High Throughput Screening* (IBC Workshop Notes), IBC, Southborough, MA, 2000. (b) Cohen, N.; Abramov, S.; Dror, Y.; Freeman, A. *Trends Biotechol.* **2001**, *19*, 507. (c) Reetz, M. T.; Jaeger, K.-E. *Topics Current Chem.* **1999**, *200*, 31. (d) Janes, L. E.; Lowendahl, A. C.; Kazlauskas, R. J. *Chem. Eur. J.* **1998**, *4*, 2324.
6. Youvan, D. C.; Coleman, W. J.; Bylina, E. J. *ACS Symp. Ser.* (this volume).
7. (a) Deckert, G. et al *Nature* **1998**, *392*, 353. (b) Callen, W.; Mathur, E. J. WO Patent 9907837A1, 2/18/1999. (c) Short, J. M. U.S. Patent 5,958,672, 9/28/1999. (d) Davis, M. C. *Trends Biotechnol.* **1998**, *16*, 102.
8. *Enzymes in Industry*; Gerhartz W., Ed.; VCH. Weinheim, Germany, 1990, Chapter 3.
9. For example, (a) Kodera, Y; et al *Biotechnol. Lett.* **1998**, *20*, 177. (b) Yang, Z.; Domach, M.; Auger, R.; Yang, F. X.; Russell, A. J. *Enzyme Microb. Technol.* **1996**, *18*, 82. (c) Inada, Y.; et al *Trends Biotechnol.* **1995**, *13*, 86. (d) Takahashi, K.; Saito, Y.; Inada, Y. *J. Am. Oil. Chem. Sci.* **1988**, 65, 911.
10. (a) Wang, P.; Sergeeva, M. V.; Lim, L.; Dordick, J. S. *Nature Biotechnol.* **1997**, *15*, 789. (b) Dordick, J. S.; Novick, S. J.; Sergeeva, M. V. *Chem. Ind. (London)*, **1998**, 17.
11. For example, (a) Bakker, M.; Van de Velde, F.; Van Rantwijk, F.; Sheldon, R. A. *Biotechnol. Bioeng.* **2000**, *70*, 342. (b) Gordon, R. K.; Bhupendra, P. WO Patent 200064957A2, 11/2/2000. (c) Ferreira-Dias, S.; Correia, A.C.; Baptista, F. O. *Bioprocess Eng.* **1999**, *21*, 517.
12. (a) Okahata, Y.; Ijiro, K. *J. Chem. Soc., Chem. Commun.***1988**, 1392. (b) Tsuzuki, W.; Okahata, Y.; Katayama, O.; Suzuki, T. *J. Chem. Soc., Perkin Trans, 1.* **1991**, 1245. (c) Tsuzuki, W.; Sasaki, T.; Suzuki, T. *J. Chem. Soc., Perkin Trans, 2.* **1991**, 1851. (d) Goto, M.; Kameyama, H.; Miyata, M.; Nakashio, F. *J. Chem. Eng. Jpn.***1993**, 26, 109.
13. Goto, M.; Kamiya, N.; Miyata, M.; Nakashio, F. *Biotechnol. Prog.* **1994** 10, 263.

14. (a) Bruno, F.F., Akkara, J.A., Ayyagari, M.,Kaplan, D.L., Gross, R.A., Swift, G., Dordick, J.S., *Macromolecules*,**1995**, 28, 8881. (b) Paradkar, V. M.; Dordick, J. S. *J. Amer. Chem. Soc.* **1994**, *116*, 5009.
15. Xie, J.; Hsieh, Y.-L. *ACS Symp. Ser.* (this volume).
16. Tanaka, K.; Yu, H. *ACS Symp. Ser.* (this volume).
17. For example, (a) Fukuda, H.; Kondo, A.; Fujita, Y.; Tanaka, A.; Ueda, M.; Noda, H. WO Patent 200179483A1, 10/25/2001. (b) Holtzapple, M. T.; Ross, M. K.; Chang, N.-S.; Chang, V. S.; Adelson, S. K.; Brazel, C. *ACS Symp. Ser.* **1997**, *666*, 130. (c) Freer, S. N.; Skory, C. D.; Bothast, R. J. *Recent Res. Dev. Microbiol.* **1998**, *2*, 201. (d) Henk, L.L.; Linden, J. C. *ACS Symp. Ser.* **1994**, *566*, 391.
18. Brooks T. A.; Ingram, L. O. *Biotechnol. Prog.* **1995**, *11*, 619.
19. (a) *Wood-Ethanol Report*; Environment Canada, January 1, 1999; web site: http://www.pyr.ec.gc.ca./ep/wet/index.html. (b) BioTimes®- The Enzyme e-zine; October 15, 2001 (Novozymes). Web: www.novozymesbiotech.com/.
20. Press Release: 11/22/1999. Web: www.dupont.com; www.genencor.com. Also cited in McCoy, M *Chem. Eng. News* **2001**, *79*, 12.
21. Hasegawa, Y.; Tokuyama, T.; Lau, P. C. K. *ACS Symp. Ser.* (this volume).
22. Steinbuchel, A. *ACS Symp. Ser.* (this volume).
23. (a) Chen, X.; Zhang, J.; Kowal, P.; Liu, Z.; Andreana, P. R.; Lu, Y.;Wang, P. G. *J. Amer. Chem. Soc.* **2001**, *123*, 8866. (b) Chen, X.; Fang, J.; Zhang, J.; Liu, Z.; Shao, J.; Kowal, P.; Andreana, P. R.; Wang, P. G. *J. Amer. Chem. Soc.* **2001**, *123*, 2081.
24. Him, J. L. K.; Chanzy, H.; Pelosi, L.; Putaux, J.; Bulone, V. *ACS Sym. Ser.* (this volume).
25. For example, (a) Watkins, K. J. *Chem. Eng. News* **2001** (February 19), *79*, 26. (b) Wu, T. D. *Trends Biotechnol.* **2001**, *19*, 479. (c) Edwards, A. M.; Arrowsmith, C. H.; Pallieres, B. *Modern Drug Discovery* **2000**, *5*, 35. (d) Shuler, M. L. *Biotechnol. Prog.* **1999**, *15*, 287.
26. Dordick, J. S.; Kim, D.; Wu, X. *ACS Symp. Ser.* (this volume).
27. Cheng, H. N.; Q.-M. Gu in *Glycochemistry: Principles, Synthesis, and Applications*; Wang, P. G.; Bertozzi, C. R., Eds.; M. Dekker, New York NY, 2001, pp. 567-579.
28. For example, (a) Knani, D.; Gutman, A.L.; Kohn. D. H. *J. Polymer. Sci. Part A: Polymer Chem.* **1993**, 31, 1221. (b) Shuai, X.; Jedlinski, Z.; Kowalczuk, M.; Rydz, J.; Tan, H. *Eur. Polymer J.* **1999**, 35, 721. (c) Dong, H.; Gui, C-S.; Li, Z-Q.; Han, S.P.; You, D.L.; Shen, J.C. *J. Polym. Sci. A, Polym. Chem.* **1999**, 37, 1265. (d) Kobayashi, S.; Takeya, K.; Suda, S.; Uyama, H. *Macromol. Chem. Phys.* **1998**, 199, 1729. (e) Klibanov, A. M. *Trends Biotechnol.* **1997**, *15*, 97.
29. For example, (a) de Bont, J. A. M. *Trends Biotechnol.* **1998**, *16*, 493. (b) Zijlstra, G. M.; de Gooijer, C. D.; van der Pol, L. A.; Tramper, J. *Enzyme*

Microb. Technol. **1996**, *19*, 2. (c) Chen, J.-P.; Lee, M.-S. *Enzyme Microb. Technol.* **1995**, *17*, 1021.

30. For example, (a) Rajagopal, A. N. *Enzyme Microb. Technol.* **1996**, *19*, 606. (b) Yamada, Y.; Kuboi, R.; Komasawa, I. *Biotechnol. Prog.* **1995**, *11*, 682.
31. For example, (a) Mesiano, A.J.; Beckman, E. J.; Russell, A.J. *Chem. Rev.* **1999**, *99*, 623. (a) Yu, H.-M. et al *J. Chin. Chem. Soc. (Taipei)* **1999**, *46*, 647. (b) Mori, T.; Okahata, Y. *Chem. Commun.(Cambridge)* **1998**, 2215.
32. (a) Wichmann, R.; Wandrey, C.; Kula, M.-R. *Biotechnol. Bioeng.* **1981**, *23*, 2789. (b) Kula, M.-R.; Wandrey, C.; In *Methods in Enzymology*, Vol. 136 (Mosbach, K.; Ed.), pp. 9-21. (c) Seelbach, K.; Kragl, U. *Enzyme Microb. Technol.* **1996**, *20*, 389.
33. (a) Woodley, J. M.; Titchener-Hooker, N.J. *Bioprocess Eng.* **1996**, *14*, 263. (b) Lilly, M. D.; Woodley, J. M. *J. Ind. Microb.* **1996**, *17*, 24. (c) Chauhan, R. P.; Woodley, J. M. *Chemtech* **1997**, *27*(6), 26.
34. For example, (a) Nidetzky, B.; Haltrich, D.; Kulbe, K. D. *Chemtech* **1996**, *26*, 31. (b) Adlercreutz, P. *Biocatalysis and Biotransformation* **1996**, *14*, 1. (c) Seelbach, K.; Riebel, B.; Hummel, W.; Kula, M.-R.; Tishkov,V.; Egorov, A.; Wandrey, C.; Kragl, U. *Tetrahedron Lett.* **1996**, *37*, 1377.
35. For example, (a) Willner, I.; Mandler, D. *Enzyme Microb. Technol.* **1989**, *11*, 467. (b) Vilker, V. L.; Reipa, V.; Mayhew, M.; Holden, M. J. *J. Amer. Oil Chem. Soc.* **1999**, *76*, 1283.
36. For example, (a) Schultz, P. G. *Proc. Nat. Acad. Sci.USA* **1998**, *95*, 14590. (b) Wentworth, P.; Janda, K. D. *Current Opin. Chem. Biol.* **1998**, *2*, 138. (c) Hilvert, D. *Top. Stereochem.* **1999**, *22*, 83.
37. (a) Margolin, A. *Trends Biotechnol.* **1996**, *14*, 223. (b) Lalonde, J. *Chemtech* **1997**, *27*, 38. (c) Zelinski, T.; Waldman, H. *Angew. Chem. Int. Ed. Engl.* **1997**, *36*, 722. (d) Lee, T. S.; Turner, M. K.; Lyle, G. J. *Biotechnol. Prog.* **2002**, *18*, 43.
38. Strauss, E. *Science* **1998**, *282*, 1406.
39. For example, (a) Wiseman, A.; Dalton, H. *Trends Biotechnol.* **1987**, *5*, 241. (b) Yu, C.; Mosbach, K. *J. Org. Chem.* **1997**, *62*, 4057. (c) Uezu, K.; Yoshida, M.; Goto, M.; Furusaki, S. *Chemtech* **1999**, *29*, 12. (d) Lu, Y.; Berry, S. M.; Pfister, T. D *Chem. Rev.* **2001**, *101*, 3047. (e) Kruger, H.-G. *Angew. Chem. Int. Ed.* **1999**, *38*, 627. (f) Yamato, K. et al *Bull. Chem. Soc. Jap.* **2000**, *73*, 903. (g) Halcrow, M. A. et al *Chem Commun.* **1998**, 2465.
40. Monti, D.; Riva, S. *Biocatalysis and Biotransformation* **2001**, *19*, 251.
41. Wandrey, C.; Laue, S. *ACS Polym. Prepr.* **2000**, *41*(2), 1851.
42. Hannink, J.M.; Cornelissen, J.J.L.M.; Farrera, J.A.; Foubert, P.; De Schryver, F.C.; Sommerdijk, N.A.J.M.; Nolte, R. J. M *Angew. Chem. Int. Ed.* **2001**, *40*, 4732.
43. Gutman, A. L.; Zuobi, K.; Boltansky, A. *Tetrahedron Lett.* **1987**, 28, 3861.
44. Brandstadt, K. F.; Saam, J. C.; Sharma, A. *ACS Symp. Ser.* (this volume).

45. Kobayashi, S.; Uyama, H. *ACS Symp. Ser.* (this volume).
46. Nomura, R.; Ueno, A.; Endo, T. *Macromolecules*. **1994**, 27, 620.
47. Bisht, K. S.; Svirkin, Y. Y.; Henderson, L. A.; Gross, R. A.; Kaplan, D. L.; Swift, G. *Macromolecules*. **1997**, 30, 7735.
48. (a) Namekawa, S.; Suda, S.; Uyama, H.; Kobayashi, S. *Int. J. Biol. Macromol.* **1999**, 25,145. (b) Uyama, H.; Kikuchi, H.; Takeya, K.; Kobayashi, S. *Acta Polym.* **1996,** 47, 357. (c) Uyama, H.; Kikuchi, H.; Takeya, K.; Kobayashi, S. *Bull Chem. Soc., Jpn.* **1995,** 68, 56. (d) Uyama, H.; Takeya, K.; Hoshi, N.; Kobayashi, S. *Macromolecules*. **1995**, 28, 7046
49. (a) Bisht, K. S.; Henderson, L. A.; Gross, R. A.; Kaplan, D. L.; Swift, G. *Macromolecules*. **1997**, 30, 2705. (b) Kumar, A, Kalra, B.; Dekhterman, A.; Gross, R.A. *Macromolecules* **2000**, 33, 6303. Kumar, A.; Kalra, B.; Dekhterman, A.; Gross, R.A. *Polymer Preprint* **2000** 41 (2), 1832.
50. (a) Namekawa, S.; Uyama, H.; Kobayashi, S. *Polym. J.* **1996**, 28, 730. (b) Kobayashi, S.; Uyama, H.; Namekawa, S.; Hayakawa, H. *Macromolecules*, **1998**, 31, 5655. (c) Dong, H.; Wang, H-da.; Cao, S-gui.; Shen, J-C. *Biotechnol Letters*. **1998**, 20, 905.
51. (a) Bisht, K. S.; Svirkin, Y. Y.; Henderson, L. A.; Gross, R. A.; Kaplan, D. L.; Swift, G. *Macromolecules*. **1997**, 30, 7735. (b) Al-Azemi, T.F.; Bisht, K.S. *Macromolecules*. **1999**, 32, 6536. (c) Kobayashi, S.; Kikuchi, H.; Uyama. H. *Macromol. Rapid Commun.***1997,** 18, 575. (d) Matsumura, S.; Tsukada, K.; Toshima, K. *Macromolecules*. **1997**, 30, 3122.
52. Bisht, K. S.; Al-Azemi, T. F. *ACS Symp. Ser.* (this volume).
53. Matsumura, S.; Tsukada, K.; Toshima, K. *Int. J. Biol. Macromol.* **1999**, 25, 161.
54. (a) Jedlinski, Z.; Kowalczuk, M.; Adamus, G.; Sikorska, W.; Rydz, J. *Int. J. Biol. Macromol.* **1999**, 25, 247. (b) Kobayashi, S.; Uyama, H.; Namekawa, S. *Polymer Degradation and Stability*. **1998**, 59, 195.
55. (a) Kumar, A.; Gross, R. A. *ACS Polym. Prepr.* **2000**, 41(2), 1863. (b) Kumar, A.; Gross, R. A. *J. Am. Chem. Soc.* **2000**, 122, 11767. (c) Kumar, A.; Kalra, B.; Gross, R. A. *ACS Symp. Ser.* (this volume).
56. For example, (a) King, E. C.; Watkins, W. J.; Blacker, A. J.; Bugg, T. D. H. *J. Polym. Sci. Part A; Polym. Chem.* **1998**, *36*, 1995. (b) Munoz-Guerra, S.; Melis, J.; Perez-Camero, G.; Bou, J.J.; Congregado, F. *ACS Polym. Prepr.* **1998**, *39*(2), 138. (c) Kunioka, M.; Goto, A. *Appl. Microbiol. Biotechnol.* **1994**, *40*, 867.
57. (a) Lee, S. Y. *Trends Biotechnol.* **1996**, *14*, 431; *Nature Biotechnol.* **1997**, *15*, 17. (b) Boswell. C. *Chem. Marketing Reporter* August 27, 2001. (c) Cho, K.-S.; Ryu, H. W.; Park, C.-H.; Goodrich, P.R. *Biotechnol. Lett.* **1997**, *19*, 7. (d) Ondrey, *Chem. Eng.* **1997**, *104*(1), 23. (e) Purushothaman, M.; Anderson, R.K.I.; Narayana, S.; Jayaraman, V. K. *Bioprocess Biosys. Eng.* **2001**, *24*, 131.

58. Lutke-Eversloh, T.; Kawada, J.; Marchessault, R. H.; Steinbuchel, A. *Biomacromolecules* **2002**, *3*, 159.
59. Kelley, A. S.; Srienc, F. *ACS Symp. Ser.* (this volume).
60. Some recent references include: (a) Uyama, H.; Maruichi, N.; Tonami, H.; Kobayashi, S. *Biomacromolecules* **2002**, *3*, 187. (b) Mita, N.; Oguchi, T.; Tawaki, S.; Uyama, H.; Kobayashi, S. *ACS Polym. Prepr.* **2000**, *41*(1), 223. (c) Tonami, H.; Uyama, H.; Kobayashi, S.; Fujita, T.; Taguchi, Y.; Osada, K. *Biomacromolecules.* **2000**, *2*, 149. (d) Uyama, H.; Kurioka, H. N.; Kobayashi, S. *Colloids and Surfaces A: Physiochem. and Eng. Aspects.* **1999**, *153*, 189. (e) Michon, T.; Wang, W.; Ferrasson. E.; Gueguen, J. *Biotechnol. and Bioeng.* **1999**, *63*, 449. (f) Michon, T.; Chenu, M.; Kellershon, N.; Desmadril, M.; Gueguen, J. *Biochemistry.* **1997**, *36*, 8504.
61. (a) Ikeda, R.; Sugihara, J.; Uyama, H.; Kobayashi, S. *Macromolecules* **1996**, 29, 8702. (b) Ikeda, R.; Uyama, H.; Kobayashi, S. *Macromolecules.* **1996,** 29, 3053.
62. Alva, K.S.; Kumar, J.; Marx, K.A.; Tripathy, S. *Macromolecules* **1997**, 30, 4024. (b) Kumar, J.; Tripathy, S.; Senecal, K.J.; Samuelson, L. *J. Am. Chem. Soc.* **1999**, *121*, 71. (c) Tripathy, S. *Chem. Eng. News.* **1999**, *77*, 68.
63. (a) Derango, A.R.; Chiang, L-C.; Dowbenko, R.; Lasch, J.G. *Biotechnol. Tech.* **1992**, *6*, 523. (b) Uyama, H.; Lohavissavapanich, C.; Ikeda, R.; Kobayashi, S. *Macromolecules* **1998**, 31, 554. (c) Emery, O.; Lalot, T.; Brigodiot, M.; Marechal, E. *J Polym. Sci.,Part A, Polym. Chem.* **1997**, 35, 3331. (d) Teixeira, D.; Lalot, T.; Brigodiot, M.; Marechal, E. *Macromolecules* **1999**, 32, 70.(e) Kalra, B.; Gross, R.A. *ACS Polym. Prepr.* **2000**, 41(1), 213. (f) Kalra, B.; Gross, R.A. *Biomacromolecules,* **2000**, 1(3), 501. (g) Kalra, B.; Gross, R. A. *ACS Polym. Prepr.* **2000,** 41(2),1828. (h) Kalra, B.; Gross, R. A. *ACS Polym. Prepr.* **2000,** 41(2), 1935.
64. Kalra, B.; Gross, R. A. *ACS Symp. Ser.* (this volume).
65. For example, (a) *Glycochemistry*; Wang, P. G.; Bertozzi, C. R. Eds.; M. Dekker, New York, 2001. (b) *Preparative Carbohydrate Chemistry*; Hanessian, S. Ed.; M. Dekker, New York, 1997. (c) *Modern Methods in Carbohydrate Synthesis*; Khan, S. H.; O'Neill, R. A. Eds.; Harwood Academic, Amsterdam, 1996. (d) *Protective Groups in Organic Synthesis, 3rd Ed.*; Greene, T. W.; Wuts, P. G. M.; Wiley, New York, 1999. (e) Barresi, F.; Hindsgaul, O. In *Modern Synthetic Methods*; Ernst, B.; Leumann, C. Eds.; VCH, Weinheim, 1995.
66. For example, (a) Wang, P. G.; Fitz, W.; Wong, C.-H. *Chemtech* **1995** (April), 22. (b) Guo, Z.; Wang, P. G. *Appl. Biochem. Biotechnol.* **1997**, *68*, 1. (c) Palcic, M. *Curr. Opin. Biotechnol.* **1999**, *10*, 616, and refs. therein.
67. Shoda, S.; Kobayashi, S. *Trends Polym. Sci.* **1997**, *5*, 109.
68. For example, (a) Brown, R. M.; Saxena. I. M. *Plant Physiol. Biochem.* **2000**, *38*, 57. (b) Embuscado, M. E.; Marks, J. S.; BeMiller, J. N. *Food*

Hydrocolloids **1994**, *8*, 407. (c) Cheng, H. N.; Takai, M.; Ekong, E. A. *Makromol. Symp.* **1999**, *140*, 145. (d) Watanabe, K.; Tabuchi, M.; Ishikawa, A.; Takemura, H.; Tsuchida, T.; Morinaga, Y.; Yoshinaga, F. *Biosci. Biotechnol. Biochem.* **1998**, *62*, 1290.

69. For example, (a) Saxena, I. M.; Brown, R. M. *J. Bacteriol.* **1995**, *177*, 5276. (b) Volman, G.; Ohana, P.; Benziman, M. *Carbohydrates in Europe* **1995** (May), 20. (c) Tajima, K.; Fujiwara, M.; Takai, M. *Macromol. Symp.* **1995**, *99*, 149. (d) Ogawa, R.; Sato, M.; Miura, Y.; Fujiwara, M.; Takai, M.; Tokura, S.; In *Cellulose: Chemical, Biochemical and Material Aspects*; Kennedy, J. F. et al. Eds.; Ellis Horwood, New York, 1993; p.35.
70. For example, (a) *Industrial Gums*, 3rd Ed; Whistler, R. L.; BeMiller, J. N., Eds.; Academic Press: San Diego, CA, 1993. (b) *Food Polysaccharides and their Applications*; Stephen, A. M., Ed.; M. Dekker; New York, NY, 1995. (c) *Application of Polymers in Foods*; Cheng, H. N.; Cote, G. L.; Baianu, I. C., Eds.; (Macromolecular Symposia 140); Wiley-VCH; Weinheim, Germany, 1999.
71. Defrees, S.; Johnson, K. WO Patent 200029603A2, 5/25/2000.
72. Xie, W.; Wang, P. G. *ACS Symp. Ser.* (this volume).
73. Cheng, H. N.; Gu, Q.-M. *ACS Symp. Ser.* (this volume).
74. Xie, J.; Hsieh, Y.-L. *ACS Symp. Ser.* (this volume).
75. Payne, G.F.; Vachoud, L.; Chen, T.; Govar, J. *ACS Symp. Ser.* (this volume)
76. Gu, Q.-M. *ACS Symp. Ser.* (this volume).
77. Hu, S.; Gao, W.; Kumar, R.; Gross, R. A.; Gu, Q.-M.; Cheng, H. N. *ACS Symp. Ser.* (this volume).
78. For example, (a) Ingvorsen, K.; Waino, M.; Rasmussen, L. H. WO Patent 0130984A2, 5/3/2001. (b) Turk, S.C.; Gerrits, N.; Smeekens, J.; Weisbeek, P.J. WO Patent 9729186A1, 8/14/1997. (a) *Cellulose Derivatives. Modification, Characterization, and Nano-structures*; Heinze, T. J.; Glasser, W. G., Eds.; (ACS Symp. Ser. 688); American Chemical Society, Washington, DC, 1998. (b) *Industrial Polysaccharides*; Yalpani, M., Ed.; Elsevier, Amsterdam, NL, 1987.
79. Gelman, R. A. *J. Appl. Polym. Sci.* **1982**, *27*, 2957.
80. For example, (a) Benen, J.A.E.; Kester, H.C.M.; Parenicova, L.; Visser, J. *Biochemistry* **2000**, *39*, 15563. (b) Whitaker, J. R. *Food Sci. Technol.(NY)* **1994**, *61*, 425.
81. For example, (a) Wong, T.Y.; Preston, L.A.; Schiller, N. L. *Annu. Rev. Microbiol.* **2000**, *54*, 289. (b) Moe, S.T.; Draget, K.I.; Skjak-Braek, G.; Smidsrod, O. In *Food Polysaccharides and Their Applications*; Stephen, A.M.; Ed.; Dekker: New York, NY, 1998, pp 245-286.
82. For example, (a) *Enzymes for Pulp and Paper Processing*; Jeffries, T. W.; Viikari, L., Eds.; (ACS Symp. Ser. 655), Amer. Chem. Soc., Washington, DC 1996. (b) Bajpai P. *Biotechnol. Prog.* **1999**, *15*, 147. (c) Call, H. P.;

Mucke, I. *J. Biotechnol.* **1997**, *53*, 163. (d) Jung, H.; Xu, F.; Li, K. *Enzyme Microb. Technol.* **2002**, *30*, 161.

83. For example, (a) Liu, S.; Suflita, J. M. *Trends Biotechnol.* **1993**, *11*, 344. (b) Renner, S. *Environ. Sci. Technol.* **1997**, *31*, 188A. (c) *Photoremediation of Soil and Water Contaminants*; Kruger, E. L.; Anderson, T.A.; Coats, J.R. (ACS Symp. Ser. 664), Washington, DC, 1997.
84. Cheng, Y.; Prud'homme, R. K. *ACS Symp. Ser.* (this volume).
85. Norwood, D. *ACS Symp. Ser.* (this volume).
86. (a) Gutman, A. L.; Zuobi, K.; Bravdo, T. *J. Org. Chem.* **1990**, 55, 3546. (b) Gutman, A. L.; Zuobi, K.; Boltansky, A. *Tetrahedron Lett.* **1987**, 28, 3861. (c) Gutman, A.L.; Oren, D.; Boltanski, T.; Bravdo, T. *Tetrahedron Lett.* **1987**, 28, 5367.
87. Kazlauskas, R. J. U.S. Patent 4,879,421, 11/7/1989.
88. Abramowicz, D. A.; Keese, C. R. U.S Patent 4,892,822, 1/9/1990.
89. (a) Gray, K. A. et al *Adv. Synth. Catal.* **2001**, *343*, 607. (b) Swanson, P. E. *Curr. Opin. Biotechnol.* **1999**, *10*, 365. (c) Swanson, P. E. U.S. Patent 5372944, 12/13/94.
90. (a) Johnson, B. F.; Amaratunga, M.; Lobos, J. H. U. S. Patent 6,114,157, 9/5/2000. (b) Amaratunga, M.; Lobos, J. H.; Johnson, B. F.; Williams, E. D. U.S. Patent 6,030,819, 2/29/2000.
91. For example, (a) Meade, T. J. U.S.Patent 5,340,913, 8/23/1994. (b) Mobley, D. P.; Shank, G. K. U.S. Patent 6,066,480, 5/23/2000. (c) Mobley, D. P.; Dietrich, D. K. U.S.Patent 5,049,496, 9/17/1991.(d) Spivack, J. L.; Salvo, J. J. U.S. Patent 5,105,025, 4/14/1992. (e) Lobos, J. H.; Leib, T. K.; Su, T.-M. U.S. Patent 5,132,228, 7/21/1992.
92. (a) Shibatani, S.; Kitagawa, M.; Tokiwa, Y.; *Biotechnol. Letts.* **1997**, 19, 511. (b) Kitagawa, M.; Tokiwa, Y. *Carbohydrate Lett.* **1997**, 2, 893. (c) Tokiwa, Y.; Kitagawa, M.; Fan, H.; Raku, T.; Hiraguri, Y.; Shibatani, S.; Kurane, R. *Biotechnol. Tech.* **1999**, 13,173.
93. Kitagawa, M.; Tokiwa, Y.; Raku, T.; Fan, H. *ACS Symp Ser.* (this volume).
94. Tokiwa, T.; Fan, H.; Raku, T.; Kitagawa, M. *ACS Symp. Ser.* (this volume).
95. Patil, D.R.; Dordick, J.S.; Rethwisch,D.G. *Macromolecules*, **1991**, 24, 3462.
96. Gruning, B.; Hills, G.; Josten, W.; Schaefer, D.; Silber, S.; Weitemeyer, C. U.S. Patent 6,288,129, 9/11/2001.
97. Namekawa, S.; Suda, S.; Uyama, H.; Kobayashi, S. *Int. J. Biol. Macromol.* **1999**, 25,145.
98. Cordova, A.; Iversen, T.; Hult, K. *Polymer.* **1999**, 40, 6709.
99. (a) Pavel, K.; Ritter, H. In *Enzyme in Polymer Synthesis*; Gross, R.A.; Kaplan, D. L.; Swift, G. Eds.; Washington, DC. American Chemical Society. **1996**. pp.200. (b) Pavel, K.; Ritter, H. *Makromol. Chemie.* **1991**, 192, 1941. (c) Geyer, U.; Klemm, D.; Pavel, K.; Ritter, H. *Makromol. Chem. Rapid Commun.* **1995**, 6, 337.

100. Kumar, R.; Gross, R. A. *ACS Symp. Ser.* (this volume).
101. For example, (a) Hansler, M.; Jakubke, H.-D. *J. Peptide Sci.* **1996**, *2*, 279. (b) Halling, P. J.; Eichhorn, U.; Kuhl, P.; Jakubke, H.-D. *Enzyme Microb. Technol.* **1995**, *17*, 601. (c) Eichhorn, U.; Bommarius, A.S.; Drauz, K.; Jakubke, H.-D. *J. Peptide Sci.* **1997**, *3*, 245. (d) *Enzymatic Peptide Synthesis*; Kullman, W.; CRC Press, Boca Raton, FL, 1987.
102. For example, (a) Jackson, D. Y.; Varnier, J.; Quan, C.; Stanley, M.; Tom, J.; Wells, J. A. *Science* **1994**, *266*, 243. (b) Waldman, H.; Sebastian, D. *Chem. Rev.* **1994**, *94*, 911. (c) Schelhaas, M.; Glomsda, S.; Hansler, M.; Jakubke, H.-D.; Waldman, H. *Angew. Chem. Int. Ed. Engl.* **1996**, *35*, 106. (d) Sears, P.; Wong, C. H. *Biotechnol. Prog.* **1996**, *12*, 423.
103. Wang, P.; Dordick, J. A. *Macromolecules* **1998** 31, 941.
104. For example, (a) Morse, D. E. *Trends Biotechnol.* **1999**, *17*, 230. (b) Tacke, R. *Angew. Chem. Int. Ed.* **1999**, *38*, 3015. (c) Cha, J. N.; Stucky, G. D.; Morse, D. E.; Deming, T. J. *Nature* **2000**, *403*, 289.
105 (a) Potera, C. *Genetic Eng. News* **2001**, *21*, 1. (b) Anon. *Chem. Eng. News* **2001** (October 29), 13.
106. (a) *Industrial Biotransformations*; Liese, A.; Seelbach, K.; Wandrey, C.; Wiley-VCH, Weinheim, 2001. (b) Schoevaart, R.; Kieboom, T. *Chem. Innovation* **2001**, *31*, 33. (c) Thayer, A. M. *Chem. Eng. News* **2001** (May 21), *79*, 27. (d) Rouhi, A. M. *Chem. Eng. News* **2002** (February 18), *80*, 86.

Development of Novel Methodologies

Chapter 2

Generation of Environmentally Compatible Polymer Libraries via Combinatorial Biocatalysis

Dae-Yun Kim, Xiaoqiu Wu, and Jonathan S. Dordick*

Department of Chemical Engineering, Rensselaer Polytechnic Institute, Troy, NY 12180

A combinatorial strategy for biocatalytic polymer synthesis has been demonstrated. Two polymer libraries were synthesized. In one, the polycondensation of esters (C_3-C_{10}) with polyhydroxylated compounds (e.g., diols, sugars, nucleic acids, and a natural steroid diol) was performed. The lipase from *Candida antarctica* in acetonitrile was capable of catalyzing the polycondensation of these monomers to give polymers with Mw's as high as 20,000 Da, including the preparation of novel sugar-containing polyesters. In the second library, soybean peroxidase was used to catalyze the oxidative polymerization of phenols to yield materials that bound and sensed metal ions. Histograms were developed from the fluorescent response of the library constituents to yield fingerprints for specific metals. These two libraries, along with their methods of preparation, provide a new paradigm for functional polymer discovery that may have applications in environmentally-benign materials and sensor arrays.

Enzymatic polymerizations (*1*) have become an effective method for the synthesis of polyphenols (*2-5*), polyesters (*6-9*), and polysaccharides (*10*). Coupled with chemoenzymatic methods (*11*), the scope of biocatalytic polymer synthesis expands even further to include poly(acrylates, acetylenes, ols) among others. In these transformations, enzymes are capable of displaying a high degree of selectivity (e.g. stereo-, regio-, and chemoselectivity), which is a critical advantage over chemical routes to polymer synthesis. Moreover, enzymes operate under conditions that favor environmentally benign and energy efficient synthesis.

To capture the full universe of biocatalytic transformations, we have chosen a path that has been used by nature to synthesize and select for unique materials. Scientists and engineers have applied combinatorial strategies for pharmaceutical and agrichemical discovery, and recently, materials, which include catalysts, dielectric materials, and polymers (*12*). We have now extended this methodology to include enzymatic polymerizations. With the application of combinatorial biocatalysis in polymer synthesis, we are able to synthesize diverse libraries of polymers, and shorten time scales for polymeric materials discovery.

MATERIALS

Materials. All materials and enzymes were purchased from Sigma or Aldrich (St. Louis, MO), unless otherwise stated. Solvents were dried over 3 Å molecular sieves for at least 24 h prior to use for the removal of residual water. The 96-deep well plates were purchased from Alltech associates, Inc (Deerfield, IL). Divinyladipate and divinylsebacate were purchased from TCI America (Portland, OR) and Polysciences, Inc. (Warrington, PA), respectively. Bis(2,2,2-trifluoroethyl)malonate (15.1 g, 39 % yield) and bis(2,2,2-trifluoroethyl)glutarate (19.2 g, 45% yield) were synthesized from a previously published method (*13*). Novozym-435 (lipase B from *Candida antarctica* immobilized on an acrylic resin) and the alkaline protease, Protex-6L, was obtained as gifts from Novo Nordisk Bioindustries (Bagsvaerd, Denmark) and Genencor International (Rochester, NY), respectively.

Enzyme-catalyzed polycondensation reactions. Enzymes (20 mg/ml except for freeze-dried Protex-6L, which was employed at a concentration of 5 mg/ml) were screened in different solvents for their ability to catalyze the transesterification and polymerization of aliphatic and aromatic diols (0.2 M) dissolved in 1.5 ml of a suitable solvent containing 0.2 M divinyladipate (DVA). A 96-deep well plate was shaken at a 45° angle at 250 rpm and 45°C for 24 h in an orbital shaker. The polymers were isolated by evaporating the reaction solvent in a vacuum oven and subsequently washed in methanol. The pellets were dried, redissolved in DMF, and filtered to remove the enzyme. The filtrate was then placed in an identical 96-deep well plate. For reactions performed in the full array, Novozym-435 (25 mg/ml) was added to a mixture of equimolar concentrations (0.2 M) of an activated diester and a hydroxyl-containing compound dissolved in 1.5 ml of acetonitrile. After 88h the plate was worked up similarly as above. The large-scale synthesis of poly(sorbitol adipate) was performed as follows: 0.2 M sorbitol and 0.2 M DVA were dissolved in 0.2 L of CH_3CN and the reaction was initiated upon addition of 5 g of *C. antarctica* lipase. After 93 h, the polymer was isolated and worked up as described above.

Enzyme-catalyzed polyphenol reactions. Soybean peroxidase (SBP) catalyzed synthesis of a polyphenol array was performed in aqueous buffer (50 mM Bis-Tris propane, pH 7.0, containing 20-33% (v/v) DMF) with five simple phenols as listed in Table 1. The reaction mixture contained 10 mM of a phenol (for homopolymers), or 5 mM each of two phenols (for copolymer), and 0.1 mg/ml SBP in a volume of 100 mL. The reaction was initiated by pumping 1 ml H_2O_2 into a phenol and SBP mixture to obtain a final concentration of 10 mM H_2O_2 over a period of 4 h. For reactions with charged phenols (e.g., *p*-hydroxyphenylacetic acid, and *p*-hydroxybenzoic acid), the reaction volume was decreased to 10 mL and the phenolic concentration was increased to 100 mM to ensure precipitate formation during polyphenol synthesis. The precipitated polymers were collected and washed thoroughly with deionized water and filtered through 0.45 μm pore size centrifuge filters from Alltech.

ANALYTICAL MEASUREMENTS

Quantitative analysis of acyl donors, sugars, and diols was performed by gas chromatography (Shimadzu GC-17A) with an AT-1 (Alltech) capillary column (30 x 0.25, 0.10 μm) and helium as the carrier gas. Determinations of polymer molecular weight (M_n and M_w) were made by gel permeation chromatography using a Shimadzu LC-10A LC system (Columbia, MD) differential refractometer (Waters) with the column calibrated using poly(ethylene glycol) standards in DMF. ^{13}C and ^{1}H -NMR spectral data were collected using a Varian Unity 500 NMR spectrometer (Palo Alto, CA).

Measurement of fluorescence response of the polymer array to metal ions. Phenolic polymers were distributed uniformly in 50 mM Bis-Tris propane buffer (pH 7.0). The effect of a metal on the fluorescence of a polyphenol was monitored by fluorescence intensities at specific maximum emission wavelengths

(MEW). The specific response of fluorescence intensity upon the perturbation of metal binding was defined as ΔF in the following equation (Eq. 1):

$$\Delta F = \frac{I_o - I}{I_o} \tag{1}$$

where I and I_0 represent emission intensities at MEWs in the presence and absence of metal ions, respectively. Each ΔF was calculated from a minimum of three independent measurements of I and I_0.

RESULTS AND DISCUSSION

Enyzme-catalyzed polycondensation reactions

Library 1: Divinyl adipate reacted with four diols (3 enzymes x 3 solvents x 4 diols)

An initial screen for combinatorial enzymatic polymerizations used several biocatalysts in different reaction media. With the advent of 96-well plates and automated instruments, screening and the entire polymer discovery process can be conducted efficiently using parallel synthesis. We first constructed a simple polymer array from one acyl donor, divinyl adipate (DVA), and four diols (ethylene glycol, 1,4-butanediol, 1,8-octanediol, and 1,4-benzenedimethanol). Upon review of the literature for enzymatic activity toward polycondensation reactions (*14-16*), three enzymes (Novozym-435, Porcine pancreatic lipase, and Protex-6L) were chosen for their abilities to catalyze the transesterification of DVA with these four diols in three solvents (acetone, acetonitrile, and *tert*-amyl alcohol).

In all cases, oligo-condensation was observed. In addition, as a control, a replicate array was constructed and in the absence of enzyme, no reaction was observed (Figure 1). Reactions catalyzed by *C. antarctica* lipase consistently displayed the highest activity in all three solvents. Porcine pancreatic lipase (PPL) and Protex-6L exhibited smaller molecular weights, which is consistent with lower conversions in step-wise polycondensation reactions. *C. antarctica* lipase catalyzed polymerizations showed M_w in the range of 330 to as high as 6,640, corresponding to a degree of polymerization (DP) of 17. The polydispersity index (PDI) ranged from 1.03 for the smallest oligomers to 1.8 for the larger polymers. Complete conversion of monomer was obtained for *C. antarctica* lipase in acetonitrile, and therefore used in subsequent studies for the construction of the next library.

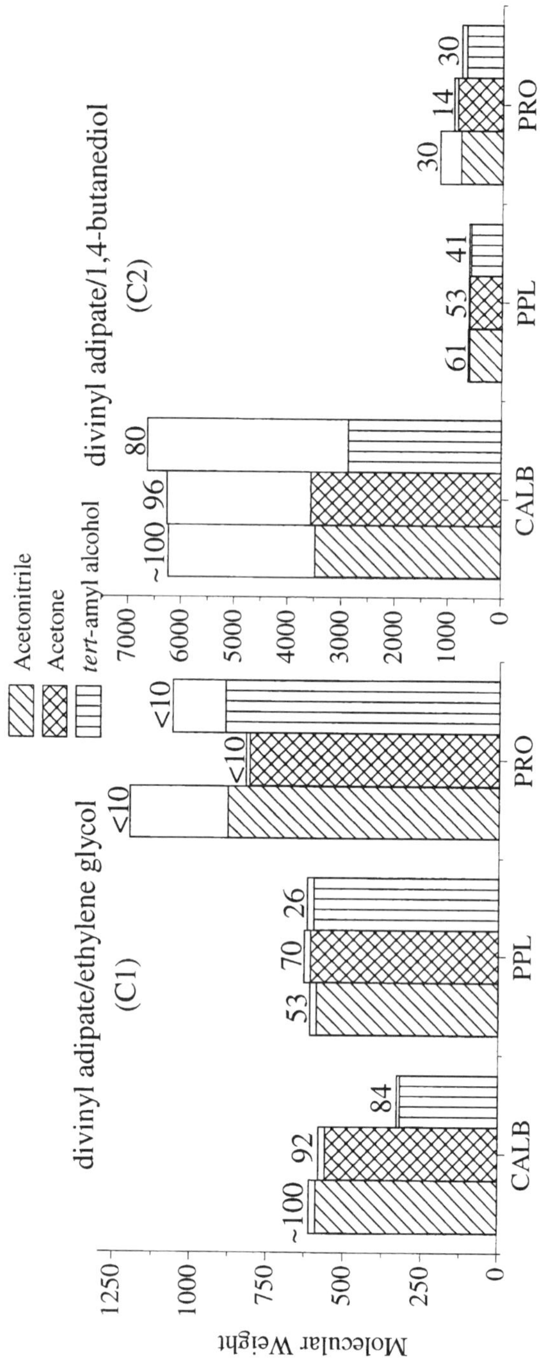

Acetonitrile
Acetone
tert-amyl alcohol
divinyl adipate/ethylene glycol
(C1)
divinyl adipate/1,4-butanediol
(C2)
Molecular Weight
1250
1000
750
500
250
0
7000
6000
5000
4000
3000
2000
1000
0
~100
92
84
53
70
26
<10
<10
<10
~100
96
80
61
53
41
30
14
30
CALB
PPL
PRO
CALB
PPL
PRO

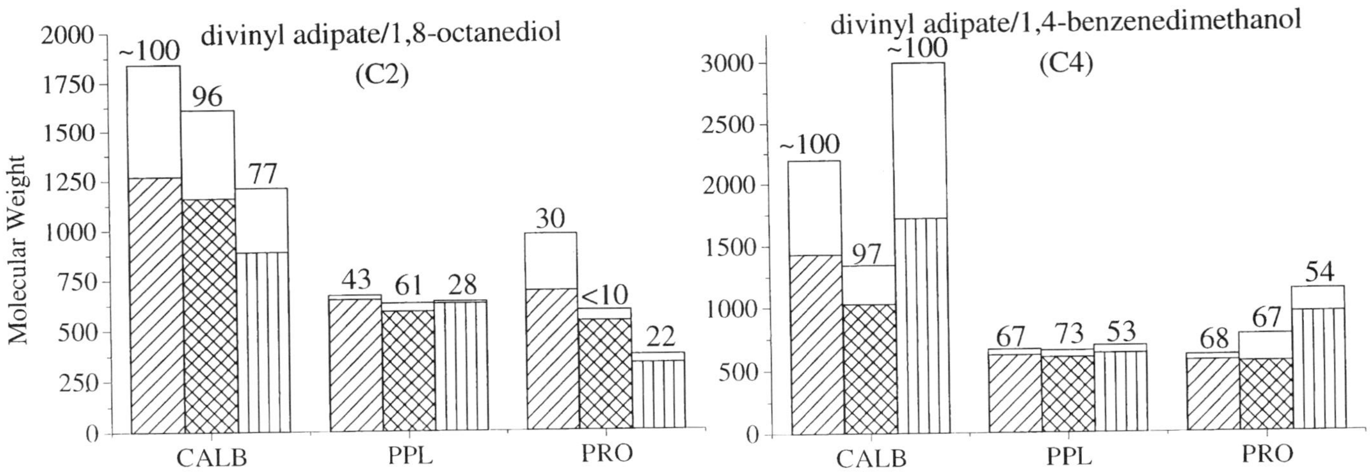

Figure 1. Library 1: Synthesized using catalysts from different sources [C. antarctica lipase (CALB), Porcine pancreatic lipase (PPL), and Protex-6L (PRO)] and by varying reaction mediums (acetonitrile, acetone, and tert-amyl alcohol). M_n=shaded portion; M_w=shaded+clear portion; Numbers refer to conversion of monomer.

Library 2: Activated esters reacted with various acyl acceptors (4 acyl donors x 12 acyl acceptors)

Biocatalyst reaction conditions were the main focus of the first library but it did not highlight the diverse number of polymers that is able to be enzymatically synthesized. To generate polymers with different architectures, we chose an additional eight diols or polyols, whose structures are shown in Figure 2. Therefore, a total of twelve acyl acceptors, which consisted of the same aliphatic and aromatic diols as in Library 1, along with carbohydrates (mono and disaccharides), nucleic acids, and a natural steroid diol, were used. We also included four straight-chain diesters as acyl donors with sizes ranging from C_3-C_{10}. This array was then duplicated with identical reaction mixtures, but without enzyme, and this half of the plate served as a necessary series of controls. Reactions were initiated with the addition of 20 mg/ml *C. antarctica* lipase. After 88h, the reactions were terminated and worked up.

A range of polymer sizes was obtained from this array with aliphatic (**1-3**) and aromatic (**4**) acyl acceptors yielding appreciable molecular weights (Figure 3). The M_w ranged from 1,540 to as high as 6,460 for the polymer synthesized from bis(2,2,2-trifluoroethyl) malonate and ethylene glycol. *C. antarctica* lipase displayed diversified polymerization activity towards the other eight acyl acceptors (sugars/nucleic acids/steroid diol), for example, rather small esters were formed with the disaccharides and all four acyl donors. Notably, polymers formed from sorbitol (**8**) with the four-acyl donors (**A-D**) displayed the highest molecular weights, which ranged from 7,000 to greater than 20,000 Da. On a weight average basis, up to 50 sorbitol molecules were incorporated into the polyester. Sugar containing polyesters are important for its interesting properties, which include a high number of hydroxyl moieties resulting in increased hydrophilicity, and may have importance in hydrogel applications (*17,18*).

The hallmark selectivity of enzymes, specifically *C. antarctica* lipase, is highlighted in this library. Polymers formed from sorbitol and mannitol showed different molecular weights, with the former exhibiting larger polymer formation These two sugar alcohols only differ in the orientation of the hydroxyl moiety at the C-2 position. Hence, *C. antarctica* lipase is able to differentiate slight variations in structure, and highlights the discriminating catalytic power of biocatalytic polymer formation over acid or base-catalysis. The enzyme-catalyzed polyester array was able to uncover the high reactivity and unusually high molecular weight for sorbitol-containing polymers indicating that this approach appears to have a role in new polymer discovery.

Detailed studies were conducted into four specific wells of the lipase-catalyzed array. Two acyl donors, DVA and divinyl sebacate (DVS), were each reacted with 1,8-octanediol and sorbitol with conversion recorded as a function of time. The reactions proceeded quickly with complete conversion of monomer

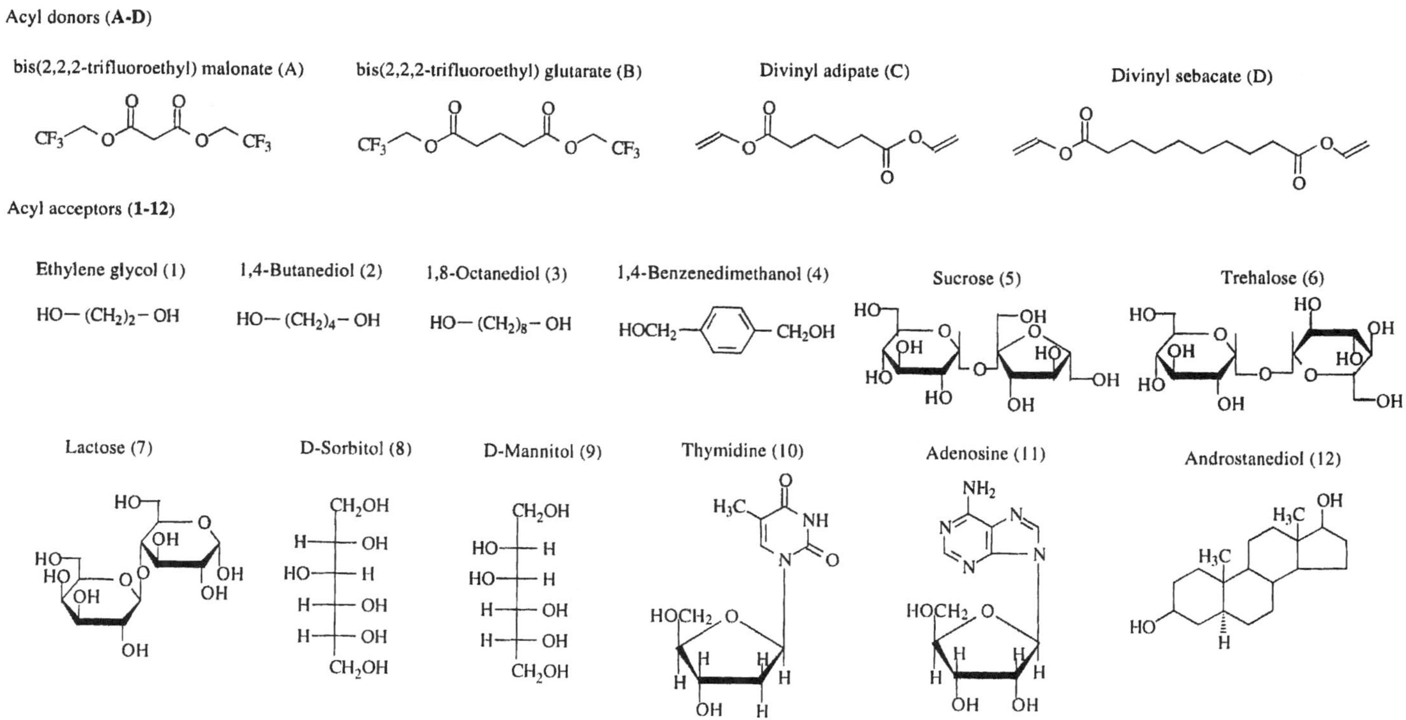

Figure 2. Array of acyl donors and acyl acceptors polycondensed for the synthesis of polymer libraries.

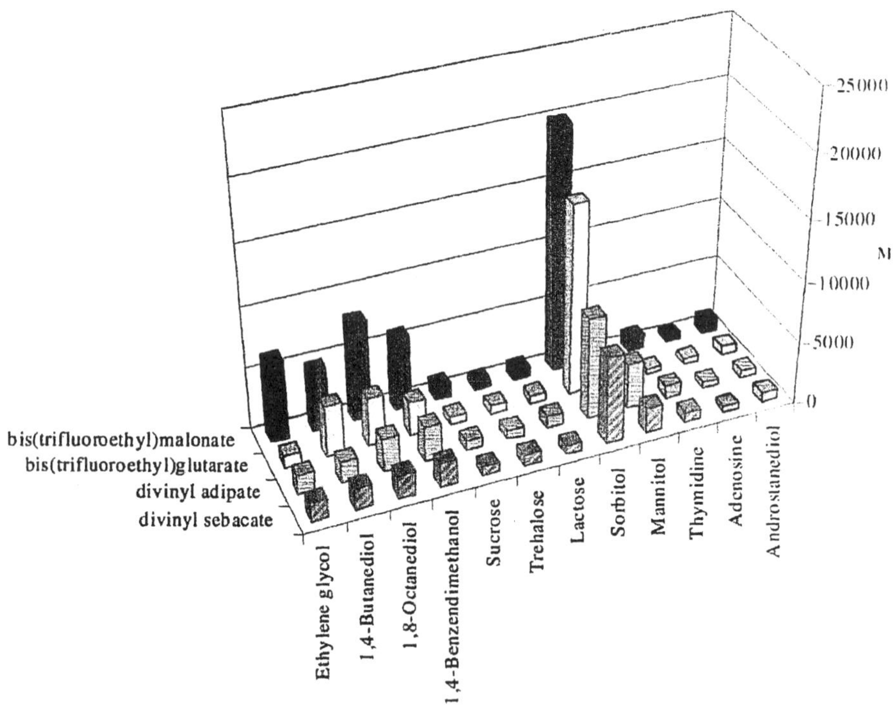

Figure 3. Library 2: Array synthesized from a diverse number of monomers (4 acyl donors x 12 acyl acceptors) to highlight the wide range application of enzymatic polymerizations.

within 25 min and 1 h for DVA/octanediol and DVS/octanediol, respectively. Monomer depletion was dependent on both acyl donor and polyol size with longer reaction times required for sorbitol. The extended monomer reaction times were 2 h and 10 h, for DVA/sorbitol and DVS/sorbitol, respectively.

Large-scale poly(sorbitol adipate) synthesis was conducted in order to investigate the regioselectivity of *C. antarctica* lipase. The polymer was isolated, as described in the materials and methods, and 5.7 grams of polymer was obtained (isolated yield of 49%). Structural analysis (^{13}C NMR) revealed the terminal carbons on the sugar polyol shifting downfield after polymerization from 62.6 to 65.8 ppm and 63.5 to 66.6 ppm for C-1 and C-6, respectively. Carbons adjacent to the primary hydroxyls indicated an upfield shift from 72.4 to 70.7 ppm and 73.8 to 71.1 ppm for C-2 and C-5, respectively. Furthermore, the chemical shift corresponding to the vinyl group on the acyl donor disappeared from the polymer spectrum completely after reaction. Therefore, as result of acylated carbons shifting downfield and adjacent carbons shifting upfield (*19*) a linear polymer was formed from the selective acylation of the terminal hydroxyls on sorbitol. In addition, Uyama and coworkers (*20*) report the synthesis of a similar polymer with sorbitol, poly (sorbitol sebacate), with an analogous conclusion of *C. antarctica* lipase regioselectively acylating the sugar alcohol at the terminal 1- and 6- positions.

Polymer molecular weights were also investigated as a function of time. The M_w for the reaction between 1,8-octanediol and DVA reached a maximum of 1,600 Da and decreased to 1,200 Da after 94 h (Figure 4). The M_w for the reaction between sorbitol and DVA reached a maximum of 21,700 Da and decreased to 13,600 Da after 94 h. These results were unusual since AA-BB polycondensation reactions proceed in a step-wise fashion, whereby the depletion of monomer occurs to produce dimers, trimers, and oligomers, which then condense to generate polymers. However, after 25 min and 2 h, for octanediol and sorbitol, repectively, the molecular weight decreased perhaps due to lower reactivity of the lipase towards higher molecular DVA/octanediol and DVA/sorbitol oligomers, relative to the monomers. Moreover, the polymer size decreases over time, well after the depletion of monomers. We speculate that a small amount of water, introduced either by the solvent and/or the biocatalyst, occurs and this results in partial depolymerization. Therefore, the lipase may catalyze the hydrolysis of polymer, resulting in ester cleavage and decreased polymer sizes.

Russell and coworkers (*21*) extensively investigated the hydrolysis and depolymerization phenomena of biocatalytic AA-BB polycondensed materials. Their discussions present insight into the hydrolysis of DVA when reacted with 1,4-butanediol. The hydrolysis of DVA during small oligomer formation is not significant, but as the polymerization progresses and as the monomer concentration decreases, the enzyme bound water is involved in hydrolysis of the vinyl end group, thereby limiting the extent of polymerization. Our results are consistent with Chaudhary et al's (*21*) that water has an important role in preventing high molecular weight formation.

Enzyme-catalyzed polyphenol reactions

Peroxidases may be thought of as the ideal combinatorial biocatalyst due to their extraordinarily broad substrate specificity. Peroxidase catalyzes the oxidative polymerization of phenols to generate phenolic polymers in the absence of formaldehyde, and peroxidase catalysis has proven to be a highly effective method to synthesize polyphenols (*5, 22, 23*). The availability of a wide variety of phenolic monomers and the broad substrate specificity of peroxidases enables a single enzyme to be used to generate a combinatorially diverse array of polyphenols (*24, 25*). Polyphenols are known to interact with metal ions, and the resulting polymer-metal complexes show altered fluorescence properties as compared with the uncomplexed polyphenols (*26, 27*). Therefore, enzymatically and combinatorially generated phenolic polymer libraries have application as a sensor array for metal ions.

Sensing for metal ions has been a rather demanding research area. Metal ions are ubiquitous in nature and also represent a significant man-made pollutant (*28, 29*). Metal ions often exist in mixtures and different metal ions may respond similarly to a given sensing material. The emerging technology of 'electronic noses' employs an array of rather non-selective sensors that utilizes a recognition technique to identify patterns of sensor elements that respond to specific analytes (*30, 31*). Such a design mimics the mammalian olfactory system, which consists of a large array of receptors (*32, 33*). Odorants interact with a combination of receptors and the pattern of receptor-odorant interactions is processed in the brain (*34*).

In library 3, we utilize sensor arrays composed of phenolic polymers generated by combinatorial catalysis using SBP and pattern recognition techniques to sense metal ions. The polyphenols prepared are more environmentally friendly than their chemical counterparts, phenol formaldehyde resins, which typically require formaldehyde during the syntheses.

Library 3: Phenolic polymer array used for metal-ion sensing

A phenolic polymer sensing array (15 members) was synthesized combinatorially from 5 phenolic monomers under the catalysis of SBP, as listed in Table 1. ΔF values of the fluorescence response of polymers upon metal binding, along with each specific MEWs when excited at 322 nm, are shown in Table 2 for 1.0 mM of Cu^{+2} and 0.2 mM of Fe^{+3}.

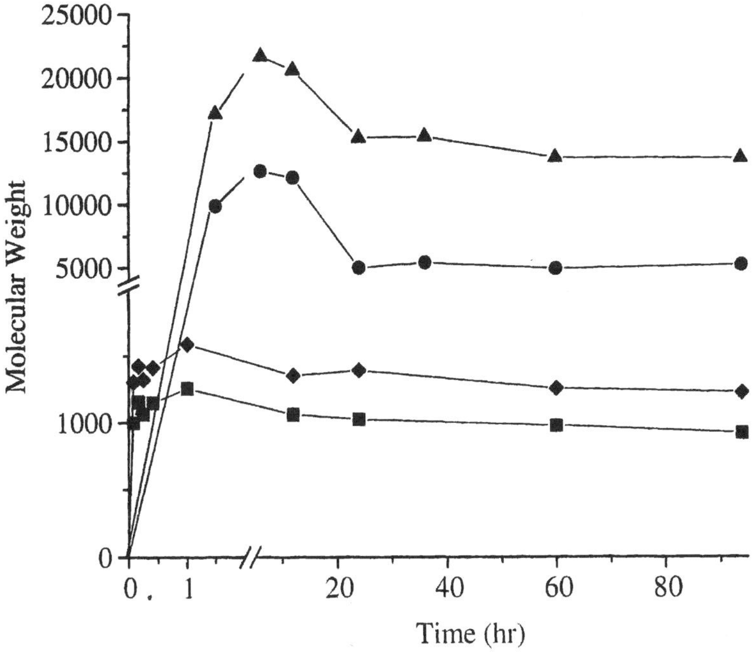

Figure 4. Molecular weight of polymers synthesized from DVA/sorbitol [M_w(▲) and M_n(●)] and DVA/octanediol [M_w(♦) and M_n(■)] as time progresses.

Table 1. Polyphenol array 2 generated of homo- and copolymers from 5 phenolic monomers.

	monomer				
monomer	M_1	M_2	M_3	M_4	M_5
M_1	P_{11}	P_{12}	P_{13}	P_{14}	P_{15}
M_2		P_{22}	P_{23}	P_{24}	P_{25}
M_3			P_{33}	P_{34}	P_{35}
M_4				P_{44}	P_{45}
M_5					P_{55}

M_1: *p*-cresol; M_2: *p*-phenylphenol; M_3: *p*-methoxyphenol; M_4: *p*-hydroxyphenyl acetic acid; M_5: *p*-hydroxybenzoic acid. All copolymers were prepared at 1:1 phenolic monomer substrate ratio (see text for details).

It is clear from Table 1 and 2 that copolymers not only have different MEW from the corresponding homopolymers, they also have different ΔF values of fluorescence response to metal ions. Every element of the 15-member array is measured for fluorescence response upon exposure to 4 different metal ions, Fe^{+3}, Cu^{+2}, Ni^{+2}, and Co^{+2}.

A statistical analysis is used to evaluate the difference of each fluorescence response. ΔZ, which represents the normalized deviation of fluorescence response from the average for each sensing element in the sensing array, is defined as follows in Eq. 2:

$$\Delta Z = \frac{\Delta F - x}{y} \qquad (2)$$

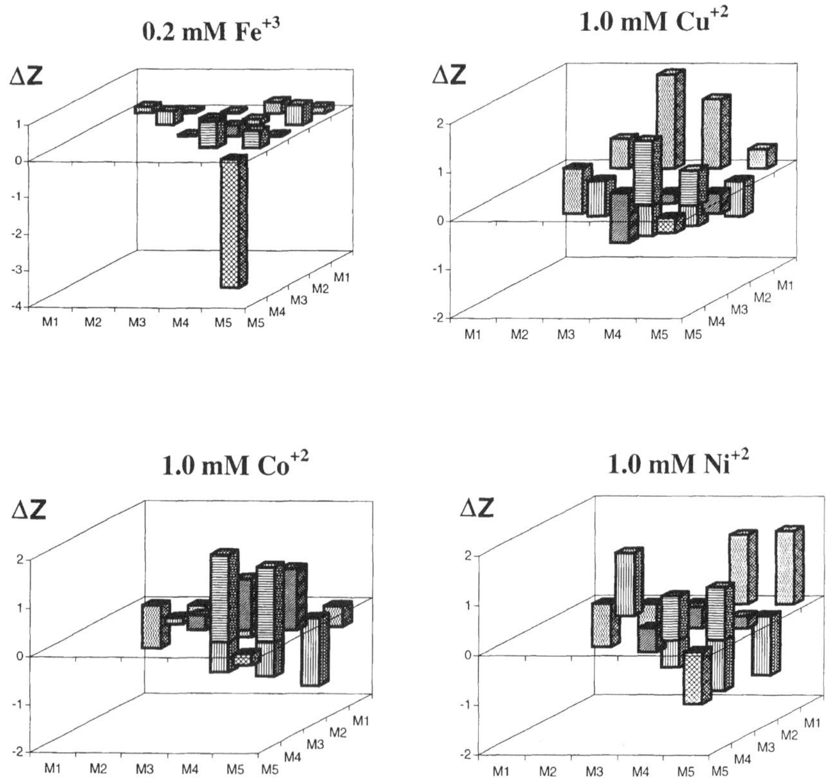

Figure 5. Histograms of 4 metal ions generated from the 15-member homo- and copolymer array, representing "fingerprints" of the metal ions and their concentrations.

Table 2. Maximum emission wavelengths (MEWs) of each element in the 15-member polyphenol sensor array II, and fluorescence intensity change ratio, ΔF, of each element to four different metal ions

	MEW	*Metal Ions*			
	(nm)	Fe^{+3} *(0.2 mM)*	Cu^{+2} *(1.0 mM)*	Ni^{+2} *(1.0 mM)*	Co^{+2} *(1.0 mM)*
P_{11}	360.8	0.52	0.14	0.03	0.03
P_{22}	360.8	0.59	0.16	0.10	0.06
P_{33}	360.8	0.47	0.13	0.04	0.09
P_{44}	408.0	0.70	0.36	0.09	0.16
P_{55}	408.0	-0.78	0.20	0.02	0.06
P_{12}	398.4	0.48	0.29	0.04	0.06
P_{13}	360.8	0.47	0.42	0.05	0.04
P_{14}	360.8	0.57	0.37	0.10	0.09
P_{55}	398.4	0.52	0.27	0.11	0.05
P_{23}	360.8	0.52	0.12	0.02	0.02
P_{24}	440.0	0.51	0.14	0.01	0.01
P_{25}	444.8	0.65	0.16	0.02	0.00
P_{34}	404.0	0.56	0.21	0.07	0.12
P_{35}	404.0	0.47	0.19	0.07	0.13
P_{45}	364.8	0.62	0.30	0.09	0.15

MEW represents maximum emission wavelength for each polymer when excited at 322 nm.

where the x and y represent the average and standard deviation. Based on ΔZ obtained for each sensing element, histograms were generated for the 15-member array using 0.2 mM or 1 mM metal ions as characteristic 'fingerprints' (Figure 5).

In conclusion, the diverse catalytic power of enzymes has been demonstrated with the generation of three unique polymer arrays. Environmentally compatible polymer libraries are synthesized by exploiting biocatalysts in a variety of reaction media, different monomers, and reaction types. This combinatorial array-based biocatalytic approach can be extended and utilized as a tool for polymer materials discovery.

Acknowledgment

The work for the generation of polycondensed arrays was sponsored by the Biotechnology Research and Development Corporation (BRDC). The polyphenolic sensor array work was supported by the Office of Naval Research (N00014-97-1-0835). Much of this work has been adopted from refs. *24, 25*, and *35*.

Literature Cited

1. *Enzymes in polymer synthesis;* Gross, R.A.; Kaplan, D.L.; Swift, G,; ACS symposium series 684; American Chemical Society: Washington, DC **1998**.
2. Akkara, J.A.; Wang, J.Z.; Yang, D.P.; Gonsalves, K.E. *Macromolecules* **2000**, 33, 2377-2382.
3. Tonami, H.; Uyama, H.; Kobayashi, S.; Kubota, M. *Macromolecular Chem. Phys.* **1999**, 200, 2365-2371.
4. Ayyagari, M.S.; Marx, K.A.; Tripathy, S.K.;Akkara, J.A.; Kaplan, D.L., *Macromolecules*, **1995,** 28, 5192-5197.
5. Dordick, J.S.; Marletta, M.A.; Klibanov A.M., *Biotechnol. Bioeng.* **1987**, 30, 31-36.
6. Binns, F; Harffey, P; Roberts, SM; Taylor, A., *J. Chem. Soc.-Perk. Trans,* **1999**, 2671-2676.
7. Geresh, S.; Gilboa Y., *Biotechnol. Bioengin.* **1990**, 36, 270-274.
8. Kumar, A.; Gross, R.A., *J. Am Chem. Soc.* **2000**, 122, 11767-11770.
9. Bisht, K.S.; Henderson, L.A.; Gross, R.A., *Macromolecules.* **1997**, 30, 2705-2711
10. Kobayashi, S.; Okamoto, E.; Wen, X.; Shoda, S, *J. Macromol. Sci.- Pure Appl. Chem* **1996**, A33, 1375-1384.
11. Dordick, J.S, *Trends Biotech.*, **1992**, 10, 287-293
12. McFarland, E.W.; Weinberg, W.H. *Tren. Biotech.* **1999**, 17, 107-115.
13. Brazwell, E.; Filos, D.; Morrow, C., *J. Poly. Sci: Part A*, **1995**, 33, 89-95.
14. Kumar, G.S.; Ghogare, A.; Mukesh, D. *J. Appl. Poly. Sci.* **1997**, 63, 35-45.
15. Anderson, E.M.; Karin, M.; Kirk, O. *Biocat. Biotransform.* **1998**, 16, 181-204.
16. Patil, D.R.; Rethwisch D.G.; Dordick, J.S. *Biotech. Bioengin.* **1991**, 37, 639-646.
17. Wulff, G.; Schmid, J.; Venhoff, T. *Macromol. Chem. Phys.* **1996**, 197, 259-274.
18. Martin, B.D.; Ampofo, S.A.; Linhardt, R.J.; Dordick, J.S. *Macromolecules.* **1992**, 25, 7081-7085.
19. Yoshimoto, K.; Itatani, Y.; Tsuda, Y. *Chem. Pharm. Bull.* **1980**, 28, 2065-2076.
20. Uyama, H.; Yaguchi, S; Kobayashi, S. *Chem. Lett.* **2000,** 7, 800-801
21. Chaudhary, A.K.; Beckman, E.J.; Russell, A.J. *Biotechnol. Bioengin.* **1997**, 55, 227-239.
22. Ryu, K.; Dordick, J.S., *Biochemistry*, **1992**, 31, 2588-2598
23. Akkara, J. A.; Senecal, K. J.; Kaplan D. L. *J. Polym. Sci., Part A: Polym. Chem.* **1991**, 29, 1561-1574.

24. Kim, J.; Wu, X.; Herman, M. R.; Dordick, J. S. *Anal. Chim. Acta.* **1998**, 370, 251-258.
25. Wu, X., Kim, J., and Dordick, J.S. *Biotech. Prog.* **2000**, 16, 513-516.
26. Alva, K. S.; Sarma, R.; Marx, K. A.; Kumar, J.; Tripathy, S. K.; Akkara, J. A.; Kaplan, D. L. *Proc. SPIE-Int. Soc. Opt. Eng.* **1997**, 3040(Smart Materials Technologies), 200-210.
27. Patel, M. N.; Sutaria, D. H.; Patel, S. D. *Angew. Makromol. Chem.* **1996**, 234, 13-20.
28. Merian, E. Metals and Their Compounds in the Environment; VCH: Weinheim, **1991**.
29. Vernet, J. –P., Ed. *Impact of Heavy Metals on the Environment;* Elsevier: New York, **1992**.
30. Muller, R. High Electronic Selectivity Obtainable with Nonselective Chemosensors. *Sens. Actuators, B.* **1991**, 4, 35-39.
31. Gardner, J. W. Detection of Vapours and Odours from a Multisensor Array Using Pattern Recognition Part 1 Principle Component and Cluster Analysis. *Sens. Actuators, B.* **1991**, 4, 109-115.
32. Breer, H. In *Handbook of Biosensors and Electronic Noses: Medicine, Food, and the Environment*; Kress-Rogers, E., Ed.; CRC press: New York, **1997**.
33. Dryer, L. Olfaction, Odorant Receptors. In *Encyclopedia of Neuroscience*; Adelman, G., Smith, B.H., Eds.; Elsevier: Amsterdam ; New York , **1999**
34. Kauer, J. *Trends Neurosci.* **1991**, 14, 79-85.
35. Kim, D.Y.; Dordick, J.S., *Biotechnol. Bioengin.* **2001**, 76, 200-206

Chapter 3

Evolving and Screening Enzymes for New Activities on Polymer Substrates

Douglas C. Youvan, William J. Coleman, and Edward J. Bylina

KAIROS Scientific Inc., 10225 Barnes Canyon Road, #A110, San Diego, CA 92121

A new digital imaging spectrophotometer and a series of colorimetric solid phase assays have been developed to screen bacterial libraries expressing mutagenized enzymes undergoing directed evolution. This high-throughput solid-phase assay system (known as 'Kcat Technology') can detect less than a 20% difference in enzyme rates within microcolonies grown at a nearly confluent density of 500 colonies per cm^2 on an assay disk. Each microcolony is analyzed simultaneously at single-pixel resolution (1.5 megapixels; 75 micron/pixel), requiring less than 100 nanoliters of substrate per measurement, a 1000-fold reduction over conventional liquid phase assays. Here we report the successful identification of variants of Agrobacterium β-glucosidase (Abg) – a glycosidase with broad substrate specificity – by simultaneously assaying two different substrates tagged with spectrally distinct chromogenic reporters. This technology should lead to the isolation of new enzyme activities that are useful in the synthesis of various substances,including specialty chemicals and pharmaceuticals.

The enzymes that are currently available for use in synthetic chemistry have evolved over millions of years to be efficient and selective catalysts for the chemical reactions taking place in living systems. However, many potential industrial applications for these catalysts involve substrates, organic solvents and other reaction conditions that were never encountered in nature. Protein engineering can be used to change the properties of these natural catalysts to suit the needs of synthetic chemistry and manufacturing. Much progress has been made by both industrial and academic laboratories in demonstrating how enzymes can be tailored for optimal performance in industrial applications by directing their evolution *in vitro*. By carefully controlling *in vitro* mutation efficiencies and screening for enhanced catalytic properties over multiple generations, researchers have developed new enzymes that are hundreds of times more active than the natural enzymes in chemical process environments.

Despite these successes, further exploitation of this enzyme engineering is severely limited by the current need to employ time-consuming liquid-phase assays for screening mutant libraries. Many companies are attempting to develop new technologies that will overcome such screening bottlenecks. Recent advances in protein engineering, genomics and combinatorial chemistry have prompted major pharmaceutical and biotechnology firms to accelerate their efforts by developing automated high-throughput liquid-phase screening systems. Conventional 96-well plates (100-200 μl per well) are typically used in liquid-phase screening. Improvements in assay miniaturization would allow acceleration in the rate of screening, reduction in the cost per assay, and conservation of compounds that are either expensive or difficult to synthesize and purify. The high-throughput screening system presented in this paper utilizes a solid-phase assay disk which may approach the limit to which these assays may be miniaturized on a 2-dimensional surface.

Feasibility of the Kcat Technology was demonstrated using the model enzyme system ('Abg') from *Agrobacterium faecalis* that hydrolyzes β-glucosides. The catalytic mechanism of the wild-type protein and several active site variants has been studied in detail (Trimbur et al., 1992; Wang et al., 1995). The enzyme is quite nonspecific, catalyzing the hydrolysis of substituted glucosides, galactosides, xylosides, fucosides, and arabinosides (Kempton & Withers, 1992). Given the versatility of this enzyme, it appeared to be an ideal candidate for directed evolution experiments. In the Abg experiments described below, we demonstrate the ability to differentiate Abg mutants differing in substrate specificity for indolyl derivatives of galactoside and glucoside. We further demonstrate the capability of characterizing these enzyme variants in solid-phase kinetics assays.

Abg activity was screened using the type of reaction shown in Figures 1 and 2. Hydrolytic cleavage of the glycosidic bond at the 3-position of the indolyl derivative generates the protonated (hydroxyl) form of indoxyl, which deprotonates and tautomerizes. In the presence of oxygen, two indoxyl molecules are then spontaneously oxidized and the C-2 carbon is deprotonated, which causes them to rapidly dimerize. The final product is an intensely colored indigo dye, which precipitates out of solution, and therefore does not diffuse away from the lysed bacterial microcolony. This is important, since absorbance is monitored as a function of time in the solid phase assay.

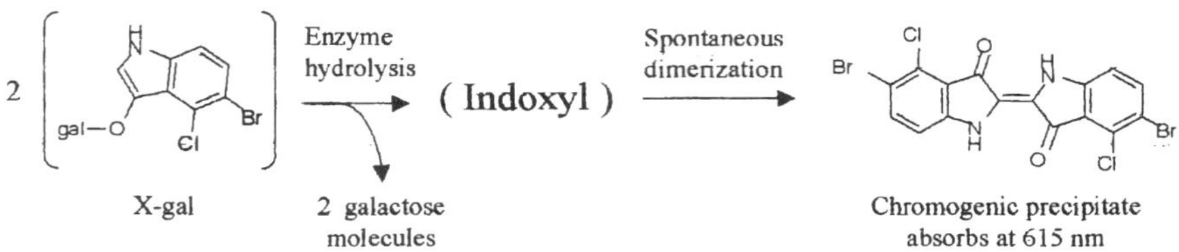

Figure 1. Enzymatic hydrolysis of an indolyl derivative used in an exemplary colorimetric solid phase assay. The classical X-gal reaction commonly used in molecular biology is shown here. The hydrolysis of other indolyl derivatives (see the next figure) follows the same pathway to form a variety of colored precipitates. Reproduced with permission from Bylina et al., 2000. Copyright 2000 Biotechnology et alia.

Indolyl derivatives	R_1	R_2	R_3	R_4	Indigo product λmax (nm)
Red	H	H	H	Cl	540
Magenta	H	H	Br	Cl	565
Iodo	H	H	I	H	575
X	H	Cl	Br	H	615
Green	CH_3	H	H	H	665
Y	H	H	H	H	680

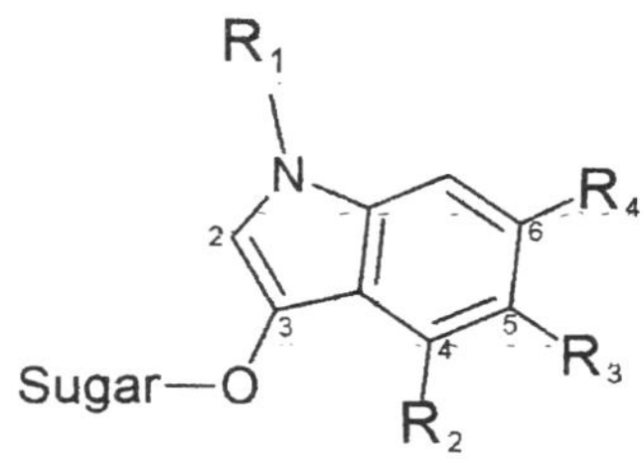

Figure 2. Indolyl derivatives and the colors of their indigo products. A variety of different colors can be produced depending on the identity of the R groups. The sugar moiety shown here can be replaced by other compounds to make the molecule a substrate for a variety of enzymes. For example, replacing the sugar with a carboxylic acid generates ester substrates for lipases and esterases. Reproduced with permission from Bylina et al., 2000. Copyright 2000 Biotechnology et alia.

Methods and Results

The active site nucleophile of Abg is the glutamic acid residue at position 358 (Withers et al., 1990). An Abg "active site phylogeny" was constructed in this region of the protein as follows. A BLAST search of the non-redundant GenBank CDS (Altschul et al., 1997) identified 20 related sequences using a search sequence comprising amino acids 350-385 of Abg. Protein sequence differences between Abg and these sequences were tabulated. Additional related protein sequences and sequence changes of active Abg mutants (Trimbur et al., 1992) were also incorporated . Figure 3 shows this phylogeny for Abg residues 361 to 373. This phylogenetic information was used to design a combinatorial cassette library. Experiments can also be based on random mutagenesis using NNN, NNK, or NNS triplets for the zeroth iteration of Recursive Ensemble Mutagenesis (REM), as previously described (Arkin and Youvan, 1992a,b; Delagrave et al., 1993; Delagrave and Youvan, 1993). In this paper, phylogeny has been used to 'bootstrap' the method – as described by Goldman et al., 1992.

A brief outline of the REM library construction follows: An Abg expression plasmid was constructed by cloning the Abg structural gene into the pQE70 expression vector (QIAgen). The Ala389 GCC codon was changed to GCA, resulting in the introduction of a BglII restriction site at this position. A doped oligonucleotide corresponding to the 90-bp NarI-BglII fragment (encoding Gly360 to Ala389 in Abg) was synthesized based on phylogenetic data entered into the 'CyberDope' computer program (KAIROS Scientific).

A REM cassette was synthesized by PCR amplification of this doped oligonucleotide:

5'CACCGAAAACGGCGYCKSCWHCRAKRWKRDGNTTSWGRAWDGCRR
GVTCMAWGACCAGCCGCGTCTCGATTATTACGCCGAACACCTCGGCA
TCGTCGCAGATCTCATCC 3'

The BglII-BsaHI digested REM cassette was ligated into the BglII-NarI digested expression vector and transformed into M15[pREP4]. The resulting library was screened using the Kcat instrument (KAIROS Scientific). Information on the Kcat instrument and the CyberDope program can be found at: www.kairos-scientific.com. More detailed methodology for performing Kcat Technology can be found in US patent 5,914,245 (Bylina et al., 1999) at: www.USPTO.gov. The software and algorithms used to sort pixels (or features) in these experiments are described elsewhere (Yang et al., 2000).

```
                                                          N
                                      Y   A               F
                              H       S   E   P       I   S       A
                  R       M   R   T   T   S   D       K   Q       Y
                  F   D   E   F   F   Q   T   N   G   N   V       E
                  V   S   S   K   I   V   F   I   K   H   T       K
                  I   G   I   E   K   E   L   Q   E   C   K   L   Q
                  L   A   F   D   D   K   I   V   D   S   R   I   H
  358                                                               374
      E   N   G   A   C   Y   N   M   G   V   E   N   G   E   V   N   D
   wt:GAAAACGGCGCCTGCTACAATATGGGCGTCGAAAACGGCGAGGTCAATGAC
  REM:GAAAACGGCGYCKSCWHCRAKRWKRDGNTTSWGRAWDGCRRGVTCMAWGAC
                  A   C   Y   N   M   G   V   E   N   G   E   V   N
                  V   A   F   D   D   K   I   V   D   S   R   I   H
                      G   I   E   K   E   L   Q   E   C   K   L   Q
                      S   S   K   I   V   F   L   K       G       K
                          T       V   R
                          N       E   M
                                  N
```

Figure 3. The combinatorial region of the cassette used to mutagenize the Abg gene. The wild-type (wt) sequence from residue 358 to 374 is given above the mixed nucleotide sequence for the combinatorial cassette. The entire sequence of this doped oligonucleotide can be found in the text. This oligonucletide was used for the first cycle of Recursive Ensemble Mutagenesis (REM). Phylogenetic data for glucosidases is given above the DNA sequences. Amino acid residues (in bold type face) were input into the CyberDope program (including wt), and the residues encoded by the REM cassette are given below the DNA sequences. In order to reduce combinatorial complexity to ~ 10^8, some of the residues in the phylogeny were omitted (not in bold type face). Reproduced with permission from Bylina et al., 2000. Copyright 2000 Biotechnology et alia.

The Graphical User Interface of the Kcat instrument and an enlarged area of the assay disk showing the spectral analysis of Abg mutagenesis on Red-gal and X-glu was used to identify microcolonies that favor either the substrate Red-galor X-glu. A high resolution 24-bit color version of this analysis can be found on-line at: http://www.kairos-scientific.com/images/ACS2000/figure4.htm. Backcoloring is used to highlight pixels in the image and absorption spectra for every pixel on the assay disk. Sorting was based on the 540 nm : 610 nm ratio. Pixels within microcolonies that favor the substrate Red-gal are sorted to the top of the contour plot and backcolored in red. Pixels within microcolonies that favor the substrate X-glu are sorted to the bottom of the contour plot and backcolored in blue. Pixels within microcolonies that favor the substrate X-glu were sorted further by catalytic velocity.

The Graphical User Interface of the Kcat instrument and an enlarged area of the assay disk was also used for showing the kinetic analysis of Abg mutants that favor the substrate X-glu. A high resolution 24-bit color version of this analysis can be found on-line at:

http://www.kairos-scientific.com/images/ACS2000/figure5.htm.

Highlighted pixels within microcolonies indicate enzyme variants with the highest velocity after sorting for X-glu specificity.

Summary

We have successfully developed a high-throughput solid-phase assay system that can be used to identify interesting enzyme variants generated by mutagenesis. This Kcat Technology can be used to distinguish small differences in both the kinetic and spectral properties of enzyme variants expressed in microcolonies. The use of Kcat Technology to evolve galactose oxidase has recently been reported (Delagrave et al., 2001).

In this study, we have used combinatorial cassette mutagenesis based on the phylogeny of homologous glycosidase genes and known point mutants in Abg to generate a recombinant library of over 10 million enzyme variants involving partial randomization of 13 amino acid residues near the active site. Different enzyme variants that show epimeric specificity – either for the hydrolysis of glucoside or galactoside – were identified by solid-phase screening. This solid phase screening technology is compatible with a variety of mutagenesis methods including error-prone mutagenesis, sequential random mutagenesis, and gene shuffling. However, using combinatorial cassette mutagenesis, this greater screening capacity enables one to survey large sequence spaces in which many residues are changed simultaneously. Isolating optimized enzymes may be accomplished through recursive cycles of cassette mutagenesis using the genetic algorithm (GA)-based technique, Recursive Ensemble Mutagenesis (REM) – emulating the natural evolution of genes by point mutation and recombination. The Abg variants identified during the screening described in this paper have been isolated and are currently being characterized.

Acknowledgement

This work was supported by NIH R44GM55470, NIH R43GM60073, and by KAIROS IR&D Funds. Material reproduced with permission from Bylina et al., 2000.

References

1. Altschul, S.F.; Madden, T.L.; Schaffer, A.A.; Zhang, J.; Zhang, Z.; Miller, W.; Lipman, D. Gapped BLAST and PSI-BLAST: a new generation of protein database search programs. *Nucleic Acids Res.* **1997**, 25, 3389-3402.
2. Arkin, A.P.; Youvan, D.C. Optimizing Nucleotide Mixtures to Encode Specific Subsets of Amino Acids for Semi-Random Mutagenesis. *Bio/Technology* **1992**, 10, 297-300.
3. Arkin, A.P.; Youvan, D.C. An Algorithm for Protein Engineering: Simulations of Recursive Ensemble Mutagenesis. *PNAS* **1992**, 89, 7811-7815.
4. Bylina, E.J.; Coleman, W.J.; Dilworth, M.R.; Silva, C.M.; Yang, M.M.; Youvan, D.C. Solid phase enzyme kinetics screening in microcolonies. U.S. Patent 5914245, 1999.
5. Bylina, E.J.; Coleman, W.J.; Grek, C.L.; Yang, M.M.; Youvan, D.C. Directed Evolution and Solid Phase Enzyme Screening. *Biotechnology et alia* <www.et-al.com> **2000**, 7, 1-6.
6. Delagrave, S.; Goldman, E.R.; Youvan, D.C. Recursive Ensemble Mutagenesis. *Protein Engineering* **1993**, 6, 327-331.
7. Delagrave; S.; Youvan, D.C. Searching Sequence Space to Engineer Proteins: Exponential Ensemble Mutagenesis. *Bio/Technology* **1993**, 11, 1548-1552.
8. Delagrave, S.; Murphy, D.M.; Rittenhouse Pruss, J.L.; Maffia, A.M.; Marrs, B.L.; Bylina, E.J.; Coleman, W.J.; Grek, C.L.; Dilworth, M.R.; Yang, M.M.; Youvan, D.C. Application of a Very High-Throughput Digital Imaging Screen to Evolve the Enzyme Galactose Oxidase. *Protein Engineering* **2001**, 14, 261-267.
9. Goldman, E.R.; Youvan, D.C. An Algorithmically Optimized Combinatorial Library Screened by Digital Imaging Spectroscopy. *Bio/Technology* **1992**, 10, 1557-1561.
10. Kempton, J.B.; Withers, S.G. Mechanism of Agrobacterium beta-glucosidase: kinetic studies. *Biochemistry* **1992**, 31, 9961-9969.
11. Trimbur, D.E.; Warren, R.A.J.; Withers, S.G. Region-directed mutagenesis of residues surrounding the active site nucleophile in β-glucosidase from Agrobacterium faecalis. *J. Biol. Chem.* **1992**, 267, 10248-10251.
12. Wang, Q.; Trimbur, D.; Graham, R.; Warren, R.A.; Withers, S.G. Identification of the acid/base catalyst in Agrobacterium faecalis beta-glucosidase by kinetic analysis of mutants. *Biochemistry* **1995**, 34, 14554-14562.
13. Withers, S.G.; Warren, R.A.J.; Street, I.P.; Rupitz, K.; Kempton, J.B.; Aebersold, R. *J. Am. Chem. Soc.* **1990**, 112, 5887-5889.
14. Yang, M.M.; Dilworth, M.R.; Youvan, D.C. Graphical User Interfaces for Single-Pixel Spectroscopy. *Biotechnology et alia* <www.et-al.com> **1999**, 5, 1-8 or *SPIE* **2000**, 3924, 108-115.
15. Youvan, D.C. Searching Sequence Space. *Bio/Technology* **1995**, 13, 722-723.

Chapter 4

Polymer-Modulated, Diffusion-Controlled Enzyme Kinetics on Monolayers

Keiji Tanaka[1,2] and Hyuk Yu[1]

[1]Department of Chemistry, University of Wisconsin at Madison, Madison, WI 53706
[2]Permanent address: Department of Applied Chemistry, Faculty of Engineering, Kyushu University, Fukuoka, Kyushu, Japan

As a model system of the hemi-leaflets of biomembranes, binary monolayers of L-α-dilauryoylphosphatidylcholine (DLPC) and poly(tert-butyl methacrylate) (PtBMA) were formed at the air/water interface by the successive addition method. First, surface pressure and lateral diffusion of a probe lipid in the monolayers, as a function of the polymer composition, were examined by the techniques of Wilhelmy plate and fluorescence recovery after photobleaching (FRAP), respectively. Then, the hydrolysis kinetics of a substrate (umbelliferone stearate) vis-a-vis lateral diffusion coefficients of the probe lipid and an interface active enzyme, lipase, were examined as a function of the polymer composition in the monolayer. The hydrolysis kinetics is shown to be closely correlated with the lateral diffusion coefficients of the probe lipid and lipase. Thus, we show quantitatively that the kinetics on the DLPC/PtBMA binary monolayers are diffusion-controlled. The same has been shown to be the case with DLPC/cholesterol binary monolayers as reported earlier.

Our focus is to find examples of surface chemical reactions modulated by surface active polymers that are diffusion-controlled; it is well known that elementary reactions such as ion pair association in dilute aqueous solutions are the typical examples of this in bulk solution.[1] Heterogeneously catalyzed surface reactions are ubiquitous in a whole host of chemical and biological processes, including in cellular signal transduction and molecular recognition.[2,3] In order to examine chemical reactions on surfaces vis-a-vis those in solutions, lateral homogeneity in macroscopic length scale is required just as in homogeneous solutions. Liquid surface is such an example; it reaches thermodynamic equilibrium easily with spontaneous capillary waves, resulting in surface roughness of 0.3-0.4 nm.[4,5] Also, all molecules on the liquid surface are at the same chemical potential, and they can be controlled to have more or less the same phase and accessibility depending on lateral pressure; such a control is not easily obtained on solid surfaces.[6] These specific advantages are taken into account here in choosing the air/water interface as a suitable system for studying surface chemical reactions.

Lipid monolayers at the air/water interface are often regarded as a well-represented model of homogeneous hemi-leaflets of lipid bilayer membranes of all biological origin.[7] Since the chemistry and kinetics on the leaflets is expected to be influenced by microdomains on biomembranes,[7] we seek to understand as a benchmark, what controls the kinetics of enzymatic catalysis on homogeneous hemi-leaflets. Hence, we examine enzyme catalysis on L-α-dilauroylphosphatidylcholine (DLPC) monolayers in "fluid state" at the air/water interface mixed with a phase compatible polymer, poly(t-butylmethacrylate)-PtBMA. Since the thickness of a monomolecular film on the liquid surface is generally a few nanometers, the system amounts to a quasi-two-dimension with a finite thickness.

We have recently studied lipase catalyzed hydrolysis reaction of umbelliferone stearate (UMB-C_{18}) on homogeneous DLPC/cholesterol binary monolayers with cholesterol fraction up to 30 mol%[8] as well as pure DLPC monolayer.[9] Cholesterol, one of three principal components of animal cell membranes,[10] was used to vary the monolayer dynamics. Lipase, which belongs to a class of lipolytic enzyme,[11] is most active at lipid/water interface due to its conformational change.[12] Combining the lateral diffusion coefficients of reactants with chemical kinetics on monolayers, it has been concluded for the first time that the enzyme catalysis on uniphasic binary monolayers of DLPC/cholesterol is entirely diffusion-controlled. The objective here is to examine the monolayer dynamics and the hydrolysis kinetics of UMB-C_{18} on binary monolayers composed of DLPC and another "viscosifier", PtBMA whose interfacial properties have been extensively studied.[13] The central question we seek to address is whether the observed cholesterol effect arises by its unique interactions with lipid molecules in the monolayer or it just comes about as a manifestation of the viscosifier effect.

Experimental

DLPC and 1-acyl-2-[12-[(7-nitro-2-1,3-benzoxadiazol-4yl)amino] dodecanoyl] phosphatidylcholine (NBD-PC), which was employed here as a fluorescent lipid probe, were obtained from Avanti Polar Lipids.. Monodisperse PtBMA with the number-average molecular weight of 10k was purchased from Polyscience, Inc. Umbelliferone stearate (UMB-C_{18}), used as a surface active substrate for the enzymatic catalysis, was synthesized by following the method published elsewhere.[9] UMB-C_{18} belongs to a class of fatty acid-phenolic esters which upon hydrolysis gives rise to a fluorogenic phenoxide, umbelliferone (UMB). Crude lipoprotein lipase LPL 200S, a mixture of glycine and lipase from *Pseudomonas cepacia*, was obtained from Amano International Enzyme Co. For lateral diffusion studies of lipase, we labeled it with fluorescein isothiocyanate (FITC), acquired from Aldrich.[8] All the details of the experimental procedures and protocol are reported in our recent publications.[8,9]

Results and Discussion

Surface pressure-area isotherms

From the surface pressure-area isotherms, which are not shown here for brevity, it is concluded that the limiting area per repeating unit of PtBMA monolayer is 0.34 nm^2 which are different from the earlier reported value of 0.27 nm^2. Also the first order derivative of the isotherm, $d\Pi/dA$, for PtBMA monolayer in this study is smaller than that reported previously by a factor of ca. 0.6. The difference could be attributed to the subphase. Here, a phosphate buffer at pH 7.0 is used instead of pure water. Since PtBMA should be stable at neutral condition, pH 7.0, it seems reasonable to envisage that the phosphate buffer used might be a "better solvent" for PtBMA monolayer than pure water. A detail analysis of the isotherms indicates that DLPC and PtBMA are miscible with negative excess Gibbs free energy of mixing. The complete miscibility was confirmed by epi-fluorescence microscopic observation of DLPC/PtBMA monolayer containing 1 mol% fluorescence dye, NBD-PC.

Lateral Diffusion

In order to gain access to the dynamic properties of DLPC/PtBMA mixed monolayers, the lateral diffusion coefficients $D^{(2)}$ of the probe lipid, NBD-PC, in the monolayers were determined with the FRAP technique. Results at each composition were reproduced in three independent trials. In every case, the fluorescence difference profile was best fit by a single exponential, indicating a single center of mass diffusion rate for the lipids ($A \approx 0.4$ nm^2) within the binary monolayers averaged over a displacement length that exceeds the cross sectional size of lipids by 4 orders of magnitude, i.e., 34 μm for the fringe spacing versus $(0.4)^{1/2}$ nm.

Figure 1 displays the surface pressure dependences of the diffusion coefficient at surface mass fractions expressed in mole fractions of the repeating unit of PtBMA. Here, the homogeneous binary monolayers were studied at different tBMA fractions up to 25 mol%, containing 1 mol% of NBD-PC. The data points on the ordinate of the plot in Figure 3, at $\pi = 0$ mN•m^{-1}, correspond to the diffusion coefficients obtained at the lift-off points of the isotherms. Once the surface pressure reaches the lift-off point from the LE/gas coexistence region and the monolayer moves into the LE state, $D^{(2)}$ decreases sharply with increasing surface pressure. Also, incorporation of tBMA into the lipid monolayer exerts an additional influence; that is, $D^{(2)}$ decreases progressively as tBMA is added. In the case of the DLPC/PtBMA monolayer with 25 mol% tBMA, $D^{(2)}$ reaches an asymptotic value of 25 μm^2•s^{-1} at $\pi = 20$ mN•m^{-1}. The collapse pressure for the DLPC/PtBMA monolayer with 25 mol% tBMA is much lower than those of other mixed monolayers employed here. The asymptotic behavior of $D^{(2)}$ with surface pressure can be seen even for the DLPC monolayer with $\Pi > 35$ mN•m^{-1}.

Since these findings are consistent with our previous results using a different viscosifier, cholesterol, of the lipid monolayer,[8,9] it can be considered that the decrement of $D^{(2)}$ with increasing surface pressure and/or tBMA fraction can be attributed to increasing the monolayer viscosity. We will shortly return to the viscosity effect.

The crux of this study is to correlate kinetics of enzyme catalyzed reaction on monolayers with its monolayer dynamics, and thus both examinations, kinetics and dynamics, must be carried out under the same condition. $D^{(2)}$ of NBD-PC and FITC labeled-lipase in the DLPC/PtBMA mixed monolayers are deduced here under the same lipase concentration in the monolayers as that for hydrolysis reaction. Also, the lateral diffusion coefficient $D^{(2)}$ of lipids or substrates is assumed to be the same as that of the probe lipid, NBD-PC. For $D^{(2)}$ measurement of lipase, the FITC labeled-lipase is used to confer fluorescence upon lipase in lieu of intact lipase which is non-fluorescent. **Figure 2** shows tBMA composition dependence of the diffusion coefficients in the DLPC/PtBMA monolayers containing lipase with surface concentration of 1.08 x 10^{-2} molec•nm^{-2}. Since the diffusion measurement is made after an equilibrium partitioning of the lipase from bulk subphase to the monolayer is reached, the surface pressure of the measurement is 25.5±0.5 mN•m^{-1}. The $D^{(2)}$ of NBD-PC decreases progressively as tBMA composition is increased. Since this trend is in good agreement with the $D^{(2)}$ retardation by tBMA in the

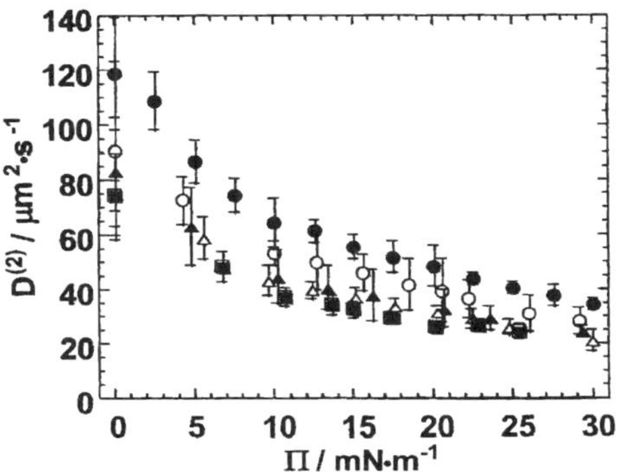

Figure 1. Lateral diffusion coefficient of a probe lipid in DLPC/PtBMA mixed monolayers as a function of surface pressure at different tBMA compositions. (Reproduced with permission from reference 5. Copyright 2002 American Chemical Society.)

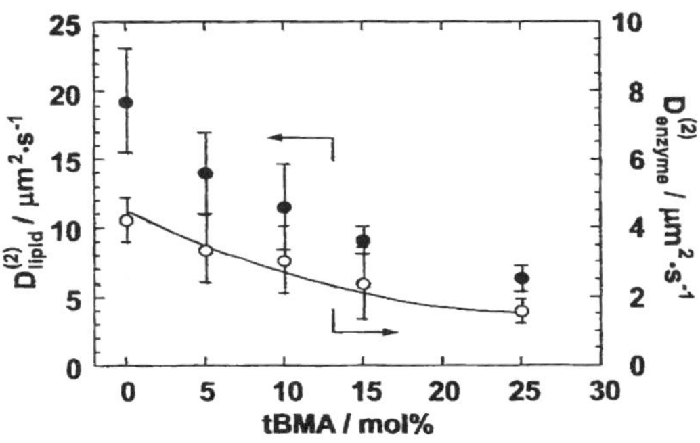

Figure 2. Lateral diffusion coefficient of NBD-PC and FITC labeled-lipase in DLPC/PtBMA mixed monolayers as a function of tBMA content: (λ), NBD-PC; (**O**), FITC-lipase. The solid curve for $D^{(2)}$ of FITC-lipase is drawn by scaling down $D^{(2)}$ of NBD-PC by a factor of 4.2.
(Reproduced with permission from reference 5. Copyright 2002 American Chemical Society.)

DLPC/PtBMA monolayers without lipase, it is most likely that a decrement of $D^{(2)}$ with the tBMA composition is attributed to the increase in the in-plane viscosity. As shown in **Figure 2**, the $D^{(2)}$ of labeled-lipase in the mixed monolayers decreases progressively in about the same fashion as the probe lipid as tBMA composition is increased. In order to account for the difference of $D^{(2)}$ between the probe lipid and the labeled-lipase at a given tBMA fraction, the HPW hydrodynamic model[15] is applied. According to the HPW model, $D^{(2)}$ of lipase can be estimated at approximately 1/3 of NBD-PC's diffusion coefficients. Although the solid curve in **Figure 2**, drawn by scaling down $D^{(2)}$ of NBD-PC, corresponds well to the retardation profile of $D^{(2)}$ for labeled-lipase obtained experimentally, the best-fit curve is obtained by a factor of 1/4.2.

Enzyme Catalyzed Hydrolysis Reaction

We now turn to discuss kinetics of the lipase catalyzed hydrolysis reaction of UMB-C_{18} on DLPC/PtBMA mixed monolayers at the air/water interface. Our kinetic model has four distinct steps: (1) the substrate adsorption from bulk phase to the monolayer, (2) substrate-enzyme encounter via in-plane diffusion, (3) hydrolysis catalysis by the lipase, and (4) spontaneous desorption of the products in the subphase.[9] Figure 7 shows the tBMA effect on initial hydrolysis reaction rate, V_0, at a fixed substrate concentration, [UMB-C_{18}] =100 nM. Measurements were carried out at a constant temperature of 296.2 ± 0.5 K, surface pressure of 25.5 ± 0.5 mN•m^{-1}, and lipase surface concentration [lipase]□ of 1.08 x 10^{-2} molec•nm^{-2}. The partitioning of the UMB-C_{18} from the bulk subphase to the monolayer was not large despite its surface activity. At [UMB-C_{18}]=100 nM, the surface concentration [UMB-C_{18}]□, estimated by accounting for the increment of total surface area to maintain the constant lateral pressure, is 1.97 x 10^{-2} molec•nm^{-2}, provided that the area per molecule of UMB-C_{18} remains at 0.24 nm^2 as determined from its surface pressure isotherm. The monolayer homogeneity at the experimental surface pressure of 25 mN•m^{-1} was confirmed by epifluorescence microscopy. The turnover number of this hydrolysis reaction on DLPC monolayers at 25 mN•m^{-1} was 166 s^{-1}, which was much larger than that in bulk solution. The V_0 decreases monotonically with the tBMA fraction in the mixed monolayer. The retardation profile of V_0 by tBMA is similar in shape to the relation of $D^{(2)}$ of reactants with tBMA composition shown in **Figure 2**. Thus, it is qualitatively obvious that V_0 is governed by the lateral mobility of reactants. This can be easily seen when V_0 is replotted as a function of the sum of $D^{(2)}$s of a lipid and lipase, as shown in the inset of **Figure 3**.

We finally come to combine the lateral diffusion coefficients of reactants with the enzymatic hydrolysis rates on the monolayers by using the two-dimensional diffusion-controlled reaction dynamics theory of Torney and McConnell.[16] For the sake of brevity, we gloss over the details of analysis and just show a comparison in **Figure 4** where the dependence on tBMA content of

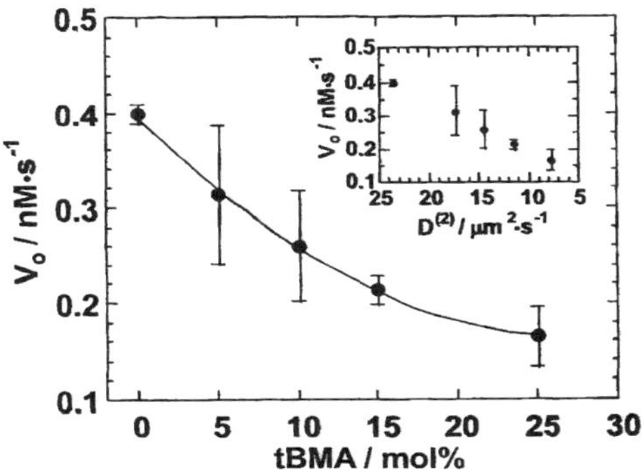

Figure 3. tBMA fraction dependence of the hydrolysis reaction rate on DLPC/PtBMA mixed monolayers. Solid curve is drawn to guide the eye. The inset is re-plotted by changing the abscissa to the sum of lateral diffusion coefficients of NBD-PC and labeled lipase at a given tBMA fraction.
(Reproduced with permission from reference 5. Copyright 2002 American Chemical Society.)

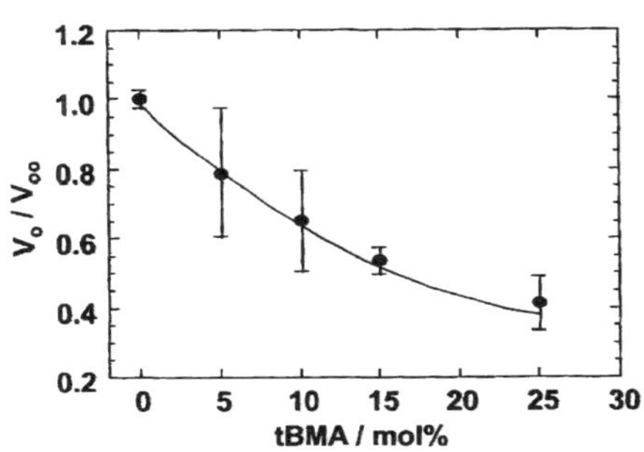

Figure 4. Normalized initial hydrolysis rates by using the data set from Figure 3. The solid line is drawn according to the two-dimensional diffusion-limited reaction dynamics theory of Torney and McConnell.
(Reproduced with permission from reference 5. Copyright 2002 American Chemical Society.)

the normalized initial hydrolysis rate is displayed where V_{oo} is in pure DLPC without tBMA. The solid line is drawn according to a single parameter in the theory, i.e., the reaction probability upon encounter □=1.0. Actually, the theoretical prediction of V_o/V_{oo} is not sensitive to the value of □. It appears clear that our results are in complete agreement with the theoretical prediction of Torney and McConnell. Thus, it seems reasonable to conclude that the substrate adsorbs to the interface, randomly diffuses to lipase and is hydrolyzed by the enzyme at the interface, with its reaction rate controlled by diffusion in the monolayer. Noting that our results in **Figure 4** are independent of □ in the theory, we claim that no fitting parameter is necessary once the normalization is allowed to focus on the diffusion retardation effect by tBMA.

Acknowledgments. This work was partially supported by the Eastman Kodak Professorship and NSF grants (DMR9203289 and DMR9711226) awarded to H.Y. We are grateful for helpful discussions with Prof. George Zografi and our colleagues Thorsteinn Adalsteinsson and Hunkyun Pak.

References

1. Rice, S. A. Chemical Kinetics, Vol. 25, Diffusion-Limitted Reactions; Elsevier: Amsterdam, **1985**.
2. Weng, G. Z.; Bhalla, U. S.; Iyengar, R. Science **1999**, 284, 92.
3. Bhalla, U. S.; Iyengar, R. Science **1999**, 283, 381.
4. Braslau, A.; Deutsch, M.; Pershan, P. S.; Weiss, A. H.; Als-Nielsen, J.; Bohr, J. Phys. Rev. Lett. **1985**, 54, 114.
5. Ma, J.; Yu, H.; Clarson, S. J.; Satija, S.; Maliszewskyj, N. submitted to Langmuir.
6. Davies, J. T.; Rideal, E. K. Interfacial Phenomena, 2nd ed.; Academic Press: San Diego, **1963**.
7. Gennis, R. B. Biomembranes, Molecular Structure and Function; Springer-Verlag: New York, **1989**; pp 36-84.
8. Tanaka, K.; Mecca, S. P.; Yu, H. *Langmuir* **2000**, *16*, 2665.
9. Tanaka, K.; Manning, P. A.; Yu, H. *Langmuir* **2000**, *16*, 2672.
10. Houslay, M. D.; Stanley, K. K. *Dynamics of Biological Membranes*; Wiley: New York, **1982**; pp 196-205.
11. Schmid, R. D.; Verger, R. *Angew. Chem. Int. Ed.* **1998**, *37*, 1608.
12. van Tilbeurgh, H.; Egloff, M.-P.; Martinez, C.; Rugani, N.; Verger, R.; Cambillau, C. *Nature* **1993**, *362*, 814.
13. Sadrzadeh, N.; Yu, H.; Zografi, G. *Langmuir* **1998**, *14*, 151.
14. Hughes, B. D.; Pailthorpe, B. A.; White, L. R., *J. Fluid. Mech.* **1981,** *110*, 349.
15. Torney, D. C.; McConnell, H. M. *Proc. R. Soc. Lond.* **1983**, *A387*, 147.

Chapter 5

Recent Developments in the Field of In Vitro Biosynthesis of Plant β-Glucans

Joséphine Lai Kee Him, Henri Chanzy, Ludovic Pelosi, Jean-Luc Putaux, and Vincent Bulone*

Centre de Recherches sur les Macromolécules Végétales, CERMAV-CNRS, B.P. 53, F-38041 Grenoble cedex 9, France

This chapter reports the use of various detergents for the preparation of vesicles and micelles from plant plasma membranes bearing β-glucan synthases. The effects of these detergents and other parameters on the *in vitro* activity of cellulose and (1→3)-β-glucan synthases are described. Morphological and structural aspects of the vesicles and micelles are presented together with the characterization of the products synthesized *in vitro* by the detergent-extracted β-glucan synthases. The possible involvement of porins for extrusion of cell wall glucans through plasma membranes as well as future developments for the study of *in vitro* biosynthesis of these polysaccharides are discussed in relation to these results.

All higher plants contain membrane-bound glucan synthases which catalyze the synthesis of (1→3)-β-glucan (callose) and (1→4)-β-glucan (cellulose) from UDP-glucose. These processive glycosyltransferases are of central importance for normal plant development. The callose synthase is also more specifically involved in the plant defense response to wound, physiological stress or infection (*1*).

Despite the importance of plant β-glucan synthases at the cellular level and the relatively simple chemical structure of their reaction products, the molecular mechanisms in which these enzymes are involved are still poorly understood. This lack of understanding contrasts with the recent progress in the identification of the genes believed to be involved in the catalysis of cell wall polysaccharide biosynthesis (*2-5*). One example is the cellulose synthase system for which molecular biology approaches have involved the use of plant mutants together with analyses of DNA sequences available in sequence data banks. These techniques have allowed important progress in the identification of putative genes without deciphering the process of cellulose synthesis and assembly into microfibrillar structures. Figure 1 summarizes the different steps believed to occur during β-glucan biosynthesis as well as the molecular aspects that still need to be clarified. For instance, the requirement of a primer to initiate polymerization of β-glucan chains remains to be determined. Although the elongation of cellulose chains has been shown to occur from the non-reducing end (*6*), no data demonstrating the existence of a similar mechanism for other plant cell wall polysaccharides like $(1\rightarrow3)$-β-glucans are currently available. Another aspect that is not clear regarding β-glucan biosynthesis concerns the mechanisms by which the polysaccharides are translocated across the plasma membrane during or after polymerization from the cytoplasmic substrate, UDP-glucose. Several models have been proposed to explain this process (*7*) but none of them has been demonstrated yet. It is also important for both fundamental and applied purposes to understand the molecular mechanisms controlling the degree of polymerization of β-glucans. The fact that the processes involved in plant β-glucan biosynthesis are not well characterized is mainly due to the difficulty in demonstrating *in vitro* the function of the products of the identified genes. Indeed, the putative glycosyltransferases corresponding to the isolated genes are not easy to express in an active form because they contain several transmembrane domains. There is therefore a need to further develop biochemical approaches to study plant β-glucan synthases in order to complement the important progress recently made using molecular genetic techniques. We have started this work on suspension cultures of cells from *Arabidopsis thaliana* and the blackberry *Rubus fruticosus*. If *A. thaliana* has the advantage of being the plant model system currently the most widely used for genomic approaches, it is however important to use in parallel a second model like *R. fruticosus* for comparative biochemical studies.

So far, the purification and characterization of glucan synthases have proved difficult, largely because of their location in plasma membranes and their inherent instability. This is particularly true for cellulose synthases which

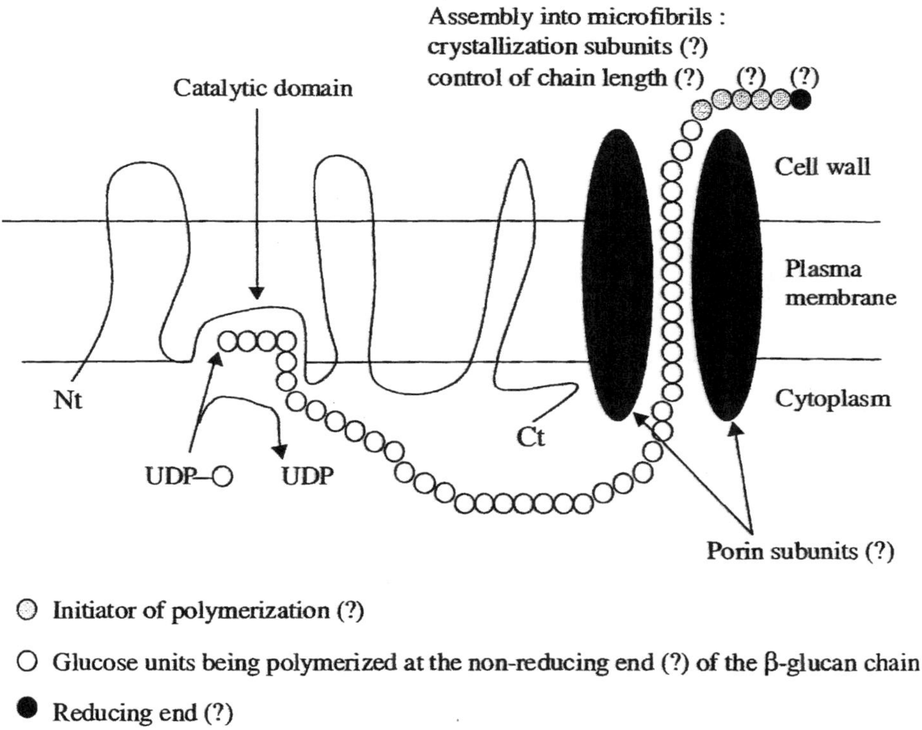

Figure 1. Hypothetical model for the biosynthesis of plant β-glucans. Question marks indicate the aspects that still have to be clarified: (a) involvement of a primer to initiate polymerization, (b) orientation of the chain being extruded, (c) mechanism of translocation across the plasma membrane, (d) involvement of crystallization subunits and (e) control of the degree of polymerization. The regulation subunits that may be involved in the biosynthetic process are not represented.

are organized as labile macromolecular complexes, usually poorly active *in vitro* after detergent extraction. On the other hand, significant progress has been made on the determination of the general biochemical properties of plant callose synthases, thanks to the use of specific detergents that preserve enzyme activities. Detergent extractions of plant (1→3)-β-glucan synthases are indeed well documented, but morphological and structural studies on the products and on the vesicles and micelles formed *in vitro* under various conditions are limited. In the present work partly based on transmission electron microscopy (TEM) and physico-chemical analyses of the polysaccharides synthesized *in vitro*, we report on these aspects, using the callose synthases from *A. thaliana* and *R. fruticosus*, incubated with or without UDP-glucose. The data presented in the next section show for instance that it should become possible to control the characteristics of the (1→3)-β-glucans synthesized *in vitro*, e.g. their degree of polymerization, depending on the conditions used to extract callose synthases from membrane preparations. Several detergents generated vesicles whereas others allowed the production of particulate structures of the micelle type. We have obtained a fraction enriched in a (1→3)-β-glucan synthase from *A. thaliana* and sequenced a porin that co-purifies with enzyme activity. The involvement of porins for cell wall polysaccharide extrusion through plasma membranes is discussed. We also describe a procedure that allows *in vitro* synthesis of cellulose from cell-free extracts of *R. fruticosus* cells. This protocol should facilitate the direct study of the cellulose synthase system using biochemical approaches.

Results and Discussion

Determination of Conditions for *In Vitro* Synthesis of β-Glucans.

(1→3)-β-Glucan Synthase Activity

Assay conditions for (1→3)-β-glucan synthases of *A. thaliana* and *R. fruticosus* were determined from enzyme preparations solubilized with CHAPS (0.5% final concentration). Our presentation is limited to the determination of assay conditions for the callose synthase from *R. fruticosus* since comparable data were obtained with enzymes from both plant systems.

Callose synthase activity was optimal in the presence of 1 mM UDP-glucose and enzymes were fully active in the range of pH 6.5 to 7.0. MOPS buffer was used at pH 6.8 for further assays. As Ca^{2+} and Mg^{2+} are usually added to assay plant callose synthases, we tested their effect on the enzymes from our plant systems. A very low activity was measured in the absence of Ca^{2+}, indicating that this cation is required for maximal activity. The optimal Ca^{2+} concentration ranged between 1.6 to 16 mM, the activity decreasing rapidly for concentrations higher than 16 mM. Mg^{2+} inhibited the callose synthase from *R. fruticosus* at concentrations as low as 0.5 mM, as opposed to other plant systems where Mg^{2+} is most often added in the assay mixtures (*8-10*). *In vitro* callose synthase activity is usually maximal in the presence of disaccharides such as cellobiose, laminaribiose or gentiobiose which are considered as allosteric activators (*8, 11*). Enzymes from our plant systems were weakly stimulated (5 to 15%) by cellobiose (20 mM) in comparison with other systems such as *Lolium multiflorum* (*8*) or *Beta vulgaris* (*11*). A strong inhibition was observed when the assay mixture contained laminarioligosaccharides having a degree of polymerization higher than 2 such as those found in laminarin. This inhibition is most likely due to the direct binding of laminarioligosaccharides in the active site in an almost irreversible manner, preventing the successive transfer of glucose units from UDP-glucose to the growing glucan chains. These data indicate that laminarioligosaccharides are not used by callose synthases as primers to initiate polymerization of (1→3)-β-glucan chains. The requirement of a possible primer for synthesis of β-glucans in general remains to be demonstrated.

In summary, the conditions giving maximal (1→3)-β-glucan synthase activity *in vitro* correspond to an assay mixture containing 50 mM MOPS buffer pH 6.8, 1 mM UDP-glucose, 8 mM Ca^{2+} and 20 mM cellobiose. These conditions were used for all experiments presented in the next sections.

Cellulose synthase activity

Considerable efforts have been made in recent years to achieve *in vitro* synthesis of cellulose from plant cell-free extracts. In most protocols in which detergent extractions of the cellulose synthase complexes were used, the treatments resulted in a loss of (1→4)-β-glucan synthase activity (*12*). Instead, high levels of callose were present in the reaction mixture after incubation with UDP-glucose (*12*). We have screened a series of detergents for their ability to preserve cellulose synthase activity and to allow *in vitro* synthesis of native cellulose. We have been able to synthesize cellulose *in vitro* from *R. fruticosus* cells when the cellulose synthase was extracted from plasma membranes with polyoxyethylene 20 cetyl ether (Brij 58; 0.05% final concentration) and incubated with 1mM UDP-glucose in 50 mM MOPS buffer pH 6.8. Eight mM Mg^{2+} was added in the reaction mixture to inhibit (1→3)-β-glucan synthase activity and favor cellulose synthesis. Enzyme activity was found to be higher in the presence of 20 mM cellobiose, the role of which (activator or primer) remains to be determined for cellulose biosynthesis. Figure 2 clearly shows the morphological difference between cellulose coming from the cell walls and contaminating plasma membrane preparations and the cellulose synthesized *in vitro*. The *in vitro* product was most often associated with granular structures that may correspond to rosette-like particles responsible for the synthesis of cellulose. Similar structures were observed in a fraction from *Vigna radiata* obtained by native PAGE and containing cellulose synthase activity (*13*). Interestingly, the microfibrils of (1→3)-β-glucan synthesized *in vitro* by a callose synthase from *R. fruticosus* solubilized with CHAPS (0.5%) also presented granular structures at their end (not shown). These granular structures exhibited a smaller size than those associated with the cellulose microfibrils and may correspond to the complexes responsible for callose synthesis.

Characterization of the β-glucans synthesized *in vitro*.

Several reactions were performed to accumulate up to 50 mg of the (1→3)-β-glucans synthesized *in vitro* using each of the various detergents that preserve enzyme activity. The polysaccharides were characterized using a battery of physico-chemical techniques. For example, the data obtained with the polymer synthesized by a callose synthase from *R. fruticosus* extracted with CHAPS are summarized in Table I. They unequivocally demonstrate that the product synthesized *in vitro* under these conditions is a linear (1→3)-β-glucan characterized by a degree of polymerization higher than 100 and a low degree of crystallinity associated with a microfibrillar structure (Figure 4A).

So far, the cellulose synthesized *in vitro* by enzyme extracts prepared with Brij 58 was characterized using biochemical techniques only, and more product is currently being accumulated in order to obtain a complete chemical and physico-chemical characterization of the polymer. Cellulose was synthesized with ^{14}C-labelled UDP-glucose and up to 50% of the radioactive polymer was hydrolyzed after treatment with a mixture of the specific cellobiohydrolases I (Cel7a) and II (Cel6A) from *Humicola insolens* cloned and expressed in *Aspergillus oryzae* (not shown). 50% of the reaction product that was resistant

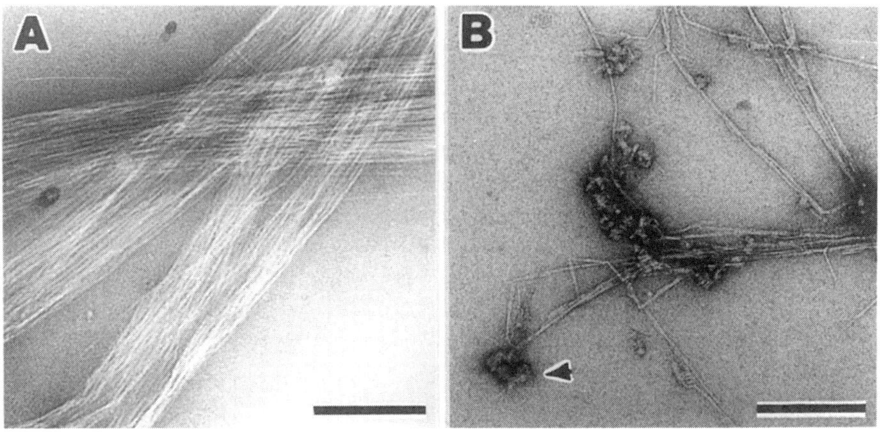

Figure 2. TEM micrographs (negative staining using 2% uranyl acetate) of cellulose from the cell walls of R. fruticosus *cells (A), and of the cellulose synthesized* in vitro *after extraction of* R. fruticosus *cellulose synthase with Brij 58 (B). Granular structures that may correspond to rosette-like particles responsible for the synthesis of cellulose are associated with the microfibrils (B, arrowhead). Bars=200 nm.*

to Cel7A and Cel6A was however sensitive to laminarinase, indicating that it corresponded to (1→3)-β-glucan.

Table I. Characterization of the (1→3)-β-glucan synthesized *in vitro* by the CHAPS-extracted callose synthase from *R. fruticosus*

Solubility	Insoluble in water Soluble in DMSO Soluble in NaOH
Infrared spectroscopy	Absorption band at 889 cm^{-1} (β linkages)
Methylation* (GC/MS)	~100% (1→3)-linked glucose Degree of polymerization > 100
^{13}C-NMR spectroscopy in DMSO solution	Resonance signals at 60.9, 68.5, 72.9, 76.4, 86.2 and 103.1 ppm assigned to carbons -6, -4, -2, -5, -3 and -1 of (1→3)-β-glucan
X-ray diffraction	Low degree of crystallinity
Reaction with the aniline blue fluorochrome	Strong UV-fluorescence induced
Transmission electron microscopy	Microfibrillar morphology

* The permethylated derivative corresponding to the non-reducing end of the polysaccharide (1,5-di-*O*-acetyl-2,3,4,6-tetra-*O*-methyl-D-glucitol) was not detectable by GC/MS, indicating that the polymer has a degree of polymerization higher than 100.

Effect of detergents on (1→3)-β-glucan synthesis.

We have studied a series of detergents for their ability to extract callose synthases and have examined their effect on the morphology of the vesicles or micelles produced *in vitro*. In addition, we have studied the effect of some of the detergents on the characteristics of the (1→3)-β-glucans synthesized *in vitro*. Only a few examples of the data obtained are presented here. A very different morphology characteristic of either vesicles or micelles was observed when membrane preparations from a given plant species were incubated in the presence of various detergents, e.g. CHAPS or octyl-β-D-glucoside, as shown in Figure 3, A and B. Strong differences were also observed when the same detergent was used to obtain enzyme preparations from membrane fractions of our two plant models, *A. thaliana* and *R. fruticosus* (not shown). The size of the microfibrillar (1→3)-β-glucans synthesized *in vitro* was affected by the nature of the detergent used for enzyme solubilization (Figure 3, C and D). Differences in the size of the product were also observed when the same detergent was used to extract plant enzymes from both origins (compare Figure 3C with Figure 4A and Figure 3D with Figure 4B). These observations and the data obtained from enzyme kinetics (not shown) suggest that the general organization of the glucan synthase complexes is influenced by the detergent used for protein extraction. In addition, the morphology and size (number of associated chains, degree of polymerization) of the microfibrils synthesized *in vitro* might be altered as a result of modifications induced by detergents on enzyme complexes.

The results presented in this section suggest that, depending on the conditions used for enzyme extraction (e.g. detergent used), it should become possible to control some of the physico-chemical properties of the β-glucans synthesized *in vitro* such as, for example, their degree of polymerization, their morphology and crystallinity. It will however be necessary to study the lipid environment of the glucan synthase complexes in plasma membranes in order to establish for various detergents a relation between the molecular mechanisms occurring during enzyme extractions and the characteristics of the reaction products.

Characterization of a (1→3)-β-glucan synthase from *A. thaliana*.

We have obtained a preparation enriched in (1→3)-β-glucan synthase activity by ultracentrifugation of CHAPS-extracted enzymes on linear glycerol gradients (density 1.02 to 1.11). After a 4-h centrifugation at 150,000 g, (1→3)-β-glucan synthase activity sedimented as a relatively sharp peak ahead of the bulk of proteins. The callose synthase was purified 15 times from the CHAPS-extracted enzyme preparation. SDS-PAGE analysis of the fraction with the highest specific activity contained major proteins of 30, 50, 55, 63 and 66 kDa (not shown). The occurrence of these proteins in the different fractions correlated with the distribution of enzyme activity in the gradient, indicating that these proteins are the most likely candidates for participation in (1→3)-β-glucan synthase activity. Other groups have related the purification of several plant callose synthases to the presence of 30-32- (*14-17*) and 50-55-

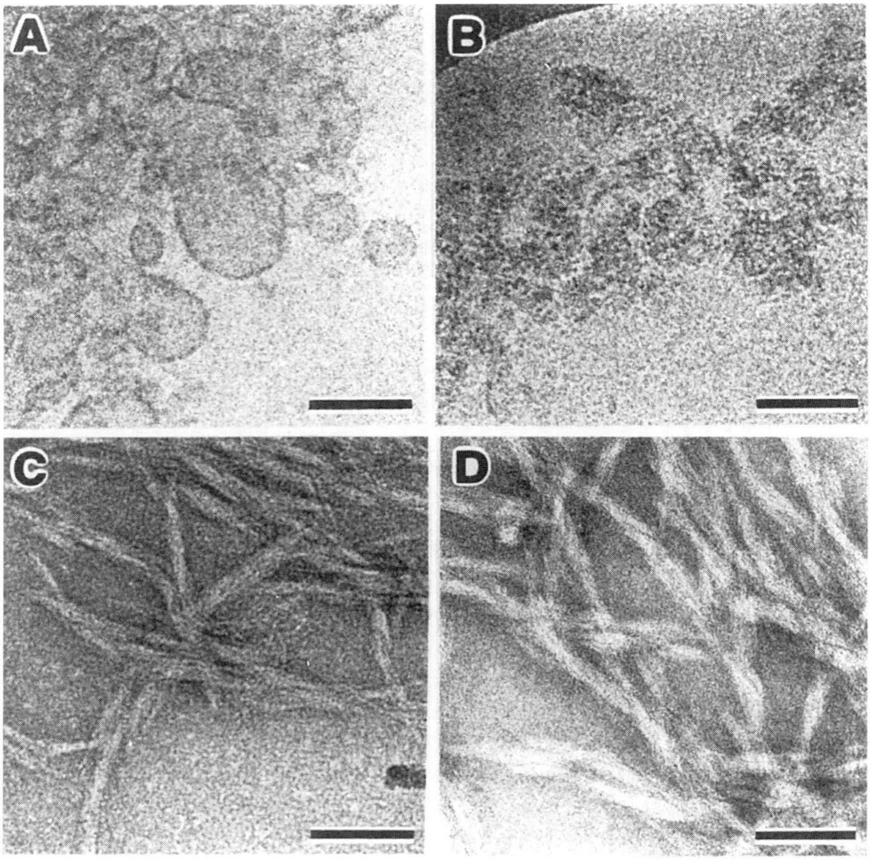

Figure 3. A and B: Cryo-TEM in vitreous ice of particulate structures obtained after incubation of A. thaliana *plasma membranes in the presence of CHAPS (A) or octyl-β-D-glucopyranoside (B). The structures shown in A have a morphology characteristic of vesicles whereas those shown in B seem to correspond to true micelles. C and D: TEM micrographs (negative staining using 2% uranyl acetate) of the (1→3)-β-glucans synthesized* in vitro *by the preparations obtained in A (C) and in B (D). Bars=100 nm.*

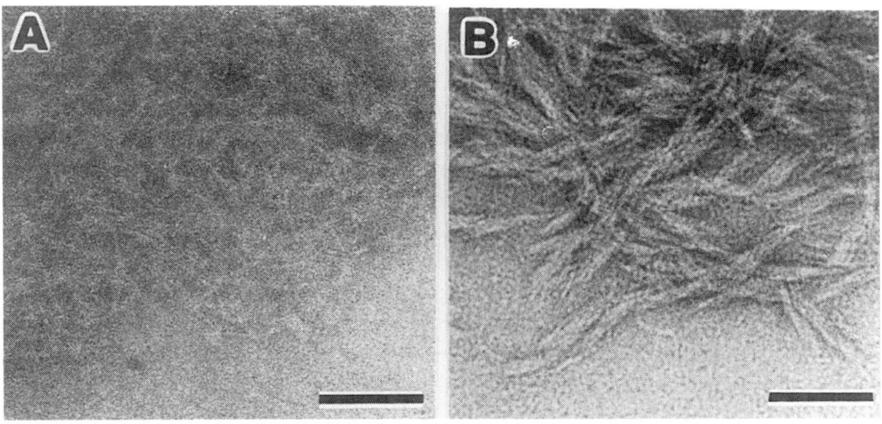

Figure 4. TEM micrographs (negative staining using 2% uranyl acetate) of the (1→3)-β-glucans synthesized in vitro *by the callose synthase from* R. fruticosus *cells, after extraction of membrane proteins with CHAPS (A) or octyl-β-D-glucopyranoside (B). Bars=100 nm.*

(*14, 18-20*) kDa polypeptides in the enriched preparations. We have started to sequence each of the 30-, 50-, 55-, 63- and 66-kDa proteins after digestion with trypsin and purification of the resulting peptides by reverse-phase HPLC. Several sequences were obtained from the 30-kDa protein. They gave 100% identity with fragments of a voltage-dependent ion channel protein (porin) from *A. thaliana* which is an early marker of the hypersensitive response to bacterial infection (*21*). Interestingly, callose synthase produces (1→3)-β-glucan chains from the cytosoluble substrate UDP-glucose in response to wound, physiological stress or microbial infection, and it has been shown that this membrane-bound enzyme can be stimulated by a variation of the membrane potential (*22*). The function of porins in higher plants is not clear, but suggested functions include the transport of water and an increasing spectrum of solutes. The co-purification of a porin with callose synthase raises questions concerning a possible role in the translocation across the membrane of nascent β-glucans. The involvement of a porin for extrusion of β-glucans through plasma membranes has previously been suggested by Qi *et al.* (*23*) who sequenced a channel protein from a fraction that contains a high callose synthase activity. Other models proposed for cellulose translocation suggest that the transmembrane helices of the putative cellulose synthase catalytic subunit would interact to form a central channel through which the glucan could be secreted (*24*). Polysaccharide extrusion across the plasma membrane is still not explained but different mechanisms requiring several associated proteins or directly involving the catalytic subunit(s) may exist for cellulose and callose.

It will be possible to describe one (or several) integrated model(s) for plant callose and cellulose biosynthesis only when a complete biochemical characterization of β-glucan synthase complexes is available. The next important step is therefore to further develop biochemical studies to better understand the mechanisms that regulate and control the various steps involved in the process of callose and cellulose synthesis (i.e. the polymerization, the extrusion and the assembly of the polysaccharide chains into supramolecular structures).

Acknowledgements

The authors are grateful to the late Dr. M. Schülein from Novo Nordisk (Denmark) for the generous gift of cellobiohydrolases Cel7A and Cel6A, and to Dr. J. Gagnon and Mr. J.P. Andrieu ("Institut de Biologie Structurale", Grenoble, France) for help with peptide sequencing. This work was supported by a Grant-in-Aid to V. B. from Hercules Incorporated (Wilmington, DE, USA).

References

1. Stone, B. A.; Clarke, A. E. *Chemistry and Biology of (1→3)-β-Glucans;* La Trobe University Press: Melbourne, Australia, 1992; pp 365-429.
2. Pear, J. R.; Kawagoe, Y.; Schreckengost, W. E.; Delmer, D. P.; Stalker, D. M. *Proc. Natl. Acad. Sci. USA* **1996,** *93,* 12637-12642.
3. Arioli, T.; Peng, L. C.; Betzner, A. S.; Burn, J.; Wittke, W.; Herth, W.; Camilleri, C.; Höfte, H.; Plazinski, J.; Birch, R.; Cork, A.; Glover, J.; Redmond, J.; Williamson, R. E. *Science* **1998,** *279,* 717-720.
4. Turner, S. R.; Somerville, C. R. *Plant Cell* **1997,** *9,* 689-701.
5. Wu, L.; Joshi, C. P.; Chiang, V. L. *Plant Physiol.* **1998,** *117,* 1125.
6. Koyama, M.; Helbert, W.; Imai, T.; Sugiyama, J.; Henrissat, B. *Proc. Natl. Acad. Sci. USA* **1997,** *94,* 9091-9095.
7. Brown, R. M. Jr.; Saxena, I. M. *Plant Physiol. Biochem.* **2000,** *38,* 57-67.
8. Ng, K.; Johnson, E.; Stone, B. A. *Plant Physiol.* **1996,** *111,* 1227-1231.
9. Eiberger, L. E.; Wasserman, B. P. *Plant Physiol.* **1987,** *83,* 982-987.
10. Read, S. M.; Delmer, D. P. *Plant Physiol.* **1987,** *85,* 1008-1015.
11. Morrow, D. L.; Lucas, W. J. *Plant Physiol.* **1986,** *81,* 171-176.
12. Delmer, D. P. *Annu. Rev. Plant Physiol.* **1987,** *38,* 259-290.
13. Kudlicka, K.; Brown, R. M. Jr. *Plant Physiol.* **1997,** *115,* 643-656.
14. Wu, A.; Harriman, R. W.; Frost, D. J.; Read, S. M.; Wasserman, B. P. *Plant Physiol.* **1991,** *97,* 684-692.
15. Fredrikson, K.; Kjellbom, P.; Larsson, C. *Physiol. Plant.* **1991,** *81,* 289-294.
16. Kuribayashi, I.; Morita, T.; Mitsui, T.; Igaue, I. *Biosci. Biotech. Biochem.* **1993,** *57,* 682-684.
17. Bulone, V.; Fincher, G. B.; Stone, B. A. *Plant J.* **1995,** *8,* 213-225.
18. Dhugga, K. S.; Ray, P. M. *FEBS Lett.* **1991,** *278,* 283-286.
19. Frost, D. J.; Read, S. M.; Drake, R. R.; Haley, B. E.; Wasserman, B. P. *J. Biol. Chem.* **1990,** *265,* 2162-2167.
20. Lawson, S. G.; Mason, T. L.; Sabin, R. D.; Sloan, M. E.; Drake, R. R.; Haley, B. E.; Wasserman, B. P. *Plant Physiol.* **1989,** *90,* 101-108.
21. Lacomme, C.; Roby, D. *FEBS Lett.* **1999,** *459,* 149-153.
22. Bacic, A.; Delmer, D. P. *Planta* **1981,** *152,* 346-351.
23. Qi, X.; Tai, C. Y.; Wasserman, B. P. *Plant Physiol.* **1995,** *108,* 387-392.
24. Delmer, D. P. *Annu. Rev. Plant Physiol. Plant Mol. Biol.* **1999,** *50,* 245-276.

Synthesis of Monomers and Macromers

Chapter 6

Cyclohexanol Biodegradation Genes: A Pathway of Opportunities

Hiroaki Iwaki[1,2], Yoshie Hasegawa[1], Masahiro Teraoka[1], Tai Tokuyama[1], Laetitia Bernard[2], and Peter C. K. Lau[2]

[1]Department of Biotechnology, Faculty of Engineering and High Technology Research Center, Kansai University, Suita, Osaka 564–8680, Japan
[2]Biotechnology Research Institute, National Research Council Canada, Montréal, Québec H4P 2R2, Canada

We have now determined the complete gene sequence of the cyclohexanol (*chn*) degradation pathway in *Acinetobacter* sp. NCIMB 9871 as well as the putative genes for the β-oxidation of adipic acid to acetyl-CoA and succinyl-CoA. In addition, a new insertion sequence, potentially useful in strain characterization, was identified. Knowledge of the nucleotide sequence of the *chn* genes was used to construct clones of *Escherichia coli* that would overproduce the requisite biocatalysts: a flavin monooxygenase (ChnB; cyclohexanone 1,2-monooxygenase [CHMO]), a ring-opening hydrolase (ChnC; ε-caprolactone hydrolase) and three oxido-reductases (ChnA, cyclohexanol dehydrogenase; ChnD, 6-hydroxyhexanoate dehydrogenase; and ChnE, 6-oxohexanoate dehydrogenase). Besides the well known application of CHMO as a Baeyer-Villiger biocatalyst that carries out stereoselective oxidations of a wide variety of ketones to the corresponding lactones, potential applications of the Chn biocatalysts in the development of "green" bioprocesses such as an "enviro-compatible" synthesis of adipic acid are discussed.

Biotechnology offers a powerful and versatile enabling technology for industrial sustainability by delivery of clean industrial products and processes (*1*). The use of microorganisms as a source of biocatalysts in the production of chemical intermediates constitutes an important step towards the "greening" of industry (*2*). World-wide, there is a growing interest and expanding list of industrial examples of biocatalysts (*3*, *4*).

We are interested in the regulation and molecular dissection of the metabolic pathways used by bacteria to degrade cycloaliphatic compounds (eg. cyclohexanol and cyclohexanone) and cyclohexylamine (used widely as an insecticide and antiseptic) with the goals of developing a knowledge-based bioresource and broad spectrum of cloned biocatalysts for value-added production and possible bioprocesses (*5*, *6*). One of the organisms chosen for study is *Acinetobacter* sp. NCIMB 9871, a strain best known for the cyclohexanone-inducible production of cyclohexanone 1,2-monooxygenase (CHMO, ChnB), a 61-kDa monomeric flavoprotein which carries out a formal Baeyer-Villiger organic reaction in the formation of chiral lactones from the respective ketones (*7*, *8*, *9*). Moreover, the biochemical pathway of cyclohexanol degradation in the strain NCIMB 9871 is well established (Figure 1). Recently, we have described the gene locus of the NCIMB 9871 chromosome that encodes a transcriptional activator (ChnR) for the expression of ChnB and the ChnE enzymes in the configuration of *chnBER* (*5*). In the present study, we completed the cloning and sequence of a 31-kb (kilo-basepairs) genetic locus that spans the entire cyclohexanol oxidation pathway and the surrounding regions. Several plasmid derivatives were made thereof with the aim of overproducing the requisite Chn biocatalysts in *E. coli* and to characterize the cloned entities for their potential applications.

Materials and Methods

Chemicals

All chemicals were purchased from Aldrich Chemical Co. (Milwaukee, Wis.) or Sigma and were the purest commercially available. The digoxigenin-11-UTP labelling system for the preparation of DNA probes (Figure 1) was purchased from Boehringer Mannheim GmbH (Germany) and used according to the manufacturer's instructions.

Biological materials, genetic manipulations and sequence analysis

The bacterial strains and plasmids used in the work are shown in Table I. Unless otherwise stated, all culture conditions and media, recombinant DNA methods, automated DNA sequencing and the softwares used in gene analysis were as previously described (*5*). The nucleotide sequences determined in this study have been submitted to DDBJ and provided accession numbers AB026668 and AB026669.

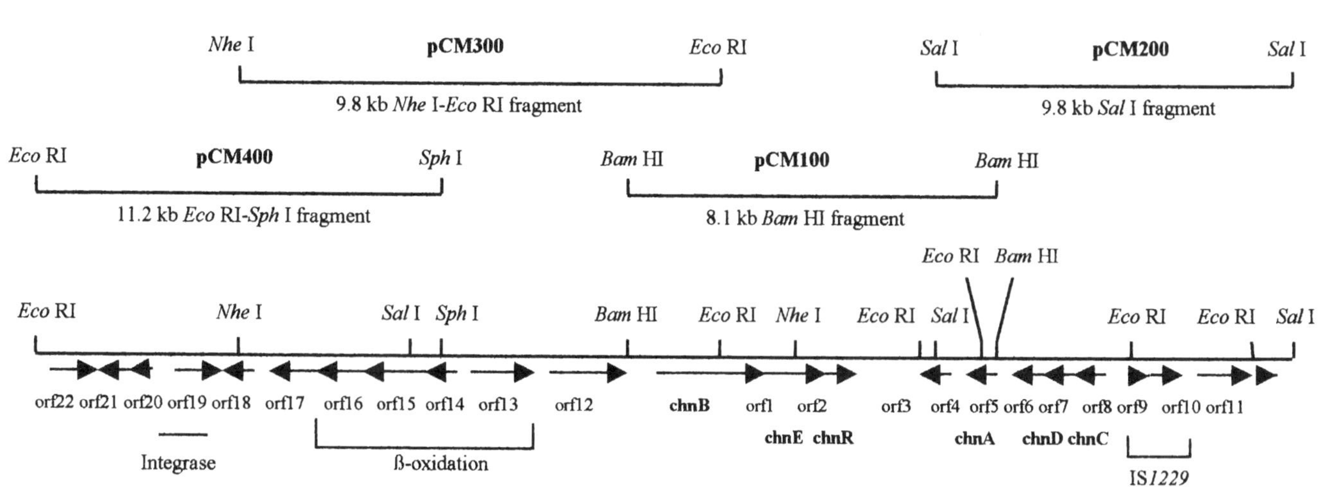

Figure 1. The biochemical steps of cyclohexanol (chn) degradation and their genes and surrounding region in Acinetobacter *sp. strain NCIMB 9871. The directions of the arrows indicate divergent or convergent transcriptions of the respective genes or open reading frames (orfs). The chn-encoded enzymes are: ChnA, cyclohexanol dehydrogenase; ChnB, cyclohexanone 1,2 monooxygenase (CHMO); ChnC, ε-caprolactone hydrolase; ChnD, 6-hydroxyhexanoate dehydrogenase; and ChnE, 6-oxohexanoate dehydrogenase. ChnR, transcriptional activator (5). Further oxidation of adipic acid to acety-CoA and succinyl-CoA is presumed to proceed via β-oxidation. The cloned restriction fragments of the NCIMB 9871 DNA in* E. coli *plasmids are as indicated.*

Subcloning in *E. coli*

In addition to the previously constructed ChnE expression clone (pCM130) in *E. coli* JM109 (*5*), the following pairs of oligos with the desired cloning sites (underlined sequences) were synthesized and used to amplify the *chnA*, *chnB*, *chnC* and *chnD* genes, respectively. Conditions of PCR amplification were as previously described (*5*).

chnA: 5'-TCCCCCGGGTTGTCGCTTGGCAGACG and
5'-TCCCCCGGGGGTATCTTCTACTCATTT;

chnB: 5'- GCACTGCAGTCCGGAATGGAGATTATCATGTCAC and
5'-GGCTGCAGCGTACGTTCGAATTAGGCATTGGCAGGTTG;

chnC: 5'-TCCCCCGGGATTTATTCCTTAAATATC and
5'-AAAACTGCAGTTTTATACGCCGTGTGGGC;

chnD: 5'-CGGAATTCATGCACTGTTACTGCGTG and
5'-CGGAATTCTTGATTTTGATTTGGGTAT

Table I. Bacterial strains and plasmids used in this study

Strains	*Characteristics*	*Reference or source*
Acinetobacter sp. NCIMB 9871	grows on cyclohexanol	NCIMB
E. coli JM109	Host for recombinant plasmids	(*10*)
E. coli DH5α	Host for recombinant plasmids	*(11)*
E. coli XL-1blue	Host for recombinant plasmids	(*12*)
Plasmids		
pUC18	Ap^r	*(10)*
pSD80	Expression vector, Ap^r	*(13)*
pCM100	8.1-kb *Bam*HI fragment in pUC18	*(5)*
pCM130	1.4-kb *Eco*RI*-*Pst*I* fragment (*chnE*) in pSD80	*(5)*
pCM200	9.8-kb *Sal*I fragment in pUC18	This study
pCM300	9.8-kb *Eco*RI-*Nhe*I frragment in pUC18	This study
pCM400	11.2-kb *Nhe*I-*Eco*RI fragment in pUC18	This study
pCM204	0.8-kb *Sma*I* fragment (*chnA*) in pSD80	This study
pCM075	1.6-kb *Pst*I* fragment (*chnB*) in pSD80	This study
pCM206	1.1-kb *Eco*RI* fragment (*chnD*) in pSD80	This study
pCM207	0.9-kb *Sma*I*-*Pst*I* fragment (*chnC*) in pSD80	This sutdy

* indicate restriction sites introduced by PCR design

In each case, the amplified DNA fragment was purified from an agarose gel, digested with the appropriate restriction enzyme(s) and cloned in the respective linearized pSD80 vector as indicated in Table I. Each cloned insert was sequenced to confirm that the PCR amplified fragments were free of mutation. With the exception of *chnB* in *E. coli* strain DH5α, other *chn* genes were cloned in JM109.

Enzyme assays and protein expression

Induction of the various pSD80 derivatives by isopropyl-β-D-thiogalactopyranoside (IPTG; final concentration 1 mM) and processing of the *E. coli* cell extracts for the various enzyme activity assays were as described for *E. coli*(pCM130) *(5)*. As in the case of 6-oxohexanoate dehydrogenase (ChnE), cyclohexanol dehydrogenase (ChnA) and 6-hydroxyhexanoate dehydrogenase (ChnD) activities were assayed following the increase in absorbance at 340 nm except NAD^+ (1 μmol) was the cofactor. Beside the test extract and cofactor, the 1-ml reaction assay in the ChnA assay contains 50 μmol of glycine/NaOH buffer (pH 10.0) and 1 μmol of cyclohexanol. In the case of ChnD, this consists of 1.0 μmol of 6-hydroxyhexanoate and 50 μmol of phosphate buffer (pH 7.0). 6-Hydroxyhexanoate was prepared as described by Donoghue and Trudgill *(7)*. ε-Caprolactone hydrolase (ChnC) was assayed by measuring the rate of lactone consumption in a 2-ml reaction mixture that contains 8 μmol of ε-caprolactone, 100 μmol of phosphate buffer (pH 7.2), and test extract. The CHMO activity was determined as previously described *(5)*. One unit of activity was defined as the amount of enzyme required to convert 1 μmol of substrate in 1 min. Protein concentrations were determined by the method of Bradford using a commercial kit available from Bio-Rad. SDS-10% polyacrylamide gel electrophoresis gels were run as previously described *(5)*.

Results and Discussion

Cloning and sequencing of the *chnBER* flanking regions

Previously, we described the cloning of a 8.1-kb *Bam*HI fragment of the *Acinetobacter* sp. NCIMB 9871 chromosome that contains the *chnBER* gene cluster in a recombinant plasmid designated pCM100 (Figure 1). As the gene repertoire for the entire cyclohexanol catabolic pathway was incomplete, we sought for the adjacent DNA fragments and sequences. In brief, three additional fragments were cloned in the *E. coli* XL1-blue cells as the result of hybridization using the designated probes (Probes 1, 2 and 3; Figure 1). These new plasmid derivatives were designated pCM200, pCM300 and pCM400. Taken together the four overlapping clones spanned 31-kb. Each of the cloned DNA was sequenced in both strands and these sequences were assembled to provide a contiguous segment of 30,981 bp.

Open reading frames upstream and downstream of the *chnBER* cluster

At first, the sequence of the small 1.8-kb *SalI*-*Bam*HI fragment of pCM100 was determined. This was found to be identical to the overlapping region in the 9.8-kb *SalI* fragment of pCM200. Complete sequencing of the 9.8-kb *SalI* fragment and subsequent analysis predicted the presence of nine open reading frames (ORFs), five on one strand and four in the opposite direction. All predicted initiator codons are preceded by an appropriately positioned consensus ribosome-binding site. Table II lists the homology of the ORFs with proteins in the GenBank and associated protein databases *(14)*. As well, this table contains eleven other predicted ORFs found in the 11.2-kb *Eco*RI-*SphI* fragment in clone pCM400 and in the 9.8-kb *NheI*-*Eco*RI bridging fragment in the plasmid pCM300. Those ORFs that are of special interest in this study are elaborated below.

Amino acid sequence features of ChnA (ORF4), ChnC (ORF7) and ChnD (ORF6)

The predicted amino acid sequence of ORF4 was found to belong to the short-chain dehydrogenase/reductase (SDR) family of proteins (*15*). Notable sequence features of ORF4 are the conservation of a coenzyme-binding motif (GlyXXXGlyXGly) at positions 14-20 and potential active-site residues, Ser144, Tyr157 and Lys161.

The deduced amino acid sequence of ORF6 revealed homology to the zinc containing alcohol dehydrogenase family of enzymes, the highest scoring candidate in the database being a NAD^+-dependent alcohol dehydrogenase of *Sulfolobus solfataricus* (37% identity). A notable sequence motif present in ORF6 and related proteins is the coenzyme-binding motif (GlyXGlyXXGly at positions 183-188). Asp207 is also conserved, a residue expected to be essential for the binding of NAD^+ *(22)*. The catalytic and structural zinc-binding residues of ORF6 are presumably Cys38, His72, and Cys158, and Cys(s) at 102, 105, 108 and 116, respectively.

The deduced amino acid sequence of ORF7 contains a pentapeptide, Gly-Asp-Ser-Ala-Gly at positions 148-152, a feature resembling the conserved GlyXSerXGly motif found in the family of hydrolases where Ser is a catalytic residue (*16*). As in the family of lipases/esterases/hydrolases, it is possible that Ser150 together with Asp249 and His275 forms the catalytic triad for ORF7 .

Table II. Homology of the ORFs with proteins in the databases (*14*)*

ORF	*Amino Acids*	*Homologous Protein*	*Source Species*	*Identity (%)*	*Accession Number*
ChnB	543	steroid monooxygenase	*Rhodococcus rhodochrous*	42	AB010439
ChnE	477	aldehyde dehydrogenase TsaD2	*Comamonas testosteroni*	42	AF305549
ChnR	313	regulatory protein GdhBR	*Pantoea citrea*	25	AF102513
Orf3	324	pilin gene inverting protein PivNM-2	*Neisseria meningitidis*	33	AE002525
Orf4	247	putative oxidoreductase	*Streptomyces coelicolor*	56	AL356592
Orf5	299	unknown protein	*Pseudomonas* sp.	29	AF043544
Orf6	352	NAD(+)-dependent alcohol dehydrogenase	*Sulfolobus solfataricus*	37	S51211
Orf7	300	putative esterase	*Streptomyces coelicolor*	40	AL049841
Orf8	92	putative transposase A	*Pantoea agglomerans*	63	AF327445
Orf9	290	putative transposase B	*Pantoea agglomerans*	68	AF327445
Orf10	449	sequestrin	*Plasmodium falciparum*	19	U09839
Orf11	79	putative prophage regulatory protein	*Escherichia coli*	45	AE005270
Orf12	574	electron transfer flavoprotein-ubiquinone oxidoreductase homolog	*Acinetobacter calcoaceticus* ADP1	91	U87258
Orf13	384	probable acyl-CoA dehydrogenase	*Pseudomonas aeruginosa*	76	AE004590
Orf14	258	probable enoyl-CoA hydratase	*Pseudomonas aeruginosa*	60	AE004590
Orf15	505	probable 3-hydroxyacyl-CoA dehydrogenase	*Pseudomonas aeruginosa*	57	AE004590
Orf16	400	probable acyl-CoA thiolase	*Pseudomonas aeruginosa*	65	AE004779
Orf17	405	conserved hypothetical protein	*Pseudomonas aeruginosa*	57	AE004482
Orf18	282	probable transcriptional regulator	*Pseudomonas aeruginosa*	60	AE004590
Orf19	425	phage integrase	*Mesorhizobium loti*	39	NC_002678
Orf20	193	probable transcriptional regulator	*Pseudomonas aeruginosa*	28	AE004728
Orf21	301	unknown conserved protein	*Bacillus halodurans*	34	AP001513
Orf22	361	hypothetical protein	*Pseudomonas aeruginosa*	51	AE004897

*The near identity of the 17-kb sequence in accession no. AF282240 reported for *Acinetobacter* sp. strain SE19 is not considered here.

The sequence identities of ORFs 4, 6 and 7 with known protein families in the databases and their motifs predict them to be ChnA, ChnD and ChnC, viz. cyclohexanol dehydrogenase, 6-hydroxyhexanoate dehydrogenase and caprolactone hydrolase, respectively. These assumptions are verified by their enzymatic activities and protein expression as described below.

Expression of ChnA, ChnB, ChnC, ChnD and ChnE in *E. coli* and activity assays.

Figure 2 examines the production of proteins in a Coomassie blue-stained SDS-polyacrylamide gel of the crude protein extracts prepared from the various *E. coli* JM109 or DH5α cells that were induced by 1.0 mM IPTG. Abundant amounts of the ChnA, ChnC and ChnE proteins were produced representing at least 10-30% of the total proteins (lanes 3, 5 and 7, respectively). The observed molecular masses of ChnA, ChnC and ChnE were found to be 25 kDa, 32 kDa and 53 kDa, values in good agreement with the DNA-predicted sizes (25.7 kDa, 33.2 kDa and 52-kDa, respectively). The overexpression of ChnB, identifiable by a 61-kDa protein was not readily produced in *E. coli* DH5α(pCM075) cells (Fig. 2; lane 4). But, in *E. coli* DH5α(pCM076) cells where the *chnB* gene was placed 3' to the *chnA* expression clone, a 61-kDa protein was readily identified. As expected, the ChnA protein was produced in a considerable amount (Figure 2B, lane 2). The appearance of a 37 kDa protein would be consistent with the size of ChnD. This was not observed in the pCM206 construct (Figure 2, lane 6) but only evident after an elaborated plasmid contruction (unpublished data).

Crude protein extracts from the various *E. coli* cells were used to determine the enzyme activities as described in Materials and Method. Table III compares the specific activities of the recombinant Chn enzymes with those of the parent *Acinetobacter* strain. Whereas the cyclohexanol dehydrogenase and ε-caprolactone hydrolase activities were much improved, a 30-33 fold increase, heterologous expression of the ChnB, ChnD and ChnE activities in the present plasmid constructs were no better than those reported for the original strain (*7*).

Table III. Specific activities of native and cloned enzymes

Enzyme	*Acinetobacter* * (U/mg)	*Clone* (U/mg)
ChnA	0.28	9.3
ChnB	0.6	0.2
ChnC	16	480
ChnD	0.03	0.08
ChnE	0.6	0.73

* According to: (*7*)

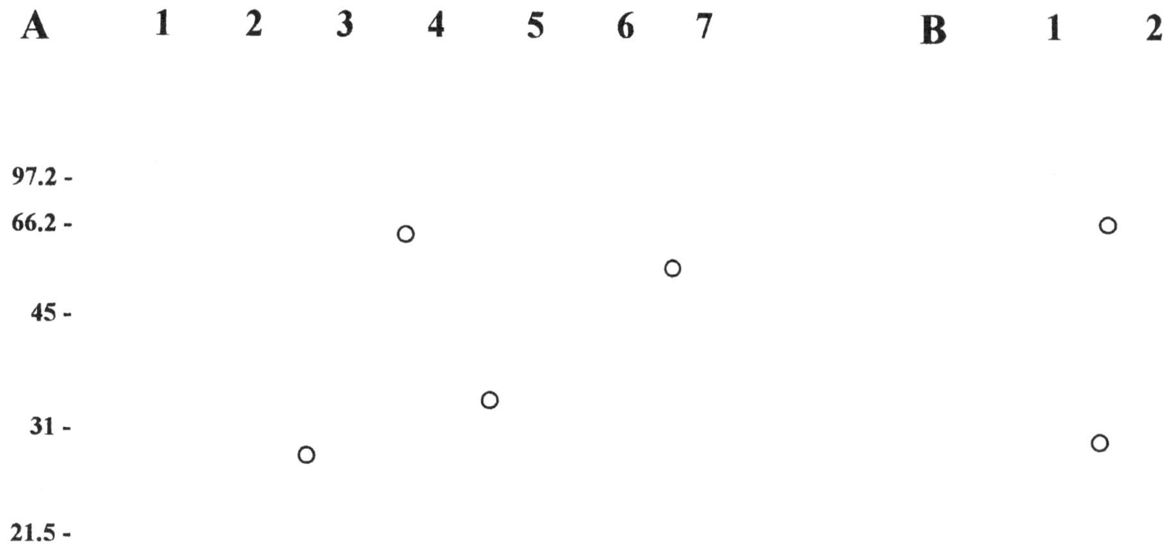

Figure. 2. SDS-Polyacrylamide (10%) gel electrophoresis of crude protein extracts from E. coli *cells induced by IPTG.* ***A.*** *lane 1, Low molecular weight standards (BioRad), as indicated alongside in kilodaltons; lane 2, pSD80 vector (control); lane 3, pCM204 (ChnA); lane 4, pCM075 (ChnB); lane 5, pCM207 (ChnC); lane 6, pCM206 (ChnD); lane 7, pCM130 (ChnE).* ***B.*** *lane 1 as in* ***A****; lane 2. pCM076 (ChnA and ChnB). The protein bands of interest are indicated by the white circles.*

"Green" opportunities and oxidations with *chn* genes

Either whole cells of *Acinetobacter* strain NCIMB 9871 or recombinant *E. coli* or yeast cells that harbor the CHMO-encoding gene have been used in a variety of asymmetric synthesis of organic compounds including drug synthesis (*9, 17, 18, 19, 20*). Besides the well known Baeyer-Villiger oxidation of ketones, CHMO is also capable of sulfoxidation, phosphoxidation, amine oxidation, boron oxidation and selenoxidation etc. Since aldehyde substrates, eg. are generally more toxic than the corresponding alcohols in bacterial cells a potentially useful strategy is to couple the *chnA* and *chnB* genes. To realize this a better understanding of the substrate specifity of ChnA is necessary since it is not known whether this enzyme is capable of accepting a wide spectrum of substrates as demonstrated by the ChnB monooxygenase. Such studies have began in our laboratory.

In addition to what has been reported for the resolution of racemic lactones *(21)* the ring-opening caprolactone hydrolase (ChnC) is anticipated to have more applications. ε-Decanolactone was found to be the next best substrate to ε-caprolactone (*21*). The availability of an overproducing clone of ChnC provides a unique opportunity to explore further substrate specificities.

An enzymatic route to the production of adipic acid from cyclohexanol is not only an intriguing possibility but an enviro-compatible one. In excess of 2.2 million metric tons per year of adipic acid (1,4-butanedicarboxylic acid) is annually produced and used in the manufacture of nylon-6,6 polyamide; other commercial uses are in products such as plasticizers, resins, plastics, foams and as a food acidulant (*22*). Current and most commercial syntheses of adipic acid use toxic benzene as a starting material and elevated pressure (800 psi) and temperature (250 °C), both are incompatible with the current global demand for energy saving process and greenhouse gas economy. Hydrogenation of benzene affords cyclohexane, which is air oxidized in the presence of metal catalysts to produce a mixture of cyclohexanone and cyclohexanol. Finally, oxidation of cyclohexanol with 60% nitric acid yields adipic acid. This terminal oxidation step is problematic since nitrous oxide, a greenhouse gas, is released. Adipic acid synthesis is estimated to account for 10% of the annual increase in atmospheric nitrous oxide (*23*). Although referred to as a laughing gas, N_2O is really no laughing matter because of its global warming effect.

As the *chn* oxidation genes are disordered and dispersed in two divergent transcripts with additional unknown ORFs (Figure 1), experiments are in progress to assemble the requisite *chnABCDE* genes under the control of a strong promoter to ensure a balanced production of the biocatalysts and a coordinated synthesis of adipic acid. Preliminary results are encouraging in at least the co-expression of ChnAB, when ChnB alone was seen not to be expressed readily (Figure 2). Further genetic manipulations including directed evolution are anticipated to make this entire biocatalytic process more attractive.

A previous green route towards adipic acid production, pioneered by Professors Draths and Frost (*24*) has led to a 1998 Presential Green Chemistry Challenge Award. Starting with D-glucose, *cis,cis*-muconic acid was produced in *E. coli* via a three-plasmid system with genes imported from *Klebsiella* and *Acinetobacter*. In the final step, catalytic hydrogenation of *cis,cis*-muconic acid, using 10% platinum on carbon, produces adipic acid. An attempt to produce adipic acid enzymatically from 1,4-dicyanobutane (adiponitrile) by a combined action of nitrile hydratase and amidase in a *Brevibacterium* sp. has also been reported (*25*).

Potential β-oxidation genes

Further metabolism of adipate has been proposed to proceed via β-oxidation to produce β-oxoadipyl-CoA which is then cleaved to acetyl-CoA and succinyl-CoA (*26*). ORF12 is most homologous to a similar-sized ORF next to the *mucK* gene in *Acinetobacter calcoaceticus* ADP1. This is a potential electron transport protein (*27*). The products of ORFs (ORF13-16) identified upstream of the *chnBER* cluster showed homology to proteins involved in fatty acid metabolism (Table II). It is interesting to note that this stretch of sequence has a high density of homologs with proteins deduced from the *Pseudomonas aeruginosa* PAO1 complete genome sequence. This includes two probable transcriptional regulators, ORF18 and ORF20. It is possible that this locus in the strain NCIMB 9871 is involved in at least the acylation and oxidation of adipic acid together with a regulatory system.

A new insertion sequence and other salient features of the *chn* gene context

Located divergently from the *chnC* gene is the presence of a new insertion sequence. This element of 1229 bp (tentatively called IS*1229*) is characterized by the following features: i) a small ORF8 and a larger ORF9 bearing 63-68% sequence identities to the putative transposases A and B of *Pantoea agglomerans* (Table 2); the two ORFs overlap at the stop and initiator codons by a one nucleotide shift; ii) a terminal inverted sequence of 25-bp; the right terminal sequence is 5'-GTGTCTAGAATTTTGGTGGCTATTC and the left terminal sequence is 5'-GTGTCTACAAAATCGGTGGCTATTC with only a 4-bp mismatch; and iii) flanking the inverted sequence is a 4-bp duplication of sequence 5'-ATAT.

It is not known at this time how many copies of IS*1229* are present in the NCIMB 9871 chromosome. But it appears unique when compared to the characteristics of the IS*1236* elements in the *Acinetobacter calcoaceticus* chromosome (*28*). Moreover, it is absent in the *chn* gene cluster recently described for *Acinetobacter* sp. strain SE19 (*29*). As in other IS elements (*30*), IS*1229* may likely be useful in strain identification.

Other notable ORFs associated with the *chn* pathway genes are: i) a "pilin gene inverting"-like sequence of ORF3; could this putative protein be responsible for the opposite orientations of the two *chn* gene clusters in NCIMB 9871?; and ii) presence of a putative phage integrase (ORF19) and a putative prophage regulatory protein (ORF11). These are indicative of recombination events in the NCIMB 9871 chromosome.

Acknowledgement

This paper is dedicated to the retirement of Professor Tokuyama on March 31, 2001. This work was supported by the Kansai University Research Grants: Grant-in-Aid for Joint Research, 1998. Funding by the Societé de secours des amis des sciences (France) and a Merit Postdoctoral Fellowship (Quebec Ministry of Education) to L. Bernard is gratefully acknowledged.

References

1. *Biotechnology for Clean Industrial Products and Processes. Towards Industrial Sustainability.* Organisation for Economic Co-operation and Development, 1998.
2. Anastas, P.T.; Williamson, T.C.; Hjeresen, D.; Breen, J.J. *Environ. Sci. & Technol.*, **1999,** 33, 116A.
3. Schmid, A.; Dordick, J.S.; Hauer, B.; Kiener, A.; Wubbolts, M.; Witholt, B. *Nature*, **2001,** *409*, 258.
4. URL http://www.ccrhq/vision/index.html.
5. Iwaki, H.; Hasegawa, Y.; Teraoka, M., Tokuyama, T.; Bergeron, H.; Lau, P.C.K. *Appl. Environ. Microbiol.* **1999,** *65*, 5158.
6. Iwaki, H.; Shimizu, M.; Tokuyama, T.; Hasegawa, Y. *Appl. Environ. Microbiol.* **1999**, 65, 2232.
7. Donoghue, N.A.; Trudgill, P.W. *Eur. J. Biochem.* **1975,** *60*, 1.
8. Willetts, A. *Trends Biotechnol.* **1997,** *15*, 55.
9. Stewart, J.D. *Curr. Org. Chem.* **1998,** *2*, 195.
10. Yanisch-Perron, C.; Vieira, J.; Messing, J. *Gene*, **1985,** *33*, 103.
11. Hanahan, D. *J. Mol. Biol.* **1983,** *166*, 557.
12. Bullock, W. O.; Fernandez, J.M.; Short, J.M. *BioTechniques*, **1987,** *5*, 376.
13. Smith, S.P.; Barber, K.R.; Dunn, S.D.; Shaw, G.S. *Biochemistry*, **1996,** *35*, 8805.
14. Altschul, S. F.; Madden, T.L.; Schaffer, A.A.; Zhang, J.; Zhang, Z.; Miller, W.; Lipman. D.J. *Nucleic Acids Res.* **1997,** *25*, 3389.

15. Jornvall, H.; Persson, B.; Krook, M.; Atrian, S.; Gonzalez-Duarte, R.; Jeffery, J.; Ghosh, D. *Biochemistry*, **1995,** *34*, 6003.
16. Olis, D.L.; Cheah, E.; Cygler, M.; Dijkstra, B.; Frolow, F.; Franken, S.M.; Harel, S.; Remington, J. Silman, I. *Protein Eng.* **1992**, *5*, 197.
17. Stewart, J.D.; Reed, K.W.; Martinez, C.A.; Zhu, J.; Chen, G.; Kayser, M.M. *J. Am. Chem. Soc.* **1998,** *120*, 3541.
18. Chen, Y.C.J.; Peoples, O.P.; Walsh, C.T. *J. Bacteriol.* **1988,** *170*, 781.
19. Roberts, S.M.; Wan, P.W.H. *J. Mol. Catalysis B:Enzymatic*, **1998,** *4*, 111.
20. Kelly, D.R.; Wan, P.W.H.; Tang, J. *Biotechnology* 2nd edition. Biotransformations I. Rehm, H.-J. Reed, R. Eds.; Wiley-VCH **1998**, vol 8a p535-587.
21. Onakunle, O.A.; Knowles, C.J.; Bunch, A.W. *Enzyme Microb. Technol.* **1997,** *21*, 245.
22. Davis, D.D.; Kemp, D.R. in Kirk-Othmer *Encyclopedia of Chemical Technology*. Kroschwitz, J.I.; Howe-Grant, M. eds (Wiley, New York, **1991**) vol 1. p466.
23. Thiemens, M.H.; Trogler, W.C. *Science*, **1991.** *251*, 932.
24. Draths, K.M.; Frost, J.W. *J. Am. Chem. Soc.* **1994,** *116*, 399.
25. Moreau, J.L.; Bigey, F.; Azza, S.; Arnaud, A.; Galzy, P. *Biocatalysis*, **1994,** *10*, 325.
26. Chapman, P.J.; Duggleby, R.G. *Biochem. J.* **1967,** *103*, 7c.
27. Williams, P.A.; Shaw, L.E. *J. Bacteriol.* **1997,** *179,* 5935.
28. Gerischer, U.; D'Argenio, D.A.; Ornston. L.N. *Microbiology.* **1996,** *142*, 1825.
29. Cheng, Q.; Thomas, S.M.; Kostichka, K.; Valentine, J.R.; Nagarajan, V. *J. Bacteriol.* **2000,** *182*, 4744.
30. Denis-Larose, C.; Labbé, D.; Bergeron, H.; Jones, A.M.; Greer, C.W.; Al-Hawari, J.; Grossman, M.J.; Sankey, B.; Lau, P.C.K. *Appl. Environ. Microbiol.* **1997,** *63,* 2915.

Chapter 7

Chemoenzymatic Synthesis of Nucleoside-Branched Poly(vinyl alcohol)

Y. Tokiwa[1,*], H. Fan[1], T. Raku[2], and M. Kitagawa[3]

[1]Research Institute of Biological Resources, National Institute of Bioscience and Human-Technology, 1–1–1 Higashi, Tsukuba, Ibaraki 305–8566, Japan
[2]Konan Chemical Ind. Company, Ltd., Osaka 569–0066, Japan
[3]Polymer Research Department, Toyobo Research Center Company, 1–1, Katata 2-Chome, Otsu, Shiga 520–0292, Japan
[*]Corresponding author: email: y.tokiwa@aist.go.jp

Transesterification reactions of hydroxyl group of ribose moiety in nucleosides with divinyladipate were catalyzed by alkaline proteases to give vinyl nucleosides, and poly(vinyl alcohol) containing nucleoside branches could be obtained by their free-radical polymerization.

Synthetic polymers containing nucleobases have attracted considerable interest. They have been a focus of intensive research as functional materials (*1*). For example, polymers containing nucleobases are useful as resins for purification of biomolecules. Although there have been several reports on polymers having nucleobases, few studies have examined polymers having nucleosides as shown in Figure 1. Hatanaka *et al.* (*2*) reported that polystyrene containing uridine showed high inhibition against sugar transferase. Inhibition activity of uridine polymer greatly increased compared with the corresponding monomer. Furthermore, Hoshino *et al.* reported the synthesis of poly(*N*-acryroyl piperidine) with NAD^+. The polymer showed reversibly soluble-insoluble property against water depending on temperature and was useful in the affinity precipitation method for purifying biochemical products (*3*). Wang *et al.* developed a new method of synthesizing nucleoside based polymers, which used phenolic group as a polymerizable group calalyzed by peroxidase (*4*). Thus nucleoside polymers have recently attractive increasing interest in the biochemical field.

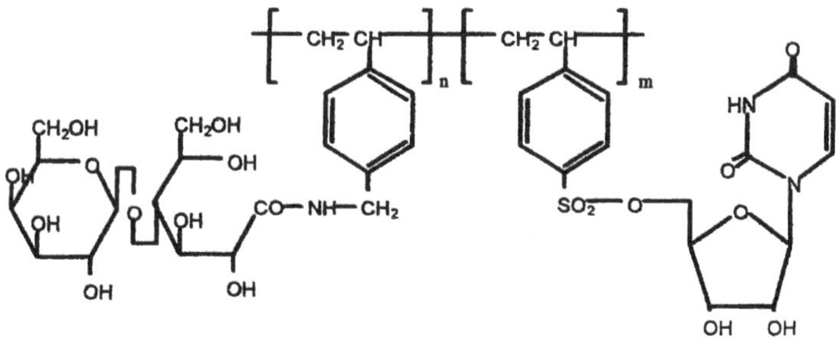

Uridine containing polymer

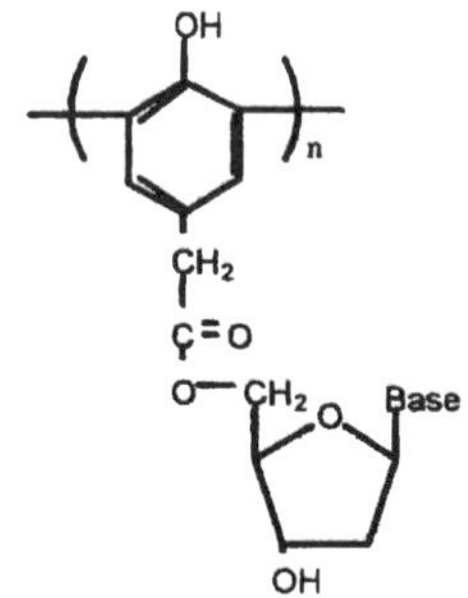

Necleoside branched polyphenol

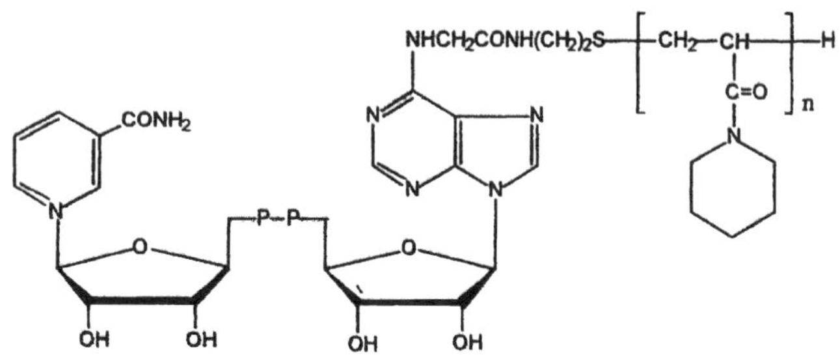

Poly(N-acriroyl piperidine) with NAD $^+$

Figure 1. Previously reported polymers containing nucleoside

In our previous studies, we investigated synthesis of biodegradable sugar-branched polymers consisting of sugar, fatty acid and poly(vinyl alcohol) by a chemoenzymatic method (*5*). By use of the same method, we examined the enzymatic synthesis of polymerizable nucleosides having ribose moiety such as vinyl uridine ester, vinyl cytidine ester, vinyl adenosine ester and vinyl guanosine ester and the polymerization of these vinyl esters (Figure 2).

Vinyl nucleoside esters were synthesized as follows: the transesterifications between nucleoside and divinyladipate were carried out by adding alkaline protease from *Streptomyces* sp. (ALP-101, Toyobo Co., Ltd., Osaka, Japan) or from *Bacillus subtilis* (Bioprase SP-10, Nagase Biochemicals Co., Ltd., Kyoto, Japan) into DMF or DMF/ DMSO (4:1 v/v) containing nucleoside, agitation for 7 days at 30°C and termination of the reaction by filtering off the enzyme. After solvents evaporation, the products were chromatographed on silica gel column with a solvent consisting chloroform / methanol as eluent, and colorless powders were obtained. The concentration of each nucleoside in the reaction mixture was measured by a high performance liquid chromatograph (HPLC) with a refractive index detector. TSK gel Amide-80 (TOSOH) was used with a mobile phase of acetonitrile/water. From the change in the value of nucleoside concentration before and after reaction, the conversion rates of each nucleoside to ester were calculated. A thin layer chromatograph (TLC) was used for qualitative analysis of nucleoside ester on silica gel plate 60F plate from Merck with a solvent consisting of ethyl acetate/methanol/water (17:3:1.5, v/v/v). The spots were developed by spraying with H_2SO_4 and heating. The position of esterification in the enzymatic synthesis reaction was established by ^{13}C NMR.

The transesterifications were catalyzed by the two kinds of alkaline protease. We found that the transesterification of the individual nucleosides differed in the disparity of chemical base structure, although all nucleosides contained ribose (Table I).

Uridine showed high conversion rate by alkaline protease from *Bacillus subtilis*. The conversion rate of uridine rapidly rose to more than 90% in 4 days. Guanosine and adenosine do not readily dissolve in DMF. Hence DMSO was added to the solvent. By addition of DMSO, the conversion rates of these nucleosides to vinyl esters were were not so low as without addition, and interestingly the only products were only 3'-esters. We found that *Streptomyces* sp. protease efficiently catalyzed a one-step acylation of secondary hydroxyl group in ribose moiety of adenosine and guanosine similarly to the synthesis of galactose whose secondary hydroxyl group of C-2 position was esterified in solvent containing 20% v/v DMSO (*9*). This positional specificity of protease compares favorably with classical chemical synthesis.

The polymerization of vinyl nucleosides were carried out in DMF and DMSO by use of azo-initiator, azobisisobutyronitrile (AIBN), and the corresponding polymers were obtained. The polymerization of vinyl nucleosides proceeded easier in DMSO than in DMF.

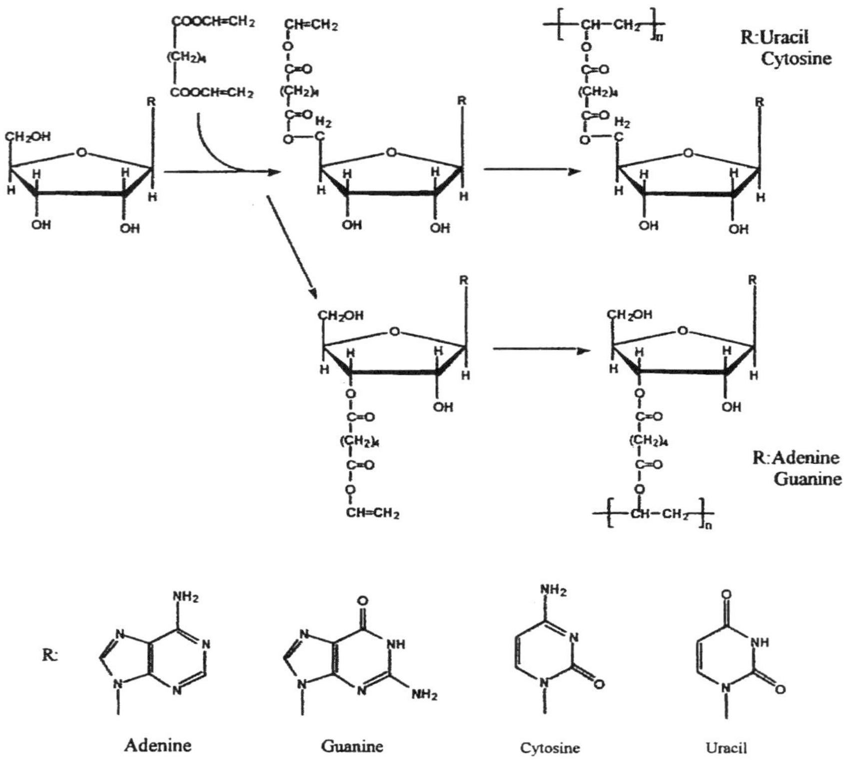

Figure 2. Enzymatic synthesis of vinyl nucleoside derivatives

Table I. Synthesis of vinyl sugar esters catalyzed by proteases from *Streptomyces* sp. or *Bacillus subtilis*

Nucleoside	*Enzyme*	*Solvent*	*Product*	*Conversion(%)*[a)]	*Ref.*
Uridine	A	DMF	5'-ester	99	6
Cytidine	A	DMF	5'-ester	75	6
Adenosine	B	DMF/DMSO	3'-ester	50	7
Guanosine	B	DMF/DMSO	3'-ester	25	6
Thymidine	B	DMF	5'-ester	95	8

a) Conversion rates were calculated from decrease of each nuclesoside's concentration measured by HPLC. A: alkaline protease from *Bacillus subtilis* (Bioprase SP-10, Nagase), B: Alkaline protease from *Streptomyces* sp. (ALP-101, Toyobo)

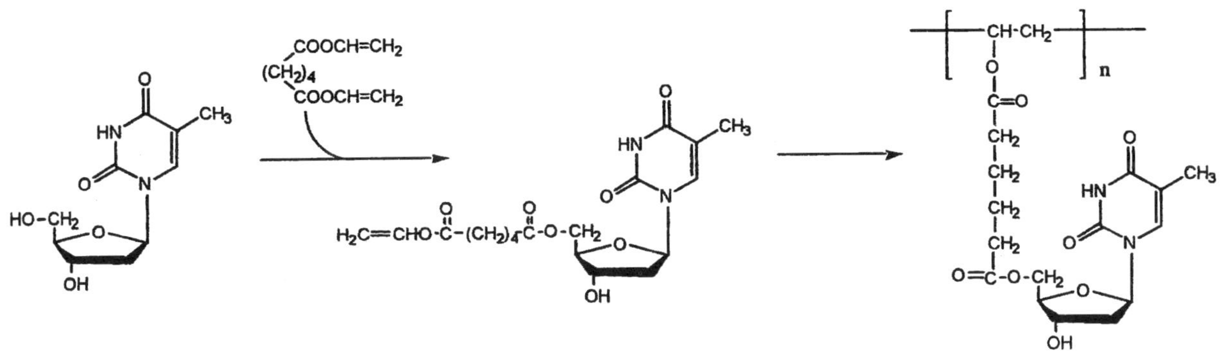

Figure 3. *Enzymatic synthesis of vinyl thymidine and its polymerization*

For nucleoside having deoxyribose moiety, thymidine, transesterification with divinyladipate catalyzed by *Streptomyces* sp. protease was examined (Figure 3). The conversion rate of thymidine increased rapidly at two days and then reached a plateau. Only one spot was detected by TLC. The product was purified by silica gel chromatography, and was characterized by ^{13}C NMR analysis. Characterization of the products by ^{13}C NMR revealed that thymidine ester was substituted at C-5' position of thymidine. Rich and Dordick (*10*) reported that effective acylation of thymidine was catalyzed by thymidine-imprinted *Bacillus licheniformis* protease and the acylation with vinyl butyrate could occur at 5'- and 3'-hydroxyl groups. We found that *Streptomyces* sp. protease selectively catalyzed acylation of 5'-hydroxyl group in thymidine and the product was only 5'-*O*-vinyladipoyl thymidine. The thymidine ester was polymerized with azo-initiator, AIBN, in DMSO, to give a polymer.

Nucleosides are constituents not only of polymeric biomolecules such as DNA and RNA, but also of monomeric biomolecules such as ATP, NAD, UDP and so on, and play important roles in a variety of biochemical reactions. Hence these polymers having nucleoside branches would be useful as functional materials in the field of biochemistry.

Acknowledgment. This work was supported by the New Energy and Industrial Technology Development Organization in Japan.

References

1. Inaki, Y.; Sugita, S.; Takemoto, T. *J. Polym. Sci., Part A: Poly. Chem.*. **1996**, 24, 3201., Inaki, Y.; Ebisutani, K.; Takemoto, T. *J. Polym. Sci., Part A: Poly. Chem.*. **1986**, 24, 3249., Cheng, C. M.; Egbe, M. I.; Grasshoff, J. M.; Guarrera, D. J.; Pai, R. P.; Warner, J. C.; Taylor, L. D. *J. Polym. Sci., Part A: Poly. Chem.* **1995**, 33, 2515.
2. Hatanaka, K; Takeshige, H; Kannno, K; Maruyama, A; Oishi, J; Kajihara, Y; Hashimoto, H. *J.Carbohydr. Chem.* **1997**, 16, 667.
3. Hoshino, K.; Koumoto, K.; Morohashi, S. The 5th Asia-Pacific Biochemical Engineering Conference, Thailand (1999).
4. Wang, P.; Dordick, J. S. *Macromolecules* **1998**, 31, 941.
5. Shibatani, S.; Kitagawa, M.; Tokiwa, Y. *Biotechnol. Lett.* **1997**, 19, 511.
6. Tokiwa, Y.; Fan, H.; Raku, T.; Kitagawa, M. *ACS Polym. Prep.* **2000**, 41, 1818.
7. Tokiwa, Y.; Kitagawa, M.; Fan, H.; Yokochi, T.; Raku, T.; Hiraguri, Y.; Shibatani, S.; Maekawa, Y.; Kashimura, N.; Kurane, R. *Biotechnol. Tech.* **1999**, 13, 563.
8. Kitagawa, M.; Fan, H.; Raku, T.; Kurane, R.; Tokiwa, Y. *Biotechnol. Lett.* **2000**, 22, 883.
9. Kitagawa, M.; Fan, H.; Raku, T.; Shibatani, S.; Maekawa, Y.; Hiraguri, Y.; Kurane, R.; Tokiwa, Y. *Biotechnol. Lett.* **1999**, 21, 355.
10. Rich, J. O.; Dordick, J. S. *J. Am. Chem. Soc.* **1997**, 119, 3245.

Chapter 8

Effect of DMSO on Regioselective Synthesis of Vinyl Sugars Catalyzed by Protease

M. Kitagawa[1], T. Raku[2], H. Fan[3], and Y. Tokiwa[3,*]

[1]Polymer Research Department, Toyobo Research Center Company, 1–1, Katata 2-Chome, Otsu, Shiga 520–0292, Japan
[2]Konan Chemical Ind. Company, Ltd., Osaka 569–0066, Japan
[3]Research Institute of Biological Resources, National Institute of Bioscience and Human-Technology, 1–1–1 Higashi, Tsukuba, Ibaraki 305–8566, Japan
[*]Corresponding author: email: y.tokiwa@aist.go.jp

Vinyl sugars involving adipic acid were synthesized by transesterification catalyzed by alkaline proteases. Hexoses (D-glucose, D-mannose, D-galactose and D-allose) were esterified with divinyladipate by alkaline protease from *Streptomyces* sp. to give corresponding 6-*O*-vinyl adipoyl sugars. In the presence of the denaturing cosolvent, DMSO, galactose and allose were selectively esterified at only the C-2 position.

Carbohydrates are useful natural resources for the development of new materials. Since they are widely used in the food industry, there are no doubts concerning toxicity and environmental compatibility. Synthesis of various types of vinyl sugar by chemical catalysts has been investigated by a number of researchers (*1*). Recently vinyl sugars have come under investigation from the viewpoint of renewable resources that could be expected to modify the antistatic properties, dyeability, adhesion, printability and biocompatibility of widely used polymers (*2*). However, previously known vinyl sugars have acrylate, methacrylate and styrene etc. as polymerizable groups and these are not biodegradable after polymerization.

Enzymatic modification of sugars offers a highly efficient process compared with conventional chemical synthesis using the process of blocking/deblocking steps (*3*). Enzymatic reaction in organic solvents to produce new materials such as surfactants (*4*) is attractive for its regio-selectivity and as an environmentally friendly method. The enzyme used for this purpose shows an overwhelming preference toward selective acylation of polyhydroxy groups of sugars.

Enzymatic Synthesis of Vinyl Sugar Esters

Prompted by an awareness of the polymerizability of vinyl esters and their use in transesterification reactions, a synthesis of a sugar branched polymer using divinyl fatty acid esters was designed as shown in Figure 1. One ester function of divinyl adipate was used in the transesterification reaction and a second vinyl ester function was used as a polymerizable group.

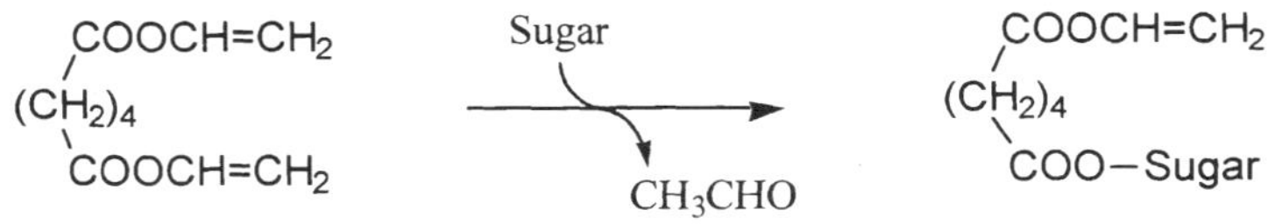

Figure 1. Vinyl sugar ester from sugar and divinyladipate

We examined the synthesis of vinyl sugar by use of alkaline protease from *Streptomyces* sp. in DMF, and could obtain vinyl sugar monomers without polymerization (*5*). Even when we used di-saccharides, vinyl sugar monomers were produced at high yield as shown in Table I. The vinyl sugars could be easily polymerized with chemical catalysts and high molecular weight polymers were produced. Furthermore, co-polymerization with some kinds of monomers was possible. Hence, vinyl sugars are useful intermediates for developing functional polymers.

During our investigations into the synthesis of vinyl sugar ester from D-glucose with divinyl adipate in DMF, which is soluble at high concentration against sugars catalyzed by *Streptomyces* sp. protease, we obtained 6-*O*-vinyladipoyl-D-glucose effectively (*6*). Primary hydroxyl group is generally more reactive than secondary hydroxyl group. On the enzymatic synthesis of sugar esters also, primary hydroxyl group was esterified preferentially, as reported by some researchers (*7*). The relation to acylation of secondary hydroxyl group of sugars, C-6 protected mono saccharide (*8*) and

6-deoxy-hexosides (*9*) catalyzed by lipase, have been investigated. But there have been few reports about acylation of secondary hydroxyl group without that of primary hydroxyl group in non-protected sugar.

Table I. Synthesis of vinyl sugar esters catalyzed by protease from *Streptomyces* sp.

	Starting sugar	*Solvent*	*Product*	*Conversion (%)[a)]*	*Ref.*
Hexose	D-Glucose	DMF	6-ester	94	10
	D-Mannose	DMF	6-ester	95	10
	D-Galactose	DMF	6-ester	94	10
	D-Allose	DMF	6-ester	92	11
	D-Glucose	DMF/DMSO	6-ester	40	10
	D-Mannose	DMF/DMSO	6-ester	68	10
	D-Galactose	DMF/DMSO	2-ester	85	10
	D-Allose	DMF/DMSO	2-ester	90	11
Pentose	D-Arabinose	DMF	5-ester	98	12
	L-Arabinose	DMF	5-ester	98	12
Di-saccharide	Maltose	DMF	6'-ester	95	13
	Trehalose	DMF	6-ester	60	13
	Sucrose	DMF	1-ester	75	14
Protected sugar	Isopropyridene-glucofuranose	DMSO	3-ester	10	15
		Toluene[b)]	3-ester	89	15

A: *Streptomyces* sp. alkaline protease, Alkaline protease from *Streptomyces* sp. (ALP-101, 20 units/mg.min) was from Toyobo Co. B: *Alcaligenes* sp. lipase, [a)] Conversion rates were calculated by HPLC. A carbohydrate analysis column, TSK gel Amide-80 (Tosoh) was used with a mobile phase of a mixture of acetonitrile/water (3:1, v/v) at a flow rate of 1.0 ml/min. The acylation of sugar were detected by ^{13}C NMR equipment. [b)] Dimethylaminopyridine was used as a chemical catalyst instead of enzyme catalyst, The reactions was initiated by adding enzyme to a solvent containing 0.25 M sugar and 1 M divinyl adipate. The suspension was stirred at 130 rpm for several days at 35 °C. The reactions were terminated by filtering off the enzyme. The DMF was evaporated. The product was separated by silica gel chromatography with an eluent consisting of chloroform-methanol (8:1). The conversion rate was calculated from the decrease in the concentration of sugar detected by HPLC with refractive index measurements.

Table I showed the various vinyl sugars obtained by transesterification of divinyladipate and various sugars in hydrophilic organic solvents catalyzed by alkaline protease.

The protease from *Streptomyces* sp. is more proteolytic active at high pH range in aqueous medium than the known alkaline protease (*16*). The *Streptomyces* sp. protease has been investigated not only for proteolysis of protein but also for transesterification reaction in organic solvents. In our previous studie, the protease was stable under reaction condition containing DMF and its transesterification activity was still more than 80% of the initial value after 5 repeated uses (*17*).

Effect of DMSO on Enzymatic Acylation

A recent study showed that substrate specificity, enantioselectivity, prochiral selectivity, regio-selectivity and chemoselectivity of enzymes turned out to depend dramatically on the nature of the solvents (*18*). We demonstrated that by addition of DMSO to DMF, galactose and allose were esterified at C-2 position selectively without esterification of C-6 position of primarily hydroxyl group catalyzed by *Streptomyces* sp. protease (*10,11*).

The conversion rates of four hexoses such as glucose, mannose, galactose and allose by use of DMF as a solvent increased with passage of time in a similar manner, and reached high conversion rates of ca. 90%. By the TLC analysis, only one spot was detected in every reaction. The enzyme activity is highly dependent on the substrate structure. Martin examined esterification of methyl glycoside catalyzed by enzyme and reported that methyl galactoside was about 2-fold more reactive than methyl mannoside (*19*). In this reaction condition there is little difference in esterification rate among various hexoses. On the other hand, when we used DMF solution containing 20% v/v DMSO as a solvent, the reaction rates of these hexoses showed different values. In the case of glucose, the conversion rate was sharply decreased by addition of DMSO. The decreases in the conversion rates of mannose and galactose in DMSO were less than those of glucose. This result implies that alkaline protease from *Streptomyces* sp. is not inactivated by DMSO but the substrate specificity of the protease in organic solvent depends on the solvents, similarly to what is observed in enzymatic synthesis of peptide (*20*).

In the case of allose, the conversion rate by use of DMF as a solvent increased with time, and reached a high conversion rate of ca. 90%. In the presence of DMSO, the conversion rate increased in the same manner. Allose reacted at a similar rate whether in the presence of DMSO or not and a main spot (Rf0.38) and minor spots (Rf 0.78, 0.84) were detected by the TLC analysis. Most enzymes are inactivated by polar solvents such as a DMSO, but DMSO is capable of dissolving sugar to a large extent. Recently a two-solvent mixture (2-methyl-2-butanol/DMSO, 4:1, v/v) was used to synthesize sucrose ester with vinyl esters catalyzed by lipase (*21*). Almarsson and Klibanov (*22*) reported that the transesterification reactions of *N*-acetyl-2-phenylalanine ethyl ester with propanole catalyzed by protease subtilisin Carlsberg can be increased more than 100-fold by the addition of denaturing organic solvents such as DMSO and enhancement of conformational flexibility has been hypothesized. These results suggest that substrate specificity of *Streptomyces* protease was also changed in the presence of DMSO.

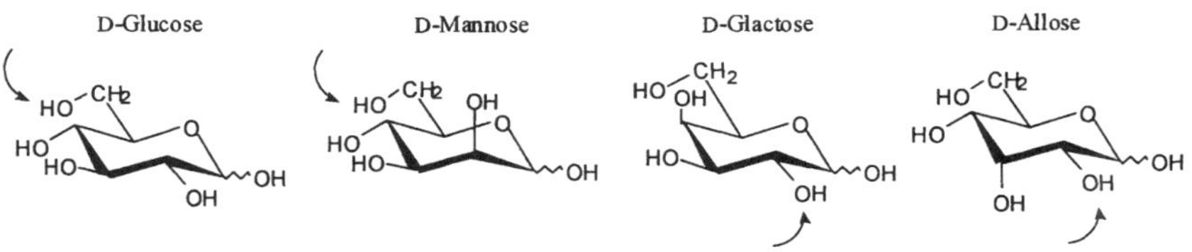

Figure 2. Regio-selective acylation of hexopyranoses in DMF/DMSO (4/1, v/v).

Therisod and Klibanov (*8*) reported the acylation of n-octyl-D-glucopyranoside, and that the enzymatically formed 2,6-glucose derivatives were chemically deblocked at C-6 to yield 2-*O*-acyl glucose. It is generally known that primary hydroxyl group of sugars is more reactive than the secondary groups by chemical or enzymatic catalysts. Why, in the presence of DMSO, galactose and allose were esterified at C-2 position preferentially compared with esterification of primary hydroxyl group is not clear (Figure 2). It is generally known that C-4 hydroxyl group of sugars has lower reactivity than other groups and that equatorial hydroxyl group has higher reactivity than axial hydroxyl group. A simple reason can not be expected to explain the result obtained for a complex polyhydroxyl compound such as sugar in this experiment. Further experiments on esterification will be needed to clarify the mechanism of the DMSO effect.

To put it briefly, we discovered that alkaline protease from *Streptomyces* sp. efficiently catalyzed a one-step acylation of secondary group in galactose and allose in DMF and DMSO in preference to esterification of primary hydroxyl group. This positional specificity of protease compares favorably with classical chemical synthesis.

Polymerization of Vinyl Sugar Esters

The vinyl sugar esters are polymerized in DMF or water with azo-initiators (*23*). After polymerization, the main chain of such a polymer is a poly(vinyl alcohol), which is well known as a biodegradable polymer (*24*). The sugar esters have many potential applications as polymeric and/or polymerizable biosurfactants in the environmental, pharmaceutical and cosmetic fields, as well as numerous others. Furthermore, the vinyl sugars co-polymerize with other monoers and improve the characteristics of the polymer (*25*)..

The vinyl sugars reported in the present paper have a reducing sugar moiety. There have been few reports of polymers containing reducing sugar. Anomeric hydroxyl group of reducing sugar is more functional than other hydroxyl groups. For example, a pendant polymer having a reducing sugar may have new applications such as DNA probe fixing material (*26*) and fixation of cell via transporter protein (*27*). Kobayashi *et al.* (*27*) reported that polystyrene or polyacrylamide containing a reducing sugar (glucose) moiety showed interaction against a glucose transporter protein of red blood cell stronger than that of the polymer having non-reducing sugar moiety. Kashimura *et al.* (*28*) further found that oxygen radical species are generated by autoxidation of reducing sugars under physiological conditions. The reactions are important in biological systems such as inactivation of viruses and cleavage of DNA. We further found that the polymer produces superoxide in aqueous solution under detection with nitrobluetetrazolium and the activity is greatly increased compared with the corresponding monomer (*29*).

The vinyl sugars have two intersting chracteristics. One is that the polymerizable group is vinyl ester having biodegradability after polmerization, and the other is that sugar moiety maintains a reducing activity. The polymers can be expected to provide functional materials developed from these two viewpoints in the near furture.

Acknowledgement. This work was supported by the New Energy and Industrial Technology Development Organization in Japan.

References

1. Whistler, R.L.; Panzer, H. P.; Goatley, J. L. *J. Org. Chem.*, **1962**, 27, 2961., Yamada, K.; Ninoda, M.; Miyamoto, T. *J. Polym. Sci. A: Polym. Chem.* **1997**, 35, 751., Kobayashi, K.; Sumitomo, H. *Macromolecules* **1980**, 13, 234., Haworth, W. N.; Gregory, H.; Wiggins, L. F. *J. Chem. Soc.* **1946**, 489., Black, W. A. P.; Colquhoun, J. A.; Dewar, E. T. *Makromol. Chem.* **1968**, 117, 210.,

Chang, D. C.; Kostelnik, R. J.; Elis, H-G. *Makromol. Chem.* **1977**, 178, 691., Iwakura, Y.; Imai, Y.; Yagi, K. *J. Polym. Sci.* **1968**, 6, 1625., Wulff, G.; Schmidt, J.; Venhoff, T. P. *Macromol. Chem. Phys.* **1996**, 197, 1285.; Klein, J.; Herzog, D. *Makromol. Chem.* **1987**, 188, 1217, Charreyre, M-T.; Boullanger, P.; Pichot, C.; Delair, T.; Mandrand, B.; Llauro, M-F. *Makromol. Chem.* **1993**, 194, 117.

2. Wulff, G.; Zhu, L.; Schmidt, H. *Macromolecules* **1997**, 30, 4533.
3. Haines, A. H. *Adv. Carbohydr. Chem. Biochem.* **1981**, 39, 13.
4. Ducret, A.; Giroux, A.; Trani, M.; Lortie, R. *J.Am. Oil Chem. Soc.* **1996**, 73, 109.
5. Shibatani, S.; Kitagawa, M.; Tokiwa, Y. *Biotechnol. Lett.* **1997**, 19, 511.
6. Kitagawa, M.; Tokiwa, Y. *J. Carbohydr. Chem.* **1998**, 17, 893.
7. Riva, S.; Chopineau, J.; Kieboom, A. P. G.; Klibanov, A. M. *J. Am. Chem. Soc.* **1988**, 110, 584.
8. Therisod, M.; Klibanov, A.M. *J. Am. Chem. Soc.* **1987**, 109, 3977.
9. Ciuffreda, P.; Colombo, D.; Ronchetti, F.; Toma, L. *J. Org. Chem.* **1999**, 55, 4187.
10. Kitagawa, M.; Fan, H.; Raku, T.; Shibatani, S.; Maekawa, Y.; Kurane, R.; Tokiwa, Y. *Biotechnol. Lett.* **1999**, 21, 355.
11. Tokiwa, Y.; Fan, H.; Raku, T.; Kitagawa, M. *Polym. Prep.* **2000**, 41, 1818.
12. Tokiwa, Y.; Kitagawa, M.; Fan, H.; Raku, T.; Hiraguri, Y.; Shibatani, S.; Kurane, R. *Biotechnol. Tech.* **1999**, 13, 173.
13. Kitagawa, M.; Chalermisrachai, P.; Fan, H.; Tokiwa, Y. *Macromol. Symp.* **1999**, 144, 247.
14. unpublished data
15. Kitagawa, M.; Raku, T.; Fan, H.; Tokiwa, Y. International Symposium of Environment and Biotechnology 2000 (Kyoto).
16. Nakanishi, T.; Matsumura, Y.; Minamiura, N.; Yamamoto, T. *Agr. Biol. Chem.* **1974**, 38, 2391.
17. Kitagawa, M.; Tokiwa, Y. *Carbohydr. Lett.* **1997**, 2, 343.
18. Wescott, C.R.; Klibanov, A.M. *Biochim. Biophys. Acta* **1994**, 1206, 1.
19. Martin, B. D.; Ampofo, S. A.; Linhardt, R. J.; Dordick, J. S. *Macromolecules* **1992**, 25, 7081.
20. Kawashiro, K.; Sugahara, H.; Sugiyama, H; Hayashi, H. *Biotechnol. Bioeng.* **1997**, 53, 26.
21. Ferrer, M.; Crices, M. A.; Bernabe, M.; Ballesteros, A.; Plou, F. J. *Biotechnol. Bioeng.* **1999**, 65, 10.
22. Almarsson, O.; Klibanov, A. M. *Biotechnol. Bioeng.* **1996**, 49, 87.
23. Kitagawa, M.; Takegami, S.; Tokiwa, Y. *Macromol. Rapid Commun.* **1998**, 19, 155.

24. Suzuki, T.; Ichihara, Y.; Yamada, M.; Tonomura, K. *Agric. Biol. Chem.* **1973**, 37, 747.
25. Raku, M.; Tokiwa, Y. *J. Appl. Poly. Sci.* **2001**, 80, 384.
26. Badey, B.; Boullanger, P.; Domard, A.; Cros, P.; Delair, T.; Pichot, C. *Makromol. Chem.* **1996**, 197, 3711.
27. Park, K-H.; Takei, R.; Goto, M.; Maruyama, A.; Kobayashi, A.; Kobayashi, K.; Akaike, T. *J. Biochem.* **1997**, 121, 997.
28. Kashimura, N.; Morita, J.; Komano, T. *Carbohydr. Res.* **1979**, 70, C3.
29. Kitagawa, M.; Tokiwa, Y. *Chem. Lett.* **1998**, 281.

Chapter 9

Enzyme-Catalyzed Synthesis of Well-Defined Macromers Built around a Sugar Core

Rajesh Kumar[1,2] and Richard A. Gross[1,*]

[1]NSF Center for Biocatalysis and Bioprocessing of Macromolecules, Department of Chemistry and Chemical Engineering, Polytechnic University, Six Metrotech Center, Brooklyn, NY 11201
[2]Current address: INSET and Center for Advanced Materials, Department of Chemistry, University of Massachusetts at Lowell, Lowell, MA 01854

By using 4-*C*-hydroxymethyl-α-D-pentofuranose as the sugar core and lipase-catalyzed transformations, a macromer was constructed with exceptional control of substituent placement around the carbohydrate core. The key synthetic transformations performed were as follows: 1) selective lipase-catalyzed acrylation along with prochiral selection of 4-*C*-hydroxymethyl-1,2-*O*-isopropylidene-α-D-pentofuranose (diastereomeric excess up to 93%); 2) the ring-opening of ε-caprolactone, ε-CL, from the remaining primary hydroxyl group to give an acryl-sugar capped macromer (M_n 11 300, M_w/M_n 1.36, initiator efficiency 50-55 %, < 5% water initiated PCL chains); 3) selective lipase-catalyzed esterification of the terminal hydroxyl of oligo(ε-CL) chains; 4) hydrolysis of the 1,2-*O*-isopropylidene group at the sugar core 5) homopolymerization of the corresponding macromer. In principle, the method developed is flexible so that it can be used to generate a wide array of unusual macromers and heteroarm stars. In the absence of biocatalytic transformation, such structural control would be extremely difficult or currently impossible to obtain.

Introduction

The synthesis and study of polymers that contain carbohydrates has captured the attention of researchers who wish to attain *i*) highly functional polymers[1] *ii*) specific biological functions, and *iii*) complex systems that fit into the category of 'smart' materials.[2,3,4] For example, such polymers have been studied for their ability to regulate interactions with lectins,[5] as pseudo-glycoproteins,[6] and as carriers for drug delivery systems.[7]

Research directed towards stars and dendrimers also reflects the aspiration of scientists to create macromolecules that function with improved efficiency and precision for catalysis, drug delivery, and much more [8,9] Hetero-arm or multi-arm star block copolymers have been prepared by living anionic and cationic polymerization.[10,11] Structure-property studies have proven that variations in the macromolecular architecture from linear to multiarm can have dramatic effects on the morphological and physical-mechanical properties of the corresponding materials.[9]

Recently, our laboratory [12,13] and others [14,15] have explored the use of *in-vitro* enzyme-catalysis for the preparation of monomers and polymers. Recent reviews and books have been published that document the rapid development of these methods.[16] Early reports that describe the synthesis of linear chains that are attached to a multifunctional initiator have been published.[17,18] The above findings provide incentive to further extend the level of control attainable during the synthesis of these products.

In this work, a general route was demonstrated that permits the efficient placement of selected structures at specific positions around a carbohydrate core. A key synthetic step that makes this possible is lipase-catalyzed diastereoselective acylation. This, and the judicious choice of a certain carbohydrate substrate, permits a level of control in the construction of hetero-arm star co-polymers that was previously non-attainable or extremely difficult to realize by traditional chemical methods where a large number of protection-deprotection steps have impeded progress.[19, 20, 21]

Results and Discussion

The lipase-catalyzed synthesis of the macromer 4'-hydroxymethylmethacryl, 4-*C*-hydroxymethyl-1,2-*O*-isopropylidene-α-D-pentofuranose (HMG, **3**, Scheme 1), its use as a multifunctional initiator to prepare a polyester arm specifically linked to the other (C-5) diastereometric center (**4**, Scheme 1), and its homopolymerization are described. Scheme-1 presents the strategy that was used for the stereoselective incorporation of the acryl group in the sugar molecule

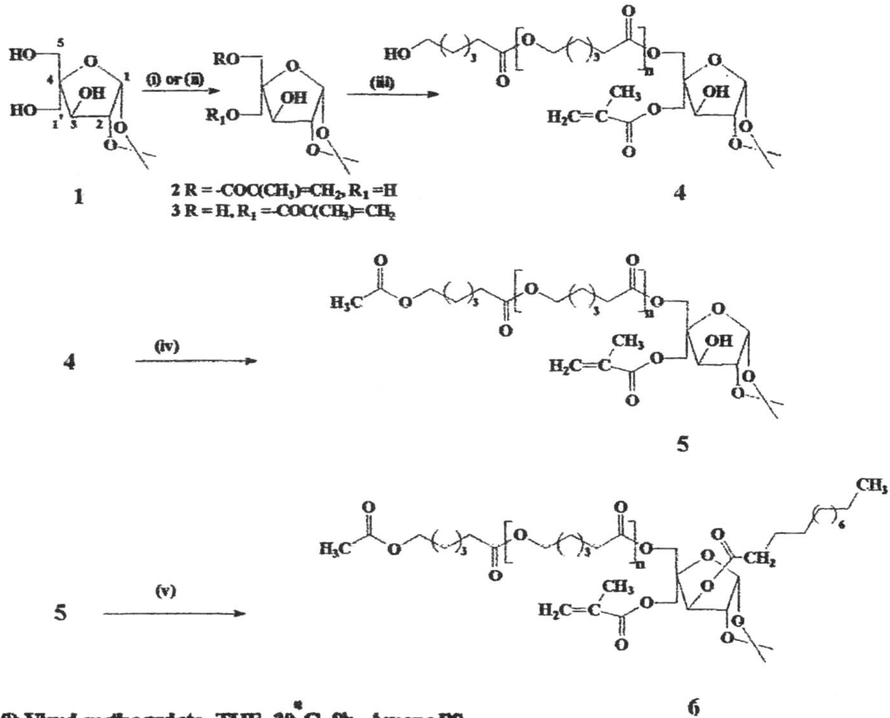

(i) Vinyl methacrylate, THF, 30°C, 8h, AmanoPS
(ii) Vinyl methacrylate, THF, 30°C, 8h, Novozyme-435
(iii) ε-CL, toluene, 60°C, 8h, Novozyme-435
(iv) Vinyl acetate, toluene, 6h, 30°C
(v) Lauric acid, DMF, DCC, DMAP, 60°C

Scheme-1. Stereoselective Acrylation followed by ROP and Selective Acylation of the PCL End- Group.

followed by initiation of ε-caprolactone (ε-CL) ring-opening polymerization to build the poly(ε=caprolactone), PCL, substituted acryl sugar macromer. 4-*C*-Hydroxymethyl-1,2-*O*-isopropylidene-α-D-pentofuranose **(1)** was synthesized by starting with diacetone glucose in a two step reaction.[22] The bishydroyxmethyl glucofuranose **1** was then subjected to acryloylation with methylacrylate in dry THF. [23] The ability of the lipases Porcine pancreatic lipase (PPL), *Candida rugosa* lipase (CRL), *PS*-30, lipases from *Pseudomonas AK*, *Pseudomonas AY* and Novozyme-435 to catalyze prochiral asymmetrization of 4-*C*-hydroxymethyl-1,2-*O*-isopropylidene-α-D-pentofuranose, **1**, was studied at 30-35°C for 8 h and the results of this work are shown in Figure 1.

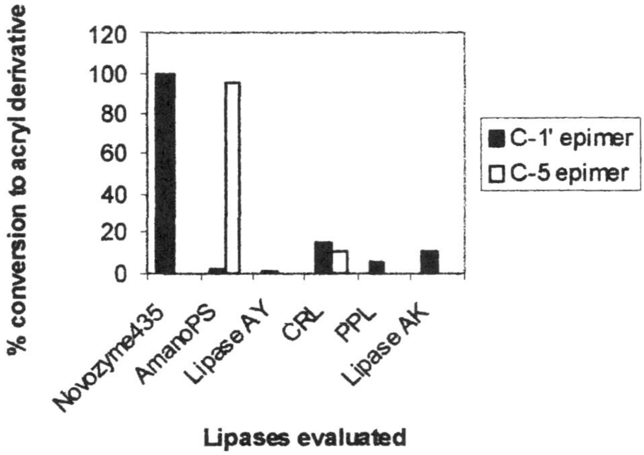

Figure 1. Ability of lipases to carry out the selective acrylation of 4-C-hydroxymethyl-1,2-*O*-isopropylidene-α-D-pentofuranose

Novozyme-435 and lipase *PS* resulted in highly diastereo-selective monoacryl derivatization using one equivalent of vinyl methacrylate as the acyl donor (Figure 1). Furthermore, for both Novozyme-435 and lipase *PS*, even when a two fold excess of vinyl methacrylate was used and the reaction was prolonged to 24 h, the major product was still the monoacryl derivative (~95-97%). The Amano *PS* catalyzed reaction preferentially placed the acryl moiety at the C-5 hydroxyl, affording **2** as the major product (*d.e. 78%*, yield = 72%,). In contrast, Novozyme-435 resulted in acrylation at the C-1' position (*d.e.* 93%, yield = 95%) giving **3** as the main product. The diastereomeric excess of the products **2** and **3** was calculated from ^{1}H NMR spectra by the difference in the integral values of the anomeric protons of the corresponding epimers at δ 5.93 and δ 6.01. Assignments of ^{1}H NMR signals to the epimer formed was done by detailed spectral analyses. It was observed that the protons of both C-1'and C-5 methylene groups of **1** appeared as a multiplet, one of the corresponding methylene protons shifted downfield on acrylation, i.e. in compound **3** the C-1' methylene protons appeared as a double doublet at δ 4.32 (J= 11.76 & 11.47) while in compound **2** the C-5 methylene protons appeared as a double doublet at δ 4.34 (J= 11.7). Furthermore the anomeric proton C-1 of **2** for major diastereomer appeared at higher δ values (in $CDCl_3$, 300 MHz) at δ 6.01 (J= 4.12) and for minor diastereomer at δ 5.90 compared to the corresponding protons in **3** the major diastereomer appeared at δ 5.93 (J= 4.12). This clearly indicates that there is reversal of selectivity during enzyme catalyzed acrylation. Also the position of the acryl group was unequivocally established by the significant NOE effect observed in H-3 on irradiation of H-1' in compound **3**.

The sugar acryl derivative **3** was studied as a multifunctional initiator for ε-CL polymerization. Based on previous work in our laboratory[12-13] and elsewhere[14-15], Novozyme-435 was chosen as the selective catalyst for this ring-opening polymerization (Scheme-1). Recent work by us [24] showed that Novozyme-435 catalysis of ε-CL polymerizations is accelerated when it is performed in low-polarity organic media. Thus, the acryl sugar **3** initiated ring-opening polymerization of ε-CL was performed in toluene. The ^{1}H NMR spectrum (see figure 2) of Product **4** from an 8 h ring-opening polymerization of ε-CL initiated by **3** and catalyzed by Novozyme-435 indicates that **4** consists of the acryl sugar moiety that is linked to the carboxyl terminal of a PCL chain (M_n 11300, M_w/M_n 1.36). Product **4** was separated from non-reacted acryl sugar by precipitating the polymer into methanol. Analysis of Product **4** by ^{1}H NMR and ^{13}C NMR revealed that the reaction was highly regioselective at the C-5 hydroxyl. Derivatization of **4** (after purification by precipitation in methanol) with oxalyl chloride revealed that <5% of the consumed ε-CL formed chains with a carboxyl terminus. The homopolymer PCL results from competitive transacylation and/or initiation by water instead of by **3**. [13] The remaining 95% of the consumed ε-CL was incorporated into chains that are exclusively attached by an ester group to the C-5 primary hydroxyl moiety of the acryl sugar **3**. From our previous work, [12] it can be inferred that the average number of repeat units, DP, of the PCL chain attached to the acryl sugar **3** can be controlled by regulating the feed ratio of initiator **3** to ε-CL.

For the attachment of a different substituent (e.g. bioactive, photolytically reactive) at another chosen site of the sugar core, it was first necessary to end-cap the terminal ω-hydroxyl group of the PCL chain of **4**. Reaction of the PCL hydroxyl chain-end group was performed using Novozyme-435 as the catalyst, vinyl acetate as an irreversible acylating group, at 30°C, in toluene as the solvent. The reaction progress was monitored by ^{1}H-NMR. At a 5:1 molar ratio of vinyl acetate to **4**, selective acetylation of the ε-hydroxyloligo(ε-CL) terminal group was observed within 6 h. Analysis of the product by ^{1}H-NMR showed < 2% acylation at the secondary hydroxyl group of the sugar. The triplet due to the ω-hydroxyl methylene protons (-CH_2OH, at 3.66 ppm) of PCL chains disappeared, and a new singlet at 2.05 ppm due to the corresponding acetyl (-COCH_3) chain end group protons was observed. Since unreacted vinyl acetate and the by-product acetaldehyde are volatile, they were easily removed *in vacuo* during product isolation. The ^{13}C-NMR spectrum of Product **5**, when compared to that for **4**, showed a downfield shift of 2.30 ppm for the signal due to the ω-hydroxymethylene carbon of the PCL segment. Also, a resonance due to the acetyl group carbon was seen at 21.06 ppm. All other carbon resonances assigned to **4** showed no substantial change after acetylation which excludes transesterification between the C-3 secondary hydroxyl of **4** and ester groups of the PCL main chain. Molecular weight analysis of **5** (M_n 10,900 g/mol, M_w/M_n= 1.36) showed no chain degradation during the end-capping reaction. Chemical esterification of **5** with lauric acid to form **6** (see Scheme 1)

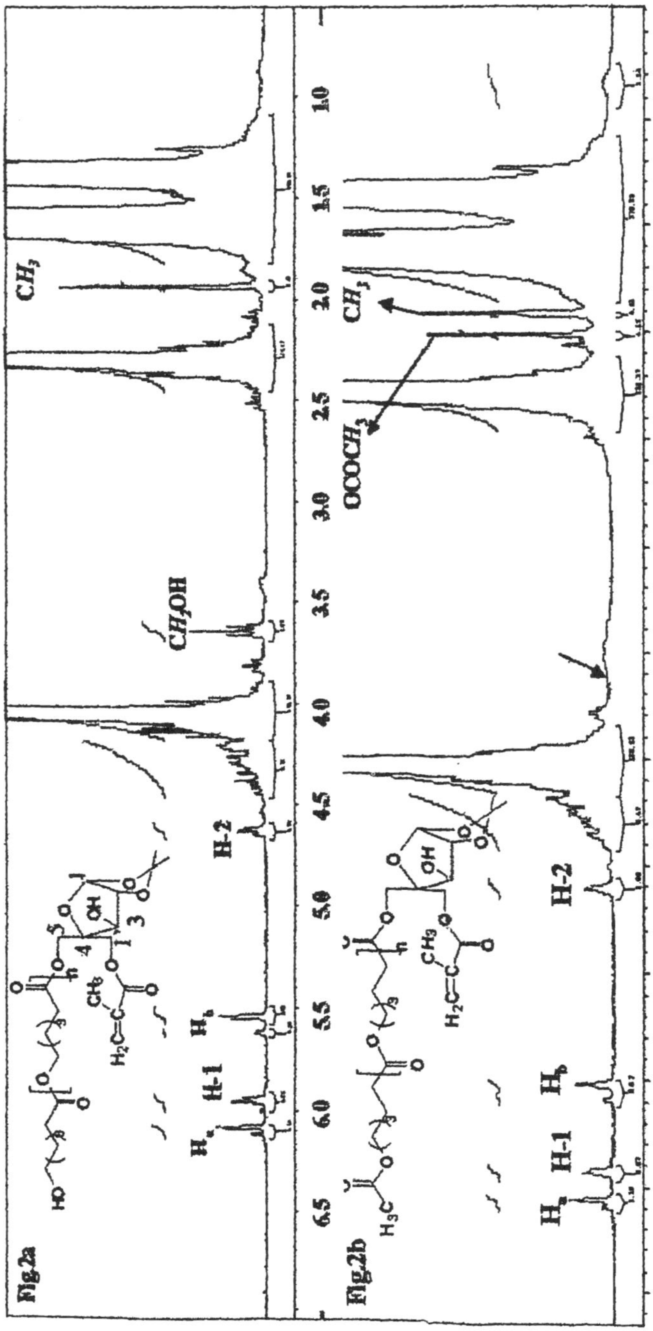
Fig.2a
Fig.2b
CH3
CH2OH
H-2
Hb
H-1
Ha
OCOCH3
6.5
6.0
5.5
5.0
4.5
4.0
3.5
3.0
2.5
2.0
1.5
1.0

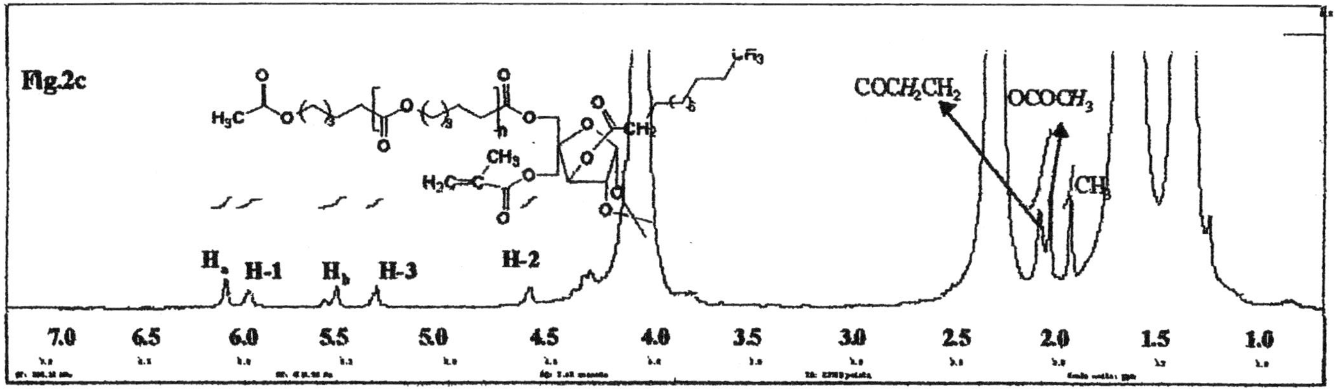

Figure 2. ^{1}H-NMR spectrum of (a) poly(ε-caprolactone) substituted acryl sugar (4), (b) End-capped product (5) (c) Chemically acetylated product (6

further confirmed that acetylation at the C-3 position results in a signal at δ 5.2 due to the corresponding *CH*-OAc proton. Study of ^{1}H-NMR spectrum shows that, for **5**, a signal at 5.2 ppm was not observed.

EXPERIMENTAL

Materials: ε-Caprolactone (CL) was procured Union Carbide Co. was purified by distillation from calcium hydride under reduced pressure. All the other chemicals used were purchased from Aldrich and used without further purification. Toluene and tetrahydrofuran (THF) were dried by distillation from melted sodium. 4-C-Hydroxymethyl-1,2-*O*-(1-methylethylidene)-3-*O*-(phenylmethyl)-α-D-pentofuranose was prepared and purified according to a literature procedure.[22] Novozyme-435 was the generous gift from Novozym Denmark.

Instrumental Methods: NMR Spectroscopy: Proton (^{1}H) NMR spectra were recorded on a Bruker spectrometer at 300 MHz, the chemical shifts in parts per million (ppm) are reported downfield from 0.00 ppm using tetramethylsilane (TMS) as an internal referece. The concerntration used was 4% w/v in chloroform-d. Carbon-13 (^{13}C) NMR spectra were recorded at 75.5 MHz on a Bruker spectrometer in 20% w/v chloroform-d solutions at 77.0 ppm.

The ketal group of **5** was successfully hydrolyzed to form 7 by using a trifluoroacetic acid/water mixture (80:20 v/v) for five minutes. The molecular weight averages of products **5** and 7 were almost identical supporting that the macromer was not hydrolyzed. Product 7, thus obtained, provides two additional hydroxyl groups at defined positions around the sugar core. These groups offer further options for the development of strict control of the three dimensional arrangement of substituents around the carbohydrate core.

The homopolymerization of 7 was conducted by using AIBN as the initiator in dry DMF (Scheme-2).[25] The polymerization was continued for 24 h at 60°C, the reaction was terminated by the precipitation of the product by acetone to yield the corresponding homopolymer (**8**) in 70% yield. The ^{1}H-NMR spectrum of the homopolymer did not show resonances due to the acryloyl protons . Furthermore, study of integral values of NMR signals as well as peak positions confirmed that the expected product was formed. The ^{13}C NMR spectrum of the polymer further supported that the sugar moiety was not disturbed and that the [-(-CH-CH_2-)-] carbon backbone was formed (M_n 23000).

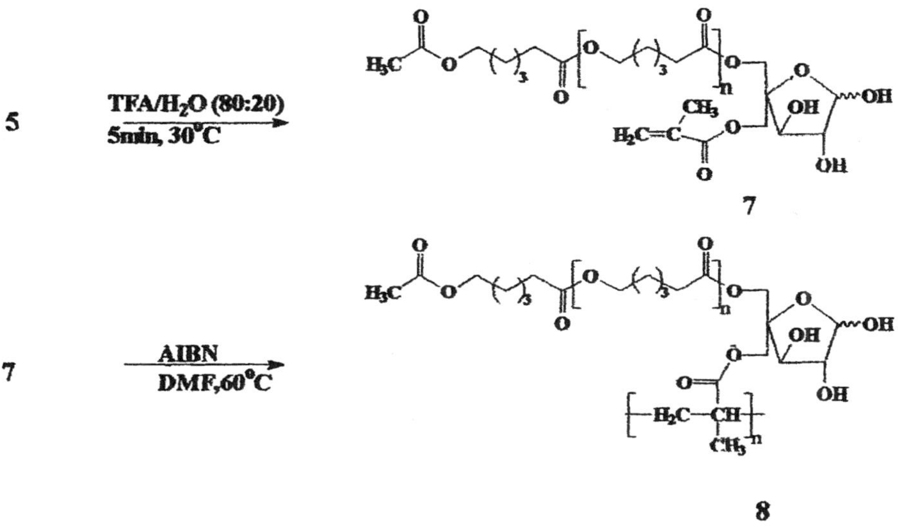

Scheme 2 : AIBN catalyzed homopolymerization of 7.

Molecular Weight Measurements: Number and weight average molecular weights (M_n and M_w, respectively) were determined by gel permeation chromatography (GPC); the GPC analysis were performed using a Waters HPLC system that includes the following: Model 510 pump, Model 717 autosampler, Model 410 refractive index detector (RI) and Styragel HR5, HR4, HR3 and HR1 columns in series. Chloroform (HPLC grade) was used as eluent at a flow rate of 1.0 mL/min. The sample concentration and injection volume were 0.2% (wt/wt) and 100μL, respectively. Molecular weights were determined by using conventional calibration generated with narrow dispersity polystyrene standards (Polymer Laboratories). Viscotec TriSEC (version 3) software was used for data processing.

General procedure for the acrylation of 1: The triol **1** (1g, 0.0045 mole) was incubated with the lipase (200 mg) in organic solvents (THF) along with equimolar amount of the acylating agent (vinyl methacrylate, 0.561g) at 30-32°C. The reaction was monitored by TLC and on completion quenched by filtering off the enzyme followed by solvent evaporation at reduced pressure. The residue was chromatographed on silica to yield the acrylated product.

Polymerization:
The enzyme-catalyzed ring opening polymerization of ε-CL by acryl sugar **3** was carried out under inert atmosphere. In a typical experiment, 100mg of HMG (3) and 100mg of Novozym-435 which had previously been dried in a vacuum desicator (0.1 mmHg, 25°C, 16 h) were transferred to an oven-dried 20 mL reaction vial. The vial was immediately stoppred with a rubber septum and purged with argon, and 0.99 mL of ε-CL (distilled at 97-98°C over CaH_2 at 10 mm of Hg) was added via syringe under argon, followed by the addition of 2mL of dry toluene. The reaction vial was then placed in a constant temperature oil bath maintained at 60°C for 8h. A control reaction was set up as described above except Novozyme-435 was not added. The reaction was quenched by removing the enzyme by vacuum filtration, the enzyme was washed 3-4 times with 5mL portions of chloroform, the filtrates were combined, and the solvent was removed in vacuo and the product was precipitated by adding methanol (25 mL) . The solid so obtained was filtered by vacuum filtration and washed 3-4 times with methanol (5 mL each time) to remove the unreacted HMG (3) and ε-caprolactone. The product was dried in vacuum oven to give 0.95 g (%-yield 87 %).

Regioselective end capping of HMG oligo(e-CL) (5): In a 20 ml oven dried reaction vial, 500mg (0.04 mmol) of HMG oligo(ε-CL) which had previously been dried in a vacuum desiccator (0.1 mmHg, 25°C, 16h) was dissolved in 5ml of dry toluene. Novozyme-435 (50 mg) dried under vacuum was added to the reaction mixture under nitrogen atmosphere. To this 20ul of vinyl acetate (0.2 mmol) was added via a syringe under nitrogen. The reaction mixture was then stirred at 30°C for 6 hours. To terminate the reaction, the enzyme was removed by vacuum filtration and washed with 3x2ml of toluene. The filtrates were combined and solvent was removed in vacuo to give 480 mg of the product.

Ketal deprotection of 5 to give 7: To 50 mg of **5** was added 0.5 ml of 80 % CF_3COOH in water and the reaction mixture was stirred for a predetermined time at room temperature. The resulting polymeric product was precipitated by adding water (10ml). The polymer was separated by filtration and washed with 5 ml of methanol, and dried in vacuum to a constant weight (35 mg).

Acknowledgements. We are grateful to the members of the NSF Center for Biocatalysis and Bioprocessing of Macromolecules at the Polytechnic University for their financial support.

References

1. Blinkovsky, A. M.; Khmelnitsky, Y. L.; Dordick, J. S. *Biotechnol. Tech.* **1994,** *8*, 33.
2. Shibatani, S.; Kitagawa, M.; Tokiwa, Y. *Biotechnol. Lett.* **1997,***19,* **511.**
3. Kitagawa, M.; Tokiwa, Y. *Carbohydr. Lett.* **1997**, 2, **343**.
4. Martin, B.D.; Ampofo, S.A.; Linhardt, R.J.; Dordick, J.S. *Macromolecules* **1992**, 25, 7081.
5. Kobayashi, K.; Kakishita, N.; Okada, M., Akaike, T. and Usui, T. J. *Carbohydr. Chem.*, **1994**,*13*, 753.
6. Nishimura, S., Matsuoka, k., Furuike, T., Ishi, S. and Kurita, K. *Macromolecules*, **1991**, *24*, 4236.
7. Chen, X., Dordick, J.S., and Rethwisch, *Macromolecules*, **1995**, *28*, 6014..
8. (a) Frechet, J.M.J. *Science*. **1994**, *263*, 1710. (b) Tomalia, D.A.; Naylor, A.M.; Goodard, W.A. *Angew. Chem., Int. Ed. Engl.* **1990**, *29*, 138.
9. (a) Ihre, H.; Hult, A.; Soderlind, E. *J.Am.Chem. Soc.* **1996**, *118*, 6388. (b) Ihre, H.; Hult, A.; Frechet, J.M.J; Gitsov, I. *Macromolecules* **1998**, *31*, 4061.
10. (a)Trollsas, M.; Hedrick, J.L.; Mecerreys, D.; Dubois, P.; Jerome, R.; Ihre, H.; Hult, A. *Macromolecules* **1997**, *30*, 8508. (b). Trollsas, M.; Hedrick, J.L. *J.Am. Chem. Soc.* **1998**, *120*, 4644.
11. (a) Hadjichristidis, N.; Tselikas, Y.; Iatrou, H.; Efstratiadis, V.; Avgeropoylos, A. *J. Macrmol. Sci., Pure Appl. Chem.* **1996,** *A33 (10),* 1447. (b) Jacob, S.; Majoros, I.; Kennedy, J.P. *Macromolecules* **1996**, *29*, 8631.
12. (a) Bisht, S.K.; Henderson, L.A.; Gross, R.A.; Kaplan, D.L.; Swift, G. *Macromolecules* **1997**, *30*, 2705. (b) Bisht, K.S.; Svirkin, Y.Y.; Henderson, L.A.; Gross, R.A.; Kaplan, D.L.; Swift, G. *Macromolecules* **1997**, *30*, 7735
13. (a) Kumar, A .; Gross, R.A. *J.Am.Chem.Soc.* **2000**, *122*, 11767; (b) Kumar, A.; Kalra, B.; Dekhterman, A. and Gross, R.A. *Macromolecules*, **2000**, *33*, 6303.
14. Cordova, A.; Hult, A.; Hult, K.; Ihre, H.; Iverson, T.; Malmsrom, E. *J.Am.Chem. Soc.* **1998**, *120*, 13521.
15. (a) Uyama, H.; Kikuchi, H.; Takeya, K.; Kobayashi, S. *Acta Polym.* **1996**, *47*, 357. (b) Kanaoka, S.; Omura, T.; Sawamoto, M.; Higashimura, T. *Macromolecules* **1992**, *25*, 6407
16. (a) Gross, R.A.; Kumar, A.; Kalra, B. *Chem. Rev.* **2001**, *101*, 2097. (b) Gross, R.A.; Kalra, B.; Kumar, A. *Appl. Microbiol. Biotechnol.* **2001**, *55*, 655. (c) Gross, R.A.; Kaplan, D.L; Swift, G. Enzymes in polymer synthesis, Eds.; ACS Symposium Series 684; American Chemical Society; Washington, DC, 1998.

17. Bisht, K.S.; Deng, F.;Gross, R.A.; Kaplan, D.L.; Swift, G. *J. Am. Chem. Soc.* **1998,** *120*, 1363.
18. Cordova, A.; Iverson, T.; Hult, K.; *Macromolecules* **1998**,*31*, 1040.
19. Wulff, G.; Schmid, J.; Venhoff, T.(1991). Carbohydrates as Organic Raw Materials, F.W. Lichtenthaler, ed. Pp. 311.
20. (a)Bollenback, G.N.; Parrish, F.W. *carbohydr. Res.* **1953**, *17*, 431.(b) Reinfeld, E.; Korn, H.F. Die Starke 1968, 20, 181
21. (a) Sugihara, J.M. *Adv. Carbohydr. Chem.* **1953**, *8*, 1 (b Ballard, J.M.; Hough, L.; Richardson, A.L. *Carbohyr. Res.* **1980**, *83*, 138. (c) Haines, A.H. *Adv. Carbohydr. Chem. Biochem.* **1981**, *39*, 13.
22. Youssefyeh, R.D.; Verheyden, J.P.H.; Moffatt, J.G. *J. Org. Chem.* **1979**, *44*, 1301.
23. Sharma, S.K.; Roy, S.; Kumar, R.; Parmar, V.S. *Tetrahedron Lett.* **1999**, *40*, 9145.
24. Kumar, A.; Gross, R.A. *Biomacromolecules* **2000**, *1*, 133.
25. Bisht, K.S.; Gao, Wei.; Gross, R.A. *Macromolecules* **2000**, *33*, 6208.

Polyesters and Polycarbonates

Chapter 10

In Vivo and In Vitro Metabolic Engineering of Polyhydroxyalkanoates Biosynthesis Pathways

Alexander Steinbüchel

Institut für Mikrobiologie, Westfälische Wilhelms-Universität Münster, Corrensstrasse 3, D–48149 Münster, Germany

Polyhydroxyalkanoates (PHA) represent a complex class of bacterial polyesters consisting of various hydroxyalkanoic acids that are synthesized by bacteria if a carbon source is present in excess. Poly(3-hydroxybutyrate), poly(3HB), is just one example. Most other PHAs, which consist not of 3HB, are only synthesized if pathways exist, which mediate between central intermediates of the metabolism or special precursor substrates on one side and coenzyme A thioesters of hydroxyalkanoic acids, which are the substrates of the PHA synthase catalyzing the polymerization, on the other side. Various such pathways allowing the *in vitro* synthesis of PHAs consisting of 3HB, 4HB and 4HV have been designed employing the PHA synthase from *Thiocapsa pfennigii* and various enzymes from *Clostridia.* These pathways were subsequently also established in recombinant *Escherichia coli* by expressing the genes for the respective enzymes. Besides these, other pathways were also metabolically engineered allowing the *in vivo* synthesis of further PHAs from simple carbon sources.

Poly(3-hydroxybutyric acid), poly(3HB), and other structurally related aliphatic polyesters from bacteria belonging to the class of polyhydroxyalkanoic acids (PHA) form biodegradable thermoplastic and elastomers that are currently in use or being considered for use in industry, medicine, pharmacy and agriculture. More then hundred different hydroxyalkanoic acids have been already identified as constituents of these polyesters (1). They are currently produced by microbial fermentations. In the future, production will be also possible by using agriculture transgenic plants (2).

Recently, also *in vitro* biosynthesis methods for various PHAs were established (2). One example is the recently established *in vitro* biosynthesis of poly(3HB) from 3-hydroxybutyryl-CoA (3HBCoA) employing purified PHA synthase from *Chromatium vinosum* ($PhaEC_{Cv}$). (3). Based on this study a three-step *in vitro* poly(3HB) biosynthesis system allowing the synthesis of poly(3HB) from 3-hydroxybutyrate and the recycling of CoA was established employing purified propionyl-CoA transferase from *Clostridium propionicum* and a commercially available acetyl-CoA synthetase from yeast in addition (4). Since this reaction was driven by ATP, and because only catalytic quantities of CoA were used, poly(3HB) could be produced on a semipreparative scale. With other enzymes, we were also able to synthesize *in vitro* poly(3-hydroxydecanoic acid), poly(3HD). For this, we employed the purified PHA synthase from *Pseudomonas aeruginosa* plus a commercially available acyl-CoA synthetase (5).

Results and Discussion

Active butyrate kinase (Buk) and phosphotransbutyrylase (Ptb) from *Clostridium acetobutylicum* were purified to almost homogeneity in three steps employing ammonium sulfate precipitation, hydrophobic interaction chromatography on phenyl-Sepharose and affinity chromatography on Matrex Red A from a recombinant strain of *Escherichia coli* expressing these proteins (6). The purified enzymes were then sucessfully exploited for the *in vitro* synthesis of 3-hydroxybutyryl-CoA (3HBCoA), 4-hydroxybutyryl-CoA (4HBCoA), 4-hydroxyvaleryl-CoA (4HVCoA). In addition, the PHA synthase of *Chromatium vinosum* ($PHAEC_{Cv}$) to use these CoA thioesters as substrates was investigated (7). The combination of Buk and Ptb with $PhaEC_{Cv}$ established a new system for *in vitro* synthesis of poly(3HB). In this system, 3-hydroxybutyric acid was converted to 3HBCoA by Buk and Ptb at the expense of ATP (7). This reaction was further driven by the polymerization of this intermediate to poly(3HB) by $PhaEC_{Cv}$ and also by the recycling of the released CoA by Ptb. Approximately 2.6 mg poly(3HB) could be synthesized from 7.7 mg 3-hydroxybutyric acid in a volume of only 1 ml. These three enzymes were *in vitro* also suitable to produce copolyesters containing various molar fractions of 3HB and 4HB if these hydroxyfatty acids were provided as substrates. The highest 4HB content in these copolyesters was 46 mol%.

This new pathway was also expressed *in vivo* in recombinant strains of *E. coli*. However, the *C. vinosum* PHA synthase was replaced by the PHA synthase of *Thiocapsa pfennigii* ($PhaEC_{Tp}$) because the latter enzyme exhibited a broader substrate range (8). For this, the four genes were cloned into plasmid pBR322, and the resulting hybrid plasmid pBPP1 conferred activities of all three enzymes to *E. coli* JM109. When the recombinant strain of *E. coli* was cultivated in M9 mineral salts medium containing glucose and hydroxyfatty acids as carbon sources, the cells accumulated PHAs. Homopolyesters of 3HB, 4HB or 4HV were obtained from each of the corrsponding hydroxyfatty acids. Various copolyesters of those hydroxyfatty acids were also obtained when two of these hydroxyfatty acids were fed at equal amounts (8). For example, cells fed with 3-hydroxybutyric acid and 4-hydroxybutyric acid accumulated a copolyester consisting of 88 mol% 3HB plus 12 mol% 4HB and contributing to approximately 69 % of the cell dry weight. Cells, which were fed with 3-hydroxybutyric acid and 4 hydroxyvaleric acid, accumulated a copolyester consisting of 94 mol% 3HB plus 6 mol% 4HV and contributing to 64 % of the cell dry weight. Finally, cells fed with 3-hydroxybutyric acid, 4-hydroxybutyric acid and 4-hydroxyvaleric acid accumulated a terpolyester consisting of 85 mol% 3HB, 13 mol% 4HB plus 2 mol% 4HV and contributing to 68.4 % of the cell dry weight.

Conclusions

This study demonstrated that a PHA biosynthesis pathway engineered *in vitro* employing purified enzymes is functionally active in *E. coli* and can be utilized for the *in vivo* synthesis of PHAs. *In vitro* engineering of pathways may be a useful strategy to evaluate whether the establishment of a particular pathway in a bacterium by *in vivo* metabolic engineering is feasible.

Acknowledgements

I would like to express my severe thanks to my coworkers R. Jossek, S.-J. Liu, Q. Qi, and B. H. A. Rehm who performed the experiments summarized in this lecture.

References

1. Steinbüchel, A.; Valentin , H.E.: *FEMS Microbiol. Lett.* **1995**, *128*, 219.
2. Steinbüchel, A.; Füchtenbusch, B.: *Trends in Biotechnology* **1998**, *16*, 419.
3. Jossek, R.; Reichelt, R.; Steinbüchel, A.: *Appl. Microbiol. Biotechnol.* **1998**, *49*, 28.
4. Jossek, R.; Steinbüchel, A.: *FEMS Microbiol. Lett.* **1998**, *168*, 319.
5. Qi, Q.; Steinbüchel, A.; Rehm, B. H. A.: *Appl. Microbiol. Biotechnol.* **2000**, 54, 37.
6. Cary, J. W.; Petersen, D. J.; Papoutsakis, E. T.; Bennett, G. N.: *J. Bacteriol.* **1988**, *170*, 4613.
7. Liu, S.-J.; Steinbüchel, A.: *Appl. Microbiol. Biotechnol.* **2000**, 53, 545.
8. Liu, S.-J.; Steinbüchel, A.: *Appl. Environ. Microbiol.* **2000**, *66*, 739.

Chapter 11

Controlling the Polymer Microstructure of Biodegradable Polyhydroxyalkanoates

Aaron S. Kelley[1] and Friedrich Srienc[2]

[1]Department of Chemical Engineering and Materials Science, Bio-Process Technology Institute, University of Minnesota, Minneapolis, MN 55455
[2]Department of Chemical Engineering and Materials Science, Bio-Process Technology Institute, University of Minnesota, St. Paul, MN 55108

Polyhydroxyalkanoates (PHAs) are biodegradable polyesters produced by many bacterial species. They are a natural part of the renewable carbon life cycle and are a sustainable source of plastics into the future. PHA production, however, cost about ten times as much as petroleum-derived thermoplastics. The value of the polymers can be increased by controlling the polymer's microstructure. By changing what substrates are available to the bacteria during polymer synthesis, different compositions of PHA can be synthesized. This technique can be used to synthesize core and shell latexes, giving specific properties to film and coatings. Additionally, alternating substrates at a faster rate can be used to synthesize different polymers within an individual polymer chain, thereby producing block copolymers. Block copolymers have uses as blend compatibilizers and thermoplastic elastomers. Producing PHAs with controlled microstructures gives control of the physical properties and allows for the production of high value polymers.

Polyhydroxyalkanoates (PHAs) are a class of biodegradable polyesters produced by many bacterial species. The polymer is synthesized when growth is limited by lack of a nutrient but carbon is available in excess. Polyhydroxybutyrate (PHB), the most prevalent PHA, has high crystallinity which leads to brittle failure.[1] Copolymers of PHB incorporating polyhydroxyvalerate (PHV) have proven to increase the toughness of the polymer. Blends of PHB with other biodegradable materials such as starch or polycaprolactone (PCL) have also lead to some improvements while maintaining biodegradability.[2] Phase separated blends present yet another way to improve the physical properties of brittle PHB. It has been shown that blends of PHB and polyhydroxybutyrate-co-valerate (PHBV) will phase separate with as little as 8% HV differences between the two polymers.[3]

Synthesizing polymer granules within bacteria in a "living" polymerization gives a unique opportunity to control the microstructure of the polymer. By controlling the substrate available to the bacteria, different polymer types can be synthesized within the same granule, or even within the same polymer chain. Curley *et al.* (1996) demonstrated the synthesis of "layered" granules in *Pseudomonas oleovorans* with diauxic use of substrates.[4] Kelley and Srienc (1999) showed that 'layered' granules could be synthesized by *Ralstonia eutropha* as well, consisting of a core of PHBV and a shell of PHB.[5]

Results and Discussion

In order to benefit from controlled PHA microstructures, a system had to be developed in *R. eutropha* to produce two different polymers that would phase separate. PHB and PHBV have been shown to phase separate.[3] Furthermore, hydroxyvaleric acid is only incorporated into the polymer chain if an odd chained fatty acid is present in the media. This presents a unique opportunity to control when PHBV is synthesized.

Figure 1 shows the approach used to synthesize controlled microstructures in *R. eutropha*. With an excess of fructose present in the media, control of PHBV synthesis is dependant upon valeric acid additions. By controlling when and how much valerate is added, the microstructures can be controlled. Valerate additions that remain in the media for 8-12 hours before exhaustion produce layered granules.[5] Multiple additions lasting approximately 4 hours each will produce multiple layered granules (work not shown).

If switches in synthesis conditions can be made to occur on a timescale equal to the synthesis time of individual polymer chains, block copolymers will be produced. Literature estimates of chain synthesis times vary from 10 minutes to 5 hours.[6,7]

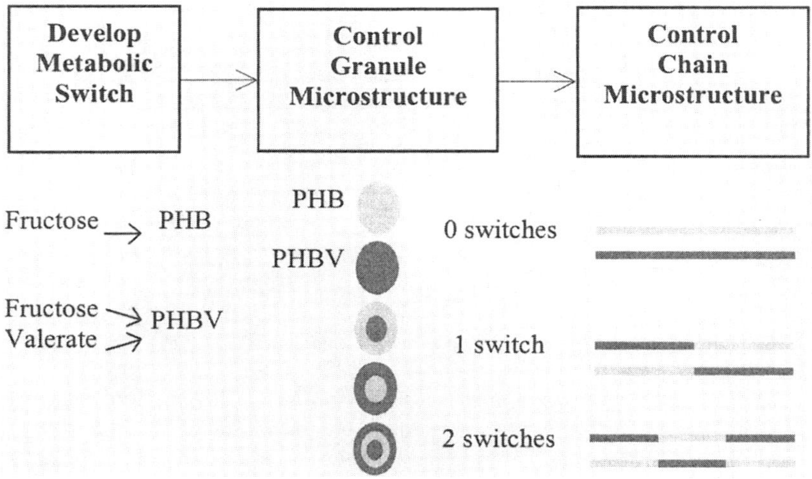

Figure 1. This figure outlines the approach to synthesizing controlled PHA microstructures. Valeric acid is added in limited amounts to media containing an excess of fructose. The number of switches and in what timeframe determine the microstructure. Switches on the order of hours produce layered granules. Switches within the synthesis time of a polymer chain produce block copolymers.

Conclusions

Polymer synthesis in microorganisms presents a unique opportunity to control microstructures to produce novel architectures. Core and shell granules should have uses as specialized latexes. Block copolymers could be used to compatabilize blends, as thermoplastic elastomers and possibly as biocompatible adhesives. Controlled deposition of materials can aid in the production of scaffolding for further nanoconstruction, presenting opportunities in the unfolding nanotechnology arena.

Acknowledgements

This work has been supported in part by the Consortium for Plant Biotechnology Research and by the National Science Foundation (BES-9708146).

References

1. Luzier, W.D. *Proc. Natl Acad. Sci. USA* **1992**, 89, 839-842.
2. Ramsay, BA, Langlade V, Carreau PJ, Ramsay JA. *Appl. Environ Microbiol.* **1993**, 59, 1242-1246.
3. Barham PJ, Organ SJ, *J. Mater. Sci.*, **1994**, 29, 1676-1679.
4. Curley JM, Lenz RW, Fuller RC. *Int. J. Biol. Macromol.*, **1996**, 19, 29-34.
5. Kelley AS, Srienc F. *Int. J. Biol. Macromol.*, **1999**, 25, 61-67.
6. Kelley AS, Jackson DE, Macosko C, Srienc F. *Poly. Degrad. Stab.*, **1998**, 59, 187-190.
7. Steinbuchel A, Aerts K, Babel W, et al., *Can. J. Microbiol.*, **1995**, 41, 94-105.

Chapter 12

Enantioselective and Regioselective Polymerization with Lipase Catalysis to Polyesters

Shiro Kobayashi and Hiroshi Uyama

Department of Materials Chemistry, Graduate School of Engineering, Kyoto University, Kyoto 606–8501, Japan

Highly enantioselective and regioselective polymerizations have been developed via lipase catalysis to give functional polyesters, which can not be obtained by conventional chemical catalysts. Ring-opening copolymerization of racemic substituted lactones with achiral lactones enantioselectively proceeded by using *Candida antarctica* lipase (lipase CA) as catalyst. In the copolymerization of racemic β-butyrolactone with 12-dodecanolide, the (*S*)-isomer was preferentially reacted to give the (*S*)-enriched optically active copolymer with enantiomeric excess of β-butyrolactone unit = 69 %. δ-Caprolactone was also enantioselectively copolymerized by the lipase catalyst to give the optically active polyester. Regioselective polymerization of divinyl sebacate and polyols has been achieved by using the lipase catalyst to give the linear reactive polymers consisting of exclusively α,ω-disubstituted unit of the polyol substrate. Lipase CA-catalyzed polymerization of divinyl sebacate and glycerol in the presence of unsaturated higher fatty acids produced a new class of crosslinkable polyesters having the unsaturated group in the side chain.

Introduction

For the last decades, an enzyme-catalyzed polymerization ("enzymatic polymerization") has been becoming important as new trend in macromolecular science (*1-7*). Enzyme catalysis has provided new synthetic strategy for useful polymers, most of which are difficult to produce by conventional chemical catalysts. *In vitro* enzymatic synthesis of polymers via non-biosynthetic (non-metabolic) pathways is recognized as a new area of precision polymer syntheses. The enzymatic polymerization also affords a great opportunity for using non-petrochemical renewable resources as starting substrates of functional polymeric materials. These methods may be useful for global sustainability without depletion of important resources. In the enzymatic polymerizations, the product polymers can be obtained under mild reaction conditions without using toxic reagents. Therefore, the enzymatic polymerization has a large potential as an environmentally friendly synthetic process of polymeric materials, providing a good example of how one may achieve "green polymer chemistry" (*6,7*).

Recently, lipase-catalyzed synthesis of polyesters has been developed as new methodology of biodegradable materials (*8,9*). Lipase is an enzyme which catalyzes the hydrolysis of fatty acid esters normally in an aqueous environment in living systems. However, lipases are sometimes stable in organic solvents and can be used as catalyst for esterifications and transesterifications. Such catalytic specificities of lipase have enabled production of functional aliphatic polyesters. The present article described our recent topics on lipase-catalyzed enantio- and regioselective polymerizations.

Enantioselective Polymerization

Optically active polymers are important functional materials having great potential for various applications such as chiral catalysts for asymmetric synthesis, packing materials of chromatographic columns for enantiomer separation, and chiral materials for the preparation of liquid crystal polymers (*10*). Enantioselective polymerizations have been investigated, in which one isomer of a racemic monomer is preferentially reacted. So far, α-olefins, methacrylates, and small-sized cyclic compounds (propylene oxide, propylene sulfide, four-membered lactones, and α-amino acid *N*-carboxylic anhydrides (5-membered)) have been reported to enantioselectively polymerize to give optically active polymers. In most cases, combinations of metal catalysts and chiral cocatalysts were used.

Lipase catalysis has been useful for enantioselective synthesis of optically active polyesters. Porcine pancreas lipase catalyzed the enantioselective polycondensation of a racemic epoxide-containing activated diester with a diol monomer to produce a highly optically active polyester (enantiomeric purity of >96%) (*11*). Racemic α-methyl-β-propiolactone was stereoselectively polymerized by *Pseudomonas cepacia* lipase (lipase PC) to give an optically active (*S*)-enriched polyester (*12*). The highest enantiomeric excess (ee) value of the polymer was 50 %. An enantioselective polymerization of β-butyrolactone (**1a**) occurred by using thermophilic lipase to give an (*R*)-enriched low-molecular weight polymer with 20-37 % ee (*13*).

In our recent study, enantioselective synthesis of optically active polyesters from four- and six-membered substituted lactones (**1a** and δ-caprolactone (**1b**), respectively) using *Candida antarctica* lipase (lipase CA) catalyst has been performed (*14*). Lipase CA was reported to efficiently catalyze the ring-opening polymerization of achiral lactones (*15-17*). At first, we examined the homopolymerization of **1a** catalyzed by lipase CA, resulting in the low polymerizability and enantioselectivity. In the bulk polymerization at 60 °C for 24 h, 59 % of **1a** was consumed to give oligomers with molecular weight less than 500. The calculated ee value of the resulting oligomer based on the optical purity of the unreacted monomer was only 4 % in (*S*)-enriched form.

On the other hand, the optically active polyesters were obtained by the lipase CA-catalyzed copolymerization of **1a** with achiral lactones (ε-caprolactone (**2a**) and 12-dodecanolide (**2b**)) (Figure 1). In the copolymerization of **1a** with **2b** (equimolar feed ratio) in diisopropyl ether at 60 °C for 4 h (entry 3 in Table 1), the gas chromatographic (GC) analysis confirmed the conversion of **1a** to be 59 % and the quantitative consumption of **2b**. The optical purity of the unreacted monomer was analyzed by using a chiral GC system. Only one peak was detected and the comparison with the authentic samples ((*R*)- and (*S*)-**1a**) showed the (*R*)-configuration of the residual monomer. Based on these data, the (*S*)-content of **1a** unit in the copolymer was estimated as 85 % (ee = 69 %). M_n and its index of the copolymer were determined by size exclusion chromatography (SEC) to be 1200 and 2.3, respectively. The specific rotation ($[\alpha]_{365}$) of the copolymer was + 4.9° (c = 6.1, chloroform).

It is well known that the catalytic site of lipase is a serine-residue and lipase-catalyzed reactions proceed via an acyl-enzyme intermediate. In the postulated mechanism of the lipase-catalyzed polymerization of lactones, the key step is the reaction of the lactone with the serine residue of lipase involving the ring-opening of the lactone to produce the acyl-enzyme intermediate (*18*). If the reaction of **1a** proceeds in a similar manner, the bond cleavage between the carbonyl carbon and oxygen atom of the lactone takes place, in which the configuration is retained. On the other hand, opening of the monomer at the alkyl-oxygen bond results in inversion or racemization of configuration. In order to determine the absolute configuration of **1a** unit in the copolymer, the

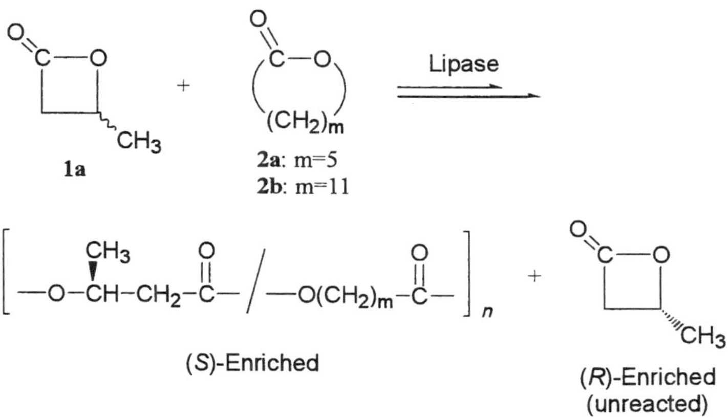

Figure 1. Lipase-catalyzed enantioselective copolymerization of β–butyrolactone with achiral lactones to optically active polyesters.

acid-catalyzed methanolysis of the copolymer and subsequent derivatization with (*R*)-(+)-phenylethylisocyanate were carried out. The configuration of methyl 3-hydroxyburyrate was found to be of (*S*)-form with optical purity of 92 % (ee = 85 %), which was fairly close to the calculated value on the basis of the conversion and optical purity of the unreacted monomer. These data support that the present copolymerization proceeds via the acyl cleavage and (*S*)-**1a** is preferentially copolymerized to give the (*S*)-enriched polyester.

Copolymerization results are summarized in Table 1. In the combination of **1a** and **2b**, the enantioselective copolymerization took place in other solvents, heptane and isooctane (entries 4 and 5), although the ee and specific rotation values were lower than those in diisopropyl ether (entry 3). Lipase CA also induced the enantioselective copolymerization of **1a** with **2a** in heptane to give the optically active polyester (entries 1 and 2).

The copolymerization of **1a** and **2b** in heptane was monitored by GC. The enzymatic conversion of **1a** proceeded more slowly than that of **2b** (Figure 2(A)). In the initial stage (after 1 h), the monomer conversion of (*S*)-**1a** was about 8.7 folds larger than that of the (*R*)-isomer, whose value was much larger than the reported one by chemical polymerization catalyst (*19*). The ee value of the polymer decreased as a function of the monomer conversion (Figure 2(B)). This is because the relative concentration of the less reactive isomer ((*R*)-isomer) increases in the monomer pool as the preferable enantiomer is incorporated in the copolymer.

Table 1. Lipase-Catalyzed Ring-Opening Copolymerization of 1 with 2[a]

Entry	*Monomers*		*Solvent*	*Time* (h)	*Conv.*[b] (%)		ee_m[c] (%)	ee_p[d] (%)	$[\alpha]_{365}$[e]	M_n[f]	M_w/M_n[f]
	1	**2**			**1**	**2**					
1	**1a**	**2a**	heptane	2	61	96	83 (*R*)	53	+12 (c = 2.4)	800	2.0
2	**1a**	**2a**	heptane	9	71	100	100 (*R*)	41	+8.2 (c = 2.6)	2300	1.5
3	**1a**	**2b**	diisopropyl ether	4	59	100	100 (*R*)	69	+4.9 (c = 6.1)	1200	2.3
4	**1a**	**2b**	heptane	6	67	100	100 (*R*)	49	+4.6 (c = 5.1)	1400	3.6
5	**1a**	**2b**	isooctane	6	67	100	100 (*R*)	69	+4.3 (c = 5.0)	1500	3.8
6	**1b**	**2a**	diisopropyl ether	4	26	100	11 (*S*)	32	-2.1 (c = 0.9)[g]	3100 [g]	2.9[g]
7	**1b**	**2a**	heptane	4	48	100	21 (*S*)	45	-3.7 (c = 2.7)[g]	2000 [g]	2.3[g]
8	**1b**	**2b**	diisopropyl ether	4	35	100	21 (*S*)	76	-1.6 (c = 1.1)[g]	6000 [g]	1.9[g]
9	**1b**	**2b**	heptane	4	44	100	42 (*S*)	53	-2.4 (c = 4.0)[g]	2600 [g]	2.2[g]

[a] Copolymerization of **1** with **2** (each 1.0 mmol) using lipase CA (50 mg) as catalyst at 60 °C in an organic solvent (5 mL). [b] Determined by GC. [c] Enantiomeric excess of unreacted **1**, determined by using chiral GC on a Chiraldex G-TA column. In parenthesis, configuration of the major unreacted monomer. [d] Enantiomeric excess of poly(**1**) unit in the copolymer, calculated on the basis of conversion and optical purity of unreacted **1**. [e] Measured in chloroform at room temperature. [f] Determined by SEC. [g] Data of methanol-insoluble part.

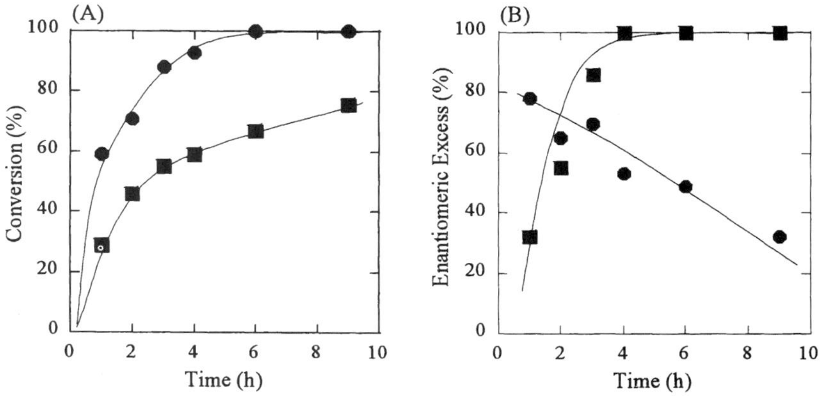

*Figure 2. Relationships between reaction time and (A) conversion of **1a** (■) and **2b** (●), and (B) enantiomeric excess of (R)-enriched unreacted **1a** (■) and (S)-enriched **1a** unit in the copolymer (●). The copolymerization of **1a** and **1b** was performed using lipase CA as catalyst in heptane at 60°C.*

δ-Caprolactone (**1b**) having three methylene units is more flexible in ring-structure than **1a**, and hence, optical resolution of **1b** will be difficult by conventional chemical catalysts. Lipase catalysis achieved the enantioselective copolymerization of δ-caprolactone with **2** to give the optically active polyester (entries 6-9) (Figure 3). In the reaction of **1b** and **2b** in diisopropyl ether, the highest ee value (76 %) was achieved (entry 8). The absolute configuration of the residual monomer was mainly the (*S*)-form, which was opposite to that of **1a**. In the esterase-catalyzed hydrolysis of **1b** in aqueous medium, the (*R*)-isomer was enantioselectively reacted, which agreed with the enantioselectivity of the present copolymerization.

Microstructural analysis of the copolymer was examined by ^{13}C NMR. A statistically binary copolymer has four different diads. Formation of random copolymers was often observed in the lipase-catalyzed copolymerization of achiral lactones in spite of the different enzymatic polymerizability (*20,21*), which indicates that the intermolecular transesterification of the polyesters frequently takes place during the copolymerization. There were three peaks due to $C(=O)OCH_2C$ or $C(=O)OCH(CH_3)C$ in the region of δ 64-68 in the ^{13}C NMR spectrum of the copolymer from **1a** and **2b** (entry 3 in Table 1); only a peak due to **1a**-**1a** homolinkage unit was not observed. These data suggest the

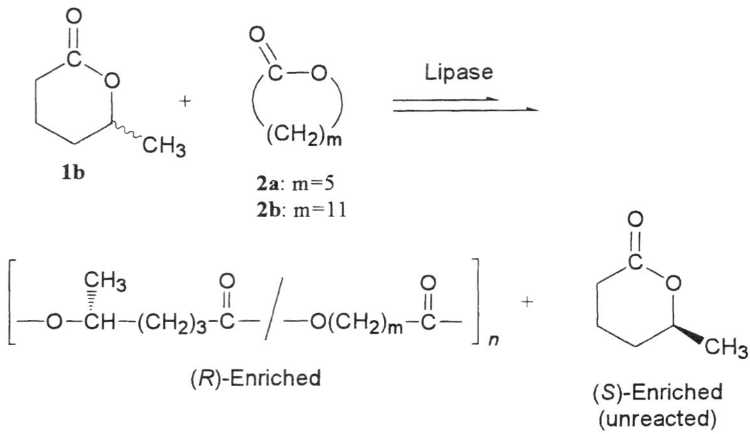

Figure 3. Enzymatic enantioselective copolymerization of δ–caprolactone with achiral lactones.

lower reactivity of the secondary alcohol derived from **1a**, *i.e.*, the conditions for the homopolymerization of **1a** without solvent brought about the low enantioselectivity. On the other hand, the terminal secondary alcohol of the polymer is preferentially reacted with a non-substituted lactone under milder reaction conditions to give polymers with higher optical purity. In the case of the copolymer from **1b** and **2b,** there was a small peak ascribed to **1b-1b** homolinkage. This may be because of the larger enzymatic reactivity of the resulting alcohol from **1b** than that from **1a**.

Regioselective Polymerization

Lipases and proteases have been used as catalyst for regioselective acylation of polyols such as sugars. Protease-catalyzed regioselective polycondensation of sucrose with bis(2,2,2-trifluoroethyl) adipate was reported to give a sugar-containing oligoester having ester linkages at the 6 and 1' positions on the sucrose *(22)*. Recently, two-step enzymatic synthesis of polyesters containing a sugar moiety in the main chain was demonstrated by utilizing the specific catalysis of lipase *(23)*. In this study, lipase-catalyzed regioselective polymerization of divinyl sebacate (**3**) with polyols was examined.

Divinyl esters have often been used as an activated diacid substrate in enzymatic synthesis of polyesters, since the leaving vinyl alcohol irreversibly tautomerizes to an aldehyde or a ketone, leading to the efficient production of polyesters under mild reaction conditions *(24,25)*. Here, we used various triols that include glycerol (**4a**), 1,2,4-butanediol (**4b**), and 1,2,6-hexanetriol (**4c**), for the regioselective polymerization with **3** in the presence of lipase catalyst in bulk (Figure 4) *(26,27)*. Polymerization results are summarized in Table 2.

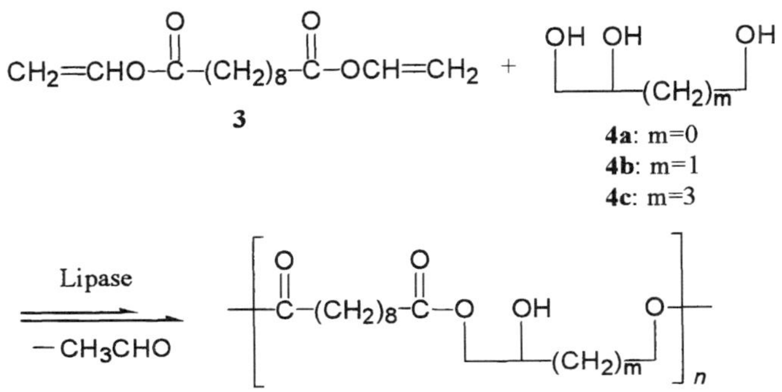

3

4a: m=0
4b: m=1
4c: m=3

Figure 4. Regioselective polymerization of divinyl sebacate and triols via lipase catalysis.

Table 2. Lipase-Catalyzed Polymerization of Divinyl Sebacate and Triols[a]

Entry	*Triol*	*Catalyst*	*Temp.* (°C)	*Yield* (%)	M_w[b] ($\times10^{-3}$)	M_w/M_n[b]
1	**4a**	lipase CA	45	45	2.2	2.2
2	**4a**	lipase CA	60	63	19	5.4
3	**4a**	lipase MM	45	63	1.3	1.9
4	**4a**	lipase MM	60	67	1.3	1.8
5	**4a**	lipase PC	45	18	1.0	1.4
6	**4a**	lipase PC	60	20	1.1	1.9
7	**4b**	lipase CA	45	71	14	3.5
8	**4b**	lipase CA	60	70	27	6.6
9	**4c**	lipase CA	30	86	1.9	1.9
10	**4c**	lipase CA	45	74	13	4.1
11	**4c**	lipase CA	60	70	12	3.4
12	**4c**	lipase MM	60	81	5.8	2.2
13	**4c**	lipase PC	60	41	1.4	1.8

[a] Polymerization of **3** with **4** (each 2.0 mmol) using lipase (100 mg) for 8 h under argon. [b] Determined by SEC.

Lipases CA, PC, and *Mucor miehei* lipase (lipase MM) catalyzed the polymerization of **3** with **4** to give the soluble polymers. Lipase CA showed the highest catalytic activity for the polymerization; in most cases, the molecular weight was beyond $1x10^4$. In the polymerization without the enzyme (control experiment), the polymeric materials were not formed, indicating that the present polymerization took place through the enzyme catalysis.

The polymer microstructure was analyzed by using various NMR techniques. Figure 5 shows expanded ^{13}C NMR spectra of the glyceride units in the polymers from **3** and **4a** obtained under the different reaction conditions. Peaks were assigned as shown in Figure 5 and unit ratio was estimated by the integrated areas of methine peaks A-D. The polymer obtained using lipase CA catalyst at 60 °C possessed four glyceride units (Figure 5(A)). From their ratio (1-substituted unit : 1,2-disubstituted unit : 1,3-disubstituted unit : trisubstituted unit = 11 : 15 : 58 : 16 (%)), the regioselectivity of acylation toward primary alcohol to secondary one is calculated as 74 % : 26 %. These data indicate the enzymatic formation of a reactive polyester consisting of 1,3-diglyceride unit as the major and a small amount of the branching unit (triglyceride) *(26-28)*. The microstructure strongly depended on the enzyme origin; the lipase MM-catalyzed polymerization produced a linear polymer consisting of 1,2- and 1,3-disubstituted units (Figure 5(B)), whereas the 1,3-disubstituted and trisubstituted units were observed in the polymer obtained using lipase PC catalyst (Figure 5(C)). There were only two peaks in the region of δ 60-75 in case of the polymer obtained by using lipase CA catalyst at 45 °C (Figure 5(D)), indicating that the regioselectivity was perfectly controlled to give the polymer consisting of exclusively 1,3-disubstituted unit. In the lipase CA-catalyzed polymerization of **3** and **4b** or **4c**, the α,ω-diacylated unit of triol was mainly formed and a similar behavior of the temperature effect was observed. The monomer feed ratio also affected the microstructure. In the polymerization of **3** and **4a** catalyzed by lipase CA at 60 °C, the linear polymer with the perfectly controlled structure was obtained in the feed ratio (**3**/**4a**) of 1.15 and 1.2.

The lipase-catalyzed regioselective polycondensation was expanded to production of polyesters bearing a sugar moiety in the main chain *(29)*. Lipase CA catalyzed the regioselective polymerization of sorbitol (**5**) with **3** in acetonitrile at 60 °C to give the polymer with M_n of $9.8x10^3$ in 64 % yield (Figure 6). The polymer was soluble in *N,N*-dimethylformamide and dimethyl sulfoxide, and insoluble in acetone, chloroform, methanol, toluene, and water. Analysis of the polymer microstructure by ^{13}C NMR showed that the polymer consisted exclusively of the acylated unit of sorbitol at 1- and 6-positions, indicating that the regioselectivity was perfectly controlled through the enzyme catalysis (Figure 7). Mannitol and *meso*-erythritol were also regioselectively polymerized with **3** to give sugar-containing polyesters.

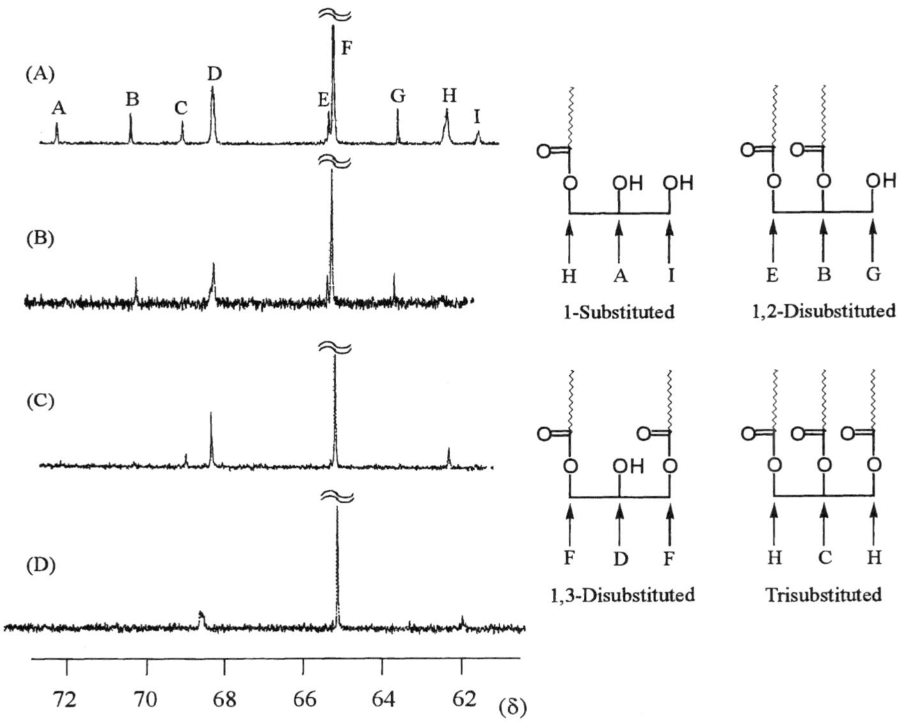

*Figure 5. Expanded ^{13}C NMR spectra of the polymers obtained from **3** and **4a** under the following conditions: (A) lipase CA catalyst at 60°C; (B) lipase MM catalyst at 60°C; (C) lipase PC catalyst at 60°C; (D) lipase CA catalyst at 45°C.*

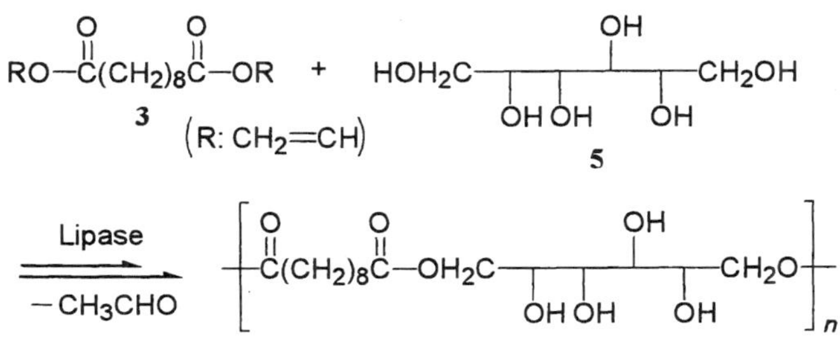

Figure 6. Lipase-catalyzed polymerization of divinyl sebacate and sorbitol.

As one possible applications of the enzymatic regioselective polymerizations, crosslinkable polyesters were synthesized by lipase CA-catalyzed polymerization of **3** and **4a** in the presence of unsaturated higher fatty acids (**6**) derived from plant oils (Figure 8) (*30*). The polymerization under reduced pressure improved the yield and molecular weight of the polyesters having the unsaturated group in the side chain. From NMR analysis of the polymer, the triglyceride unit was found to be mainly formed. The curing was examined by cobalt naphthenate catalyst or thermal treatment. Two or three unsaturated groups in the side chain were required for the hardening, yielding crosslinked transparent film. FT-IR analysis of the cured film suggests that the crosslinking mechanism of the polymer was similar to that of the oil autoxidation. The present crosslinkable polyesters are synthesized from renewable plant-based materials without use of toxic reagents under mild reaction conditions. In the curing stage, the crosslinked polymeric film is obtained in the absence of organic solvents at an ambient temperature under air. Therefore, the present method is expected as an environmentally benign process of polymer coating.

Conclusion

Highly selective polymerizations catalyzed by lipase have been achieved to produce new functional polymers. The enantioselective copolymerization of racemic substituted lactones with achiral non-substituted lactones took place through lipase CA catalysis to give the polymer with high optical purity. In the lipase-catalyzed polymerization of divinyl sebacate and triols, the triol monomer was regioselectively acylated, yielding soluble polyesters. From sorbitol, the sugar-containing polyester consisting of 1,6-diacylated sorbitol unit was obtained. The polymerization of divinyl sebacate and glycerol in the presence of unsaturated fatty acids gave the crosslinkable polyesters. The enzymatic processes for production of useful polymeric materials are highly environmentally benign since in most cases, biodegradable products are obtained from non-toxic substrates and catalysts under mild reaction conditions; therefore, the enzymatic polymerizations are expected to provide a future essential technology in chemical industry.

References

1. Kobayashi, S.; Shoda, S.; Uyama, H. *Adv. Polym. Sci.* **1995**, *121*, 1-30.
2. Kobayashi, S.; Shoda, S.; Uyama, H. In *Catalysis in Precision Polymerization*, Kobayashi, S., Ed.; John Wiley & Sons: Chichester, Chap. 8 (1997).

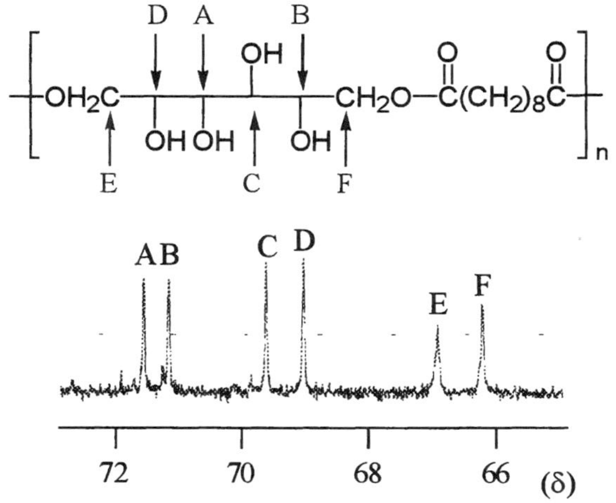

Figure 7. Expanded ^{13}C NMR spectrum of sugar-containing polyester from sorbitol.

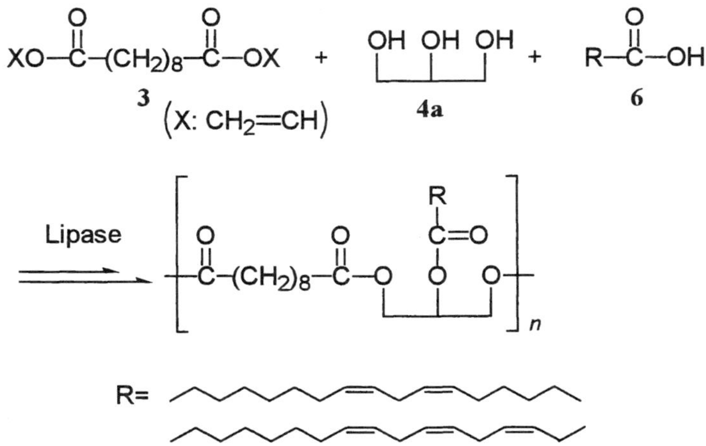

Figure 8. Enzymatic synthesis of crosslinkable polyesters.

3. Ritter, H. In *Desk Reference of Functional Polymers, Syntheses and Applications*, Arshady, R., Ed.; American Chemical Society: Washington, pp103-113 (1997).
4. Gross, R. A.; Kaplan, D. L.; Swift, G., Ed.; *ACS Symp. Ser.* **1998**, *684*.
5. Kobayashi S.; Uyama H. In *Materials Science and Technology - Synthesis of Polymers*, Schlüter A.-D., Ed.; Wiley-VCH: Weinheim, Chap 16 (1998).

6. Kobayashi, S. *J. Polym. Sci., Polym. Chem. Ed.* **1999**, *37*, 3041-3056.
7. Kobayashi, S.; Uyama, H.; Ohmae, M. *Bull. Chem. Soc. Jpn.* **2001**, *74*, 613-635; Kobayashi, S.; Uyama, H.; Kimura, S. *Chem. Rev.* **2001**, *101*, 3793-3818.
8. Kobayashi, S.; Uyama, H. In *Biopolyesters*, Babel, W.; Steinbüchel, A., Ed.; Springer-Verlag: Heidelberg, pp241-262 (2001).
9. Kobayashi, S.; Uyama, H. *Curr. Org. Chem.* in press.
10. Okamoto, Y.; Nakano, T. *Chem. Rev.* **1994**, *94*, 349-372.
11. Wallace, J. S.; Morrow, C. J. *J. Polym. Sci., Polym. Chem. Ed.* **1989**, *27*, 2553-2567.
12. Svirkin, Y. Y.; Xu, J.; Gross, R. A.; Kaplan, D. L.; Swift, G. *Macromolecules* **1996**, *29*, 4591-4597.
13. Xie, W.; Li, J.; Chen, D.; Wang, P. G. *Macromolecules* **1997**, *30*, 6997-6998.
14. Kikuchi, H.; Uyama, H.; Kobayashi, S. *Macromolecules* **2000**, *33*, 8971-8975.
15. Uyama, H.; Suda, S.; Kikuchi, H.; Kobayashi, S. *Chem. Lett.* **1997**, 1109-1110.
16. Kumar, A.; Gross, R. A. *Biomacromolecules* **2000**, *1*, 133-138.
17. Kumar, A.; Kalra, B.; Dekhterman, A.; Gross, R. A. *Macromolecules* **2000**, *33*, 6303-6309.
18. Uyama, H.; Takeya, K.; Kobayashi, S. *Bull. Chem. Soc. Jpn.* **1995**, *68*, 56-61.
19. Le Borgne, A.; Spassky, N. *Polymer* **1989**, *30*, 2312-2319.
20. Uyama, H.; Takeya, K.; Kobayashi, S. *Proc. Jpn. Acad.* **1993**, *69B*, 203-207.
21. Kobayashi, S.; Uyama, H.; Namekawa, S.; Hayakawa, H. *Macromolecules* **1998**, *31*, 5655-5659.
22. Patil, D. R.; Rethwisch, D. G.; Dordick, J. S. *Biotechnol. Bioeng.* **1991**, *37*, 639-646.
23. Park, O.-J.; Kim, D.-Y.; Dordick, J. S. *J. Polym. Chem., Polym. Chem. Ed.* **2000**, *70*, 208-216.
24. Uyama, H.; Kobayashi, S. *Chem. Lett.* **1994**, 1687-1690.
25. Uyama, H.; Yaguchi, S.; Kobayashi, S. *J. Polym. Sci. Polym. Chem. Ed.* **1999**, *37*, 2737-2745.
26. Uyama, H.; Inada, K.; Kobayashi, S. *Macromol. Rapid Commun.* **1999**, *20*, 171-174.
27. Uyama, H.; Inada, K.; Kobayashi, S. *Macromol. Biosci.* **2001**, *1*, 40-44.
28. Kline, B. J.; Beckman, E. J.; Russell, A. J. *J. Am. Chem. Soc.* **1998**, *120*, 9475-9480.
29. Uyama, H.; Klegraf, E.; Wada, S.; Kobayashi, S. *Chem. Lett.* **2000**, 800-801.
30. Tsujimoto, T.; Uyama, H.; Kobayashi, S. *Biomacromolecules* **2001**, *2*, 29-31.

Chapter 13

Enzyme-Catalyzed Direct Polyesterification

Kurt F. Brandstadt[1], John C. Saam[2], and Ajit Sharma[3]

[1]Dow Corning Corporation, P.O. Box 994, Midland, MI 48686
[2]Michigan Molecular Institute, 1910 West St. Andrews, Midland, MI 48640
[3]Central Michigan University, Mt. Pleasant, MI 48859

The ability of an inexpensive, commercially available enzyme, *porcine pancreatic* crude type II lipase, to catalyze the direct polyesterification of a primary (12-hydroxydodecanoic acid) and, to a lesser extent, a secondary (12-hydroxystearic acid) hydroxy acid under mild reaction conditions was demonstrated. *Porcine pancreatic* lipase powder was used without modification to catalyze the heterogeneous reactions in hydrophobic solvents. The enzyme was determined to selectively catalyze the esterification of the primary hydroxy acid in comparison to the secondary hydroxy acid. The removal of water formed in the reactions dramatically promoted the polyesterifications suggesting the role of reversibility. The significance of temperature, enzyme reuse and hydration, the monomer to enzyme weight ratio, surfactants, sonication, and molecular structure were evaluated. The process promises to be an environmentally compatible synthesis that would result in a 'green' material through a reduction of waste stream products, as well as an ability to recycle solvents and the catalyst.

Typically, polyester coating resins are synthesized in a one-step process at high temperatures (> 200^0C) for long periods (*1*). Although these reaction conditions favor the equilibrium of polycondensation, they also promote uncontrolled side reactions: redistribution of monomer sequences, cross-linking, and broad molecular weight distributions. Since resin properties are presently achieved on a trial and error basis, the ability to control the resin structure is essentially lost.

In 1986, the synthesis of stereoregular polyester materials with enzyme catalysts was documented (*2*). *Candida rugosa* and *Chromobacterium viscosum* lipase were used to polymerize 12-hydroxystearic acid, 12-hydroxy-*cis*-9-octadecenoic acid, 16-hydroxyhexadecanoic acid, and 12-hydroxydodecanoic acid with Mn values equal to 600-1,307 in different media including water and organic solvents. The reactions were performed at 35^0C for three days with recovery of the product and catalyst. It was observed that hydroxy acids with secondary hydroxy groups polymerized quickly to yield oligomers with wide molecular weight distributions. In contrast, hydroxy acids with primary hydroxy groups polymerized slowly to yield oligomers with narrow molecular weight distributions. Since the reported enzymatic hydroxy selectivity contrasted with classical alcohol reactivity (primary > secondary > tertiary), this advantageous observation was identified for evaluation in order to potentially promote slower esterifications with secondary alcohols.

Given the ability to synthesize structurally defined materials, the use of an enzyme to strategically catalyze a polyester material directly under mild reaction conditions was evaluated. The catalytic ability of an inexpensive, commercially available enzyme, *porcine pancreatic* crude type II lipase, to directly polymerize a primary (12-hydroxydodecanoic acid) and a secondary (12-hydroxystearic acid) hydroxy acid under mild reaction conditions was explored.

Background

Based on biological systems, enzyme-catalyzed reactions have been historically studied in aqueous media in an attempt to maintain the catalytically active tertiary configuration (*3*). The belief that a hydrophobic environment would dramatically alter the configuration and decrease the activity of the enzyme persisted and inhibited exploration for decades. Fortunately, this notion was not true. In fact, enzymes have been observed to function in two-phase, reverse micelles, one-phase, and anhydrous organic media (*4*). The use of organic solvents as media have led to the following attributes (*3,4*):

- increased enzymatic stability and activity.
- the ability to perform reactions not possible in water.
- increased solubility of a variety of substrates.
- different substrate specificities.
- decreased substrate or product inhibition.
- suppression of side reactions.
- simplified recovery of the catalyst and product.
- ability to recycle the catalyst and solvent.

Commercially available enzymes have been applied directly as a suspension, immobilized, chemically functionalized, or genetically engineered to catalyze well-defined enantio- and regioselective molecules under mild reaction conditions (*4*). Conversely, reaction times are typically long and yields may be low. Economically, the availability, specific activity, and operating lifetime influence the potential application(s) of an enzyme (*5*). In industry, enzymes have been used in large scale conversions of porcine to human insulin and in the production of high-fructose corn syrup, aspartic acid, malic acid, and guanosine 5'-phosphate (GMP) (*5*). Interestingly, given the low cost, high stability, and broad substrate specificity of lipase, Whitesides stated, "chemists comfortably use platinum as a catalyst; they will eventually use lipase (an enzyme) with no more hesitation (*5*)." The following examples illustrate successful enzymatic syntheses of novel materials.

Thermoplastic elastomers, poly(β-hydroxyalkanoates) (PHA), have been produced by bacterial fermentation under specific experimental conditions (*6*). Figure 1 illustrates the polymeric repeat unit of a PHA.

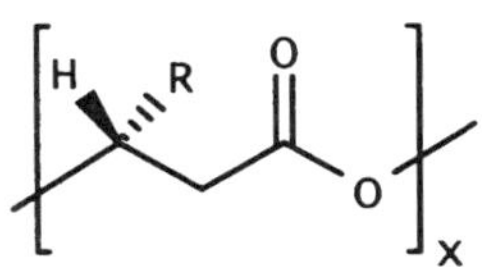

Figure 1: Stereochemical Repeat Unit of Poly(β-hydroxyalkanoate).

The thermoplastic polyester, poly(β-hydroxybutyrate) (PHB), has been accumulated in various bacteria as an intracellular polymeric carbon reserve/energy source. In the absence of an essential nutrient, select bacteria trigger the accumulation of a polymeric reserve in the presence of an excessive amount of carbon, a food source. With limited amounts of ammonium as a nutrient, 80% yields of PHB in the bacterium, *Alcaligenes eutrophus*, were reported (*6*) based on the dry weight of carbon biomass (glucose). Different carbon sources have led to the production of different PHAs. Based on microstructure, molecular weight, and the results of other physical property tests, the bacterium, *Psuedomonas oleovorans*, has produced highly reproducible PHAs from different sodium alkanoate molecules. In application, 80 to 100,000 dalton materials have been produced under mild experimental conditions in 15-20 hours. In industry, Monsanto uses this technology to produce a tough, flexible copolymer commercially sold under the Biopol brand, poly(β-hydroxybutyrate-co-β-hydroxyvalerate) (*7*). Regardless, the technology has the following limitations: (1) the polymers are restricted to the β-hydroxyalkanoate repeat unit, and (2) the polymer must be extracted from the biomass with solvent in a separate step.

In 1991, the chemoenzymatic synthesis of a novel sucrose-containing polymer was documented (*8*). Proleather, an alkaline protease from *Bacillus sp.*, was used to synthesize a regioselective monomer which was polymerized using a conventional chemical catalyst. This was an effective approach to synthesize poly(sucrose adipamide) as illustrated in Figure 2.

The research group commented that, "it occurred to us that a far more efficient approach would be to use enzymes only for the highly selective step(s) in polymer synthesis (such as monomer preparation) and to employ conventional chemical catalysts for the bulk polymer synthesis (*8*)." The monomer was prepared at 45°C for 5 days and polymerized at 35°C for 24 hours in an organic medium. Although the overall yield was low, the product demonstrated the synthetic selectivity achieved with an enzyme catalyst.

Similarly, a chemoenzymatic approach was used to synthesize optically active poly((meth)acrylic) materials as illustrated in Figure 3 (*9*).

Nucleophilic racemic alcohols (two secondary and one primary) were resolved (90-99%) in the asymmetric transesterification of (meth)acrylate esters catalyzed by lipase in neat organic solvents. Subsequently, the monomers were consumed in free-radical polymerizations resulting in materials ranging in molecular weight between 200,000 to 4,000,000 daltons.

In 1989, the viability of performing a lipase catalyzed transesterification in a nonpolar organic solvent as a means of forming optically active diester-diol (AA-BB) polyesters of high stereochemical purity was reported (*10*). In addition, an epoxy functional group was incorporated in the chiral diester as a means of demonstrating the synthetic potential of mild enzymatic polymerization conditions. The reaction is detailed in Figure 4.

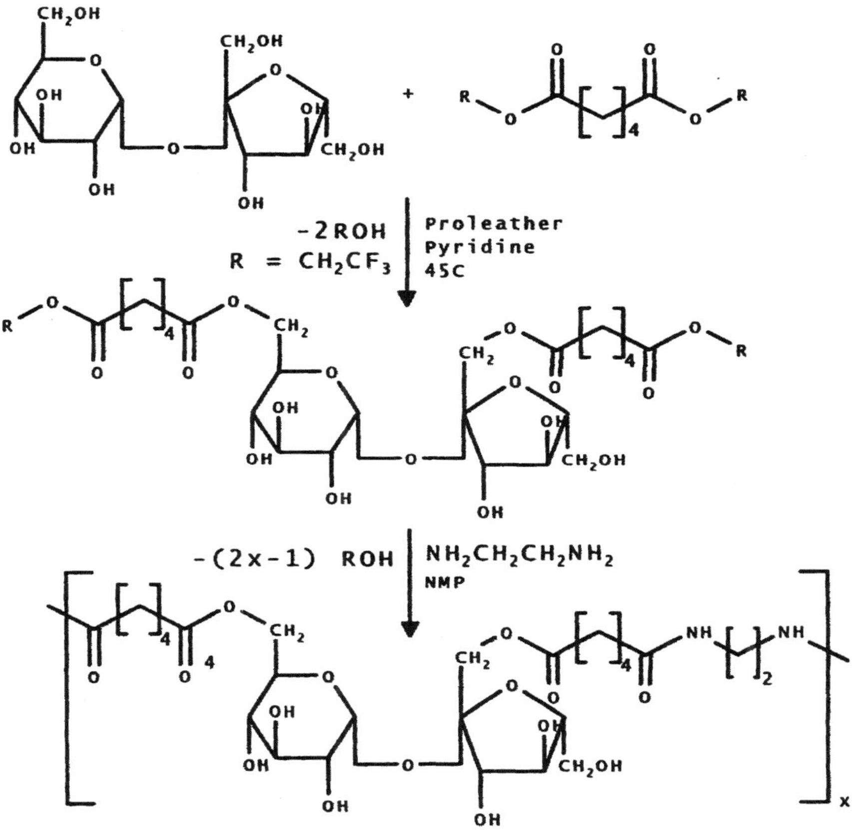

Figure 2: Chemoenzymatic Synthesis of Poly(sucrose adipamide).

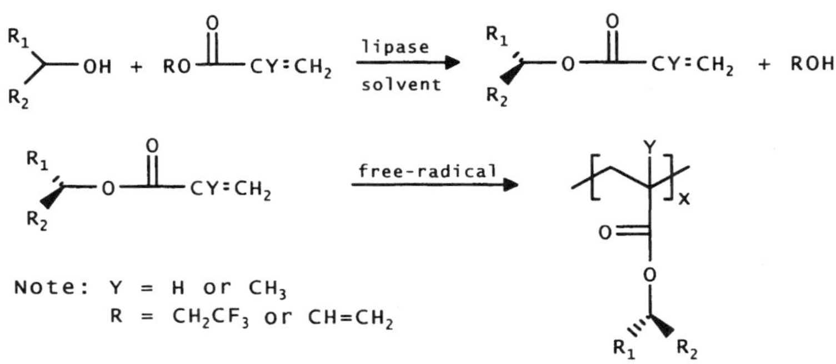

Figure 3: Chemoenzymatic Synthesis of Optically Active (Meth)acrylic Polymers.

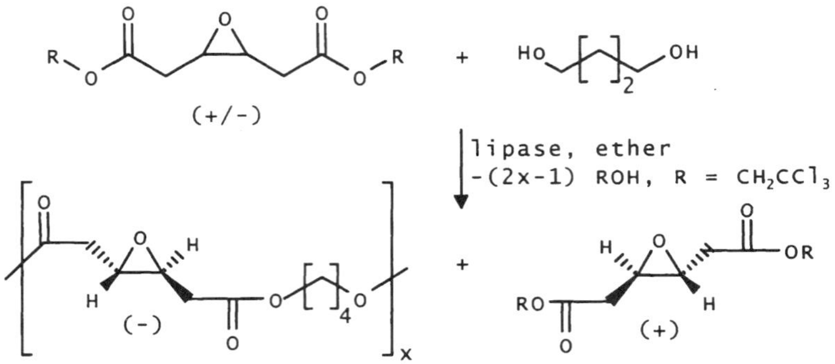

Figure 4: Enzyme-Catalyzed Diester-Diol Polyesterification.

The diester and the diol were combined in a 2:1 molar ratio under the assumption that only one enantiomer of the diester would react. Anhydrous ether was used as the reaction solvent which resulted in a slurry (i.e. the diol was partially soluble). Finally, commercial *porcine pancreatic* lipase was added to catalyze the reaction, while the slurry was stirred under an inert atmosphere. Although the diol was consumed in 6 hours, the reaction was allowed to continue for 3.5 days in an effort to promote the stepwise polymerization of the oligomeric products. Both the enzyme and the product material were recovered. Based on an endgroup analysis of the product by ^{1}H nuclear magnetic resonance spectroscopy (NMR), the product was determined to have an Mn equal to 5,300 daltons. In addition, the product had a weight-average molecular weight (Mw) of 7,900 daltons based on a gel permeation chromatographic analysis with polystyrene standards.

In 1996, the polyester, poly(ε-caprolactone), was synthesized by ring-opening polymerization of ε-caprolactone with butanol and *P. fluorescens* lipase at 65^{0}C in heptane (Figure 5) (*11*).

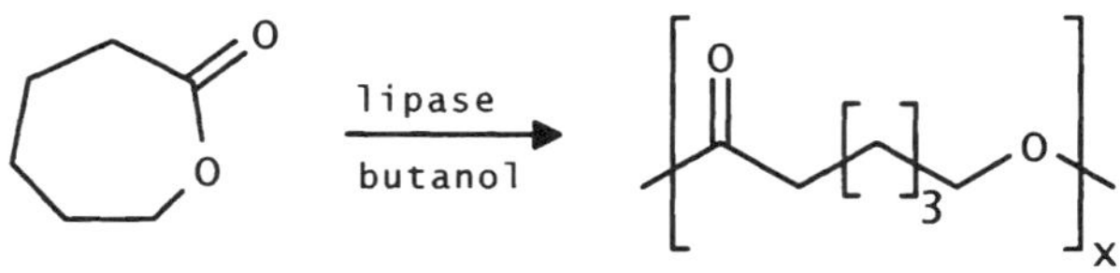

Figure 5: Ring-Opening Polymerization of ε-Caprolactone.

The absence of a leaving group was postulated to eliminate potential side reactions. The monomer conversion was 92%, forming a polyester with an Mn equal to 7,700. The polymer molecular weight increased with decreased concentrations of butanol and water. Similarly, a copolymer from two macrolactones, 11-undecanolide and 15-pentadecanolide, was produced with an Mn equal to 23,000 (*11*).

Recently, a solvent-free transesterification was performed with a commercial lipase (Novozym-435) at 50°C for 4 hours (*12*). Specifically, divinyl adipate was reacted with an equimolar amount of 1,4-butanediol. Since vinyl alcohol rapidly tautomerizes to acetaldehyde, the transesterification reaction was postulated to proceed irreversibly. The transesterification reaction is illustrated in Figure 6.

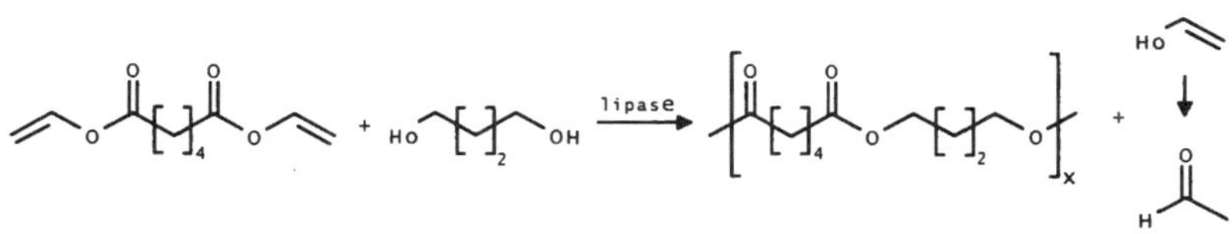

Figure 6: Solvent-Free Enzyme-Catalyzed Transesterification.

In an effort to promote the equilibrium and minimize hydrolysis, the reaction was performed in a nearly anhydrous medium. Since the commercially available enzyme was known to contain water, the ratio of catalyst to monomer was adjusted to minimize the level of pre-dried lipase. Based on the chromatographic results of the product, the reaction progressed efficiently as 98% of the monomer was consumed forming a material with an Mn equal to 23,236. Remarkably, the enzyme catalyzed the synthesis of a high molecular weight polymer in four hours.

Experimental

Materials. *Porcine pancreatic* crude type II lipase (PPL, #L3126) was purchased from Sigma (St. Louis, MO). 12-Hydroxydodecanoic acid (12-HDA, #19,878-1) and 12-hydroxystearic acid (12-HSA, #21,996-7) were obtained from Aldrich (Milwaukee, WI). The hydroxy acid monomers were recrystallized twice and dried to further purify the bifunctional monomers. Reagent grade organic solvents were used throughout the project.

Instrumentation. The extent of reaction was calculated by measuring the molecular weight of the isolated reaction products by gel permeation chromatography (GPC) with respect to polystyrene standards (Polymer Laboratories, Shropshire, UK) and a differential refractometer. The samples were prepared at 0.1% (w/v) in THF.

Solution state ^{13}C and ^{1}H NMR spectra were obtained on a Varian 400 MHz NMR spectrometer with 16 mm and 5 mm probes, respectively. The samples were prepared at approximately 50% and 5% (v/v) in deuterated chloroform with and without 0.04 M chromium(III)acetylacetonate to ensure quantitative conditions. The chemical shifts were referenced to chloroform.

'Closed' Reactions. The 'closed' reactions were performed with an approximate solvent (1 mL) to monomer weight ratio of 5:1. Typically, the monomer (0.2 g, 0.9 mmol 12-HDA, 0.7 mmol 12-HSA) to enzyme (0.1 g, 4401 units) weight ratio was formulated at 2:1. The 'closed' reactions were magnetically stirred in Parafilm® wrapped screw capped ¼ oz. glass vials at a constant temperature (+/- 1^{0}C) for 5 days. PPL was separated from the polyester by heating the solvent diluted mixture and filtering the re-solubilized product through a 0.5 μm Teflon® syringe filter. The dissolved product was washed with solvent, evaporated to dryness under a stream of nitrogen gas, and dried in a vacuum oven at 25^{0}C for ~ 14 hours to obtain a white crystalline polymer.

Open Reactions. The solvent (15 mL) and the monomer were formulated with an approximate weight ratio of 5:1. The monomer (3.0 g, 13.9 mmol 12-HDA, 10.0 mmol 12-HSA) to enzyme (1.5 g, 66015 units) weight ratio was formulated at 2:1. The open reactions were conducted in a 25 mL two-neck round bottom flask equipped with a Dean-Stark trap (filled with toluene) connected to a reflux condenser, and inlet and outlet ports for nitrogen. Additional aliquots of solvent were added to the reaction mixture to maintain constant volume. The reaction was mechanically stirred at a constant temperature (+/- 1^{0}C) for 5 days under a flow of nitrogen (10-20 cc/min.) located 1 mm over the surface of the reaction mixture. PPL was separated from the polyester by heating the solvent diluted mixture and filtering the re-solubilized product through #1 Whatman filter paper and a 0.5 μm Teflon® syringe filter. Both the product and enzyme solutions were washed with solvent, evaporated to dryness under a stream of nitrogen gas, and dried in a vacuum oven at room temperature for ~ 14 hours. The polyesters were white crystalline solids.

Results and Discussion

Feasibility. The project was initiated by catalyzing 12-HDA and 12-HSA 'closed' reactions (Experimental) with PPL in toluene under mild reaction conditions. Given the reported (*13*) necessity of water for the catalytic function of enzymes, PPL was hydrated with distilled water (~pH=6) prior to reaction. Based on the molecular weight results (Table I, entries 1-3), PPL catalyzed the direct polyesterification of the primary hydroxy acid, 12-HDA, under mild reaction conditions. The control synthesis was observed to be unreactive at 45^{0}C without PPL. This observation was confirmed by ^{13}C NMR. Comparatively, the secondary hydroxy acid did not polymerize under the defined experimental conditions.

Evaluation of Solvent. Since PPL is soluble in an aqueous medium, the relative reactivity of the primary hydroxy acid was evaluated in heterogeneous and homogeneous solutions. The 'closed' reactions were performed in hydrophobic (toluene, chloroform) and hydrophilic (THF) media at 45^{0}C. Prior to reaction, the enzyme was hydrated with distilled water (~pH = 6). Since the polymerization was successfully performed in toluene and chloroform as opposed to THF (Table I, entries 3-4), the heterogeneous poly(12-HDA) reaction mixtures were determined to promote PPL catalyzed polycondensation. The partitioning of water from the enzyme into the miscible solvent (THF) may have altered the configuration and catalytic activity of the enzyme despite the absence of diffusional controls (*14*). In comparison, 12-HSA remained completely unreactive.

Water. Since the effect of removing water to drive the equilibration reaction and adding water to catalytically activate the enzyme was unknown, untreated PPL was used to catalyze the 'closed' polyesterification reactions in toluene at 45^{0}C. Based on molecular weight measurements (Table I, entries 3, 5), the activity of the polyesterification reactions increased in the absence of added water. Notably, the secondary hydroxy acid, 12-HSA, yielded a small distribution of low molecular weight oligomers (e.g. dimer, trimer). Since the reactivity increased with an enzyme that had not been treated with water, the equilibrium could have shifted in the direction of condensation polymerization.

Since the 'closed' reactions were promoted under water starved conditions, the effect of the removal of water in a stream of nitrogen (Open Reactions) as well as the use of anhydrous sodium sulfate and three angstrom molecular sieves (*3*) as internal drying agents ('Closed' Reactions) were evaluated in the promotion of the poly(12-HDA) equilibration reaction. The reactions were performed with untreated PPL in toluene at 45^{0}C. The weights of the drying agents were formulated 1:1 with PPL. Based on the comparison of molecular weights (Table I, entries 5-7), the removal of by-produced water with a nitrogen purge shifted the equilibrium and dramatically promoted the polymerization of high molecular weight polymer. The use of internal drying agents had a variable response. Sodium sulfate slightly promoted the equilibrium reaction (Mn=3471), while the use of molecular sieves had an adverse effect (Mn=1618).

Since enzymatic reactions in water are dependent on pH coupled with an aqueous pH memory (*15*), various hydration techniques have been used to obtain catalytically optimal ionization states in organic solvents. Although PPL was reported to be catalytically optimal at pH 7, PPL was hydrated with a range of buffered aqueous solutions (pH 2-10) as well as distilled water (~pH 6) to study the enzymatic activity of hydrated PPL. The hydrated solutions were centrifuged to isolate the enzymes and remove the excess water. The hydrated enzymes were, sequentially, washed in toluene, dried under nitrogen, ground to a powder with a mortar and pestle, and dried in a vacuum oven at 25^{0}C for 14 hours. The poly(12-HDA) reactions were performed with treated PPL in toluene with stirring at 45^{0}C for 5 days in sealed glass ampoules. In comparison with untreated PPL (Table I, entries 5, 8-9), the catalytic activity of all the buffered enzymes decreased resulting in limited molecular weights. In general, the catalytic activity of the pH 2-8 buffered enzymes were similar; however, the enzyme treated at pH 10 was essentially deactivated. Upon exposure to a basic medium, the ionization of critical functional groups (e.g. the catalytic triad) and the tertiary configuration of the enzyme probably changed, thereby inhibiting polyesterification.

Temperature. In water, enzymes at elevated temperatures have been reported (*13*) to reversibly unfold and are irreversibly inactivated due to decomposition. However, in the absence of water, lipase can withstand heating in organic solvents at temperatures up to 100^{0}C. Xylene was chosen as a high boiling (140^{0}C) solvent to evaluate the catalytic activity of PPL as a function of temperature (25^{0}C - ~105^{0}C). Based on the chromatographic results of a control reaction at 70^{0}C (Mn=469), it was concluded that PPL was critical to the efficient formation of high molecular weight polyester. Based on the molecular weight results (Table I, entries 10-12), polyesterification was observed to increase throughout the temperature range, but the enzymatic activity of PPL markedly decreased between 70^{0}C and ~105^{0}C. Although PPL remained catalytically active at ~105^{0}C, the enzyme showed less activity than the activity at 25^{0}C (Mn=1551). It was postulated that PPL was approaching an extreme environment in which the enzyme began to denature and fail.

In order to study the relationship between the thermal history of the enzyme and its activity, PPL was heated at ~105^{0}C in xylene for 5 days prior to catalyzing a 'closed' poly(12-HDA) polyesterification reaction at 45^{0}C in xylene for 5 days. Based on the GPC results (Mn=525), the thermally treated enzyme was observed to be essentially deactivated as its reactivity was less than the reactivity originally obtained at ~105^{0}C. Upon denaturing and decomposing at the maximum temperature, the altered configuration of PPL remained unchanged after cooling. Thus, thermal deactivation appears to be irreversible.

Since the reactivity of the primary hydroxy acid equilibrium reaction and the enzymatic activity increased with temperature under water starved conditions, the reactivity of the secondary hydroxy acid (12-HSA) with untreated PPL was evaluated in 'closed' and open systems at 70^{0}C. Based on the molecular weight information, the removal of by-produced water resulted in an ~ 30% increase in polycondensation in comparison to the 'closed' system (Mn=871 vs. 658). However, the molecular weight distribution was still composed of low molecular weight oligomers. Therefore, it appears that PPL preferentially catalyzes the polyesterification of the primary hydroxy acid, 12-HDA.

Dispersion. The ability of an ionic (sodium dioctyl sulfosuccinate, Aerosol®) and a non-ionic surfactant (Macon-10®) to increase the reactive interfacial surface area and enzymatic activity in 'closed' reactions was assessed. The surfactants were formulated with a 1:1 weight ratio with PPL. Although the solid enzyme became completely dispersed in the presence of the ionic surfactant solution, the polyesterification reactions were completely inhibited. The largest component fraction was unreacted monomer (~85%). The equilibrium reaction was believed to be suppressed by the large amount of water in the formulation. However, the non-ionic surfactant was also observed to decrease the reactivity of PPL in the absence of water (Mn=1009). Consequently, the presence of surfactants was postulated to alter the composition of the catalytic interface and inhibit enzyme-substrate interactions.

The physical use of sonication was used to completely emulsify the heterogeneous hydroxy acid reaction mixtures in toluene in order to optimize the reactive interfacial surface area between the enzyme and the substrates. Since polyesterification was not observed in the control reaction, PPL catalyzed the polymerization reactions in the sonicated media. Based on the chromatographic results, sonication decreased the catalytic activity of PPL (Mw=5086 vs. 15210, 70^0C). Since ultrasound has been documented to decompose a variety of amino acids (*16*), the physical decomposition and alteration of the enzyme by sonication could be the cause of the decrease in catalytic activity.

Evaluation of the Monomer to Enzyme Weight Ratio. A range of monomer to enzyme weight ratios were evaluated to assess the extent of reaction as measured by Mn under similar experimental conditions. The 'closed' polyesterification reactions were performed in toluene at 45^0C. The observed Mn values are detailed in Figure 7. At 200 mg of monomer and approximately 20-40 mg (880-1760 units) of enzyme (10:1 to 5:1), the activity of PPL reached a plateau and became nearly independent of the amount of enzyme present in the heterogeneous reaction. A monomer to enzyme weight ratio from 10:1 to 5:1 appears to be satisfactory for optimal conversions and molecular weights.

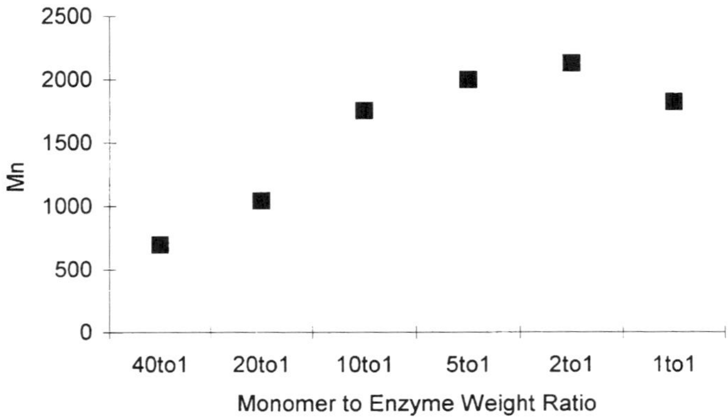

Figure 7: Evaluation of the Monomer to Enzyme Weight Ratio.

Time Study of the Condensation of 12-HDA. In order to evaluate the molecular growth of the enzyme-catalyzed polyester material with time, ten 'closed' reactions were conducted at 45^{0}C in xylene, quenched, and analyzed by GPC to evaluate the growth of the polyester over 120 hours. The molecular growth as measured by Mn and the extent of reaction (%) were plotted in Figure 8.

The Mn increased sharply in the first ten hours and continued to progress slowly as the reaction reached a plateau after 79 hours. After 120 hours, the larger oligomers condensed to form higher molecular weight molecules which increased the polydispersity of the polyester material. The enzyme-catalyzed polyesterification reaction was effectively complete after three days. Based on the results, the molecular growth of the enzyme-catalyzed polyesterification reaction was very distinct from classical step-growth kinetics.

Evaluation of the Use of Recycled Enzyme. In order to assess the catalytic activity of recovered PPL, the enzyme was reused to catalyze a 'closed' poly(12-HDA) reaction in toluene at 45^{0}C. Based on similar molecular weight measurements (Table I, entries 5, 13), recycled PPL was successfully demonstrated to catalyze the poly(12-HDA) reaction. The ability to reuse recycled PPL increases the application robustness of the enzyme.

Evaluation of Molecular Structure. Based on a ^{13}C NMR spectral analysis of poly(12-HDA), a weak spectral peak (~70 ppm) supports the presence of a small amount of an ether bond in poly(12-HDA). Such an abnormality would flip the replicated microstructure (-ABA**BB**ABA-). Since the abnormal molecule remains bifunctional, the molecular growth of the resultant polyester would not be hindered.

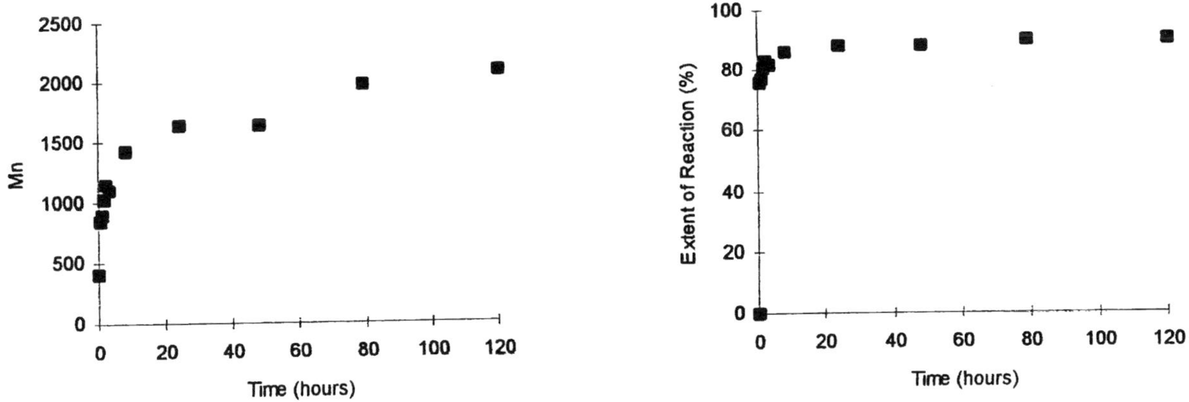

Figure 8: Five Day Study of the Condensation of 12-Hydroxydodecanoic Acid.

Table I. Poly(12-Hydroxydodecanoic Acid) Gel Permeation Chromatography Results.

Entry	*Sample Description*	*Mn*	*Mw*	*Mw/Mn*
1	1^0 hydroxy acid, 12-HDA	406	412	1.01
2	1^0 control	398	403	1.01
3	1^0, toluene, hydrated PPL	1171	2351	2.01
4	1^0, THF, hydrated PPL	398	501	1.26
5	1^0, toluene, untreated PPL	2127	5867	2.76
6	1^0, N_2 purge, crude	6264	19420	3.10
7	1^0, N_2 purge, purified	12490	20070	1.61
8	1^0, hydrated PPL, pH=7	1263	2461	1.95
9	1^0, hydrated PPL, pH=10	424	453	1.07
10	1^0, xylene, 45^0C	2308	5686	2.46
11	1^0, xylene, 70^0C	5364	15390	2.87
12	1^0, xylene, ~105^0C	1148	1933	1.68
13	1^0, toluene, recycled PPL	2533	5578	2.20

Conclusions

The ability of an inexpensive, commercially available enzyme, *porcine pancreatic* crude type II lipase, to catalyze the direct polyesterification of a primary (12-hydroxydodecanoic acid) and, to a lesser extent, a secondary (12-hydroxystearic acid) hydroxy acid under mild reaction conditions was demonstrated. The process promises to be an environmentally compatible synthesis that would result in a 'green' (*17*) material through a reduction of waste stream products, as well as an ability to recycle solvents and the catalyst.

Acknowledgments

The authors would like to thank the Michigan Molecular Institute (Midland, MI) and the Dow Corning Corporation (Midland, MI) for providing access to a laboratory, materials, and analytical equipment.

References

1. Hood, J.D.; Blount, W.W.; Sade, W.T. *Journal of Coating Technology* **1986**, 58(739), 49.
2. Matsumura, S.; Takahashi, J. *Makromol. Chem., Rapid Commun.* **1986**, 7, 369.

3. Chen, C-S.; Sih, C.J. *Angewandte Chemie, Int. Ed. Engl.* **1989**, 28, 695.
4. Guagliardi, A.; Rossi, M.; Bartolucci, S. *Chimicaoggi* **1989**, 7(5), 31.
5. Whitesides, G.M.; Wong, C-H., *Angewandte Chemie, Int. Ed. Engl.* **1985**, 24, 617.
6. Gagnon, K.D.; Lenz, R.W.; Farris, R.J.; Fuller, R.C. *Rubber Chemistry and Technology* **1992**, 65(4), 761.
7. *Chemical & Engineering News*, July 29, 1996, p 40.
8. Patil, D.R.; Dordick, J.S.; Rethwisch, D.G. *Macromolecules* **1991**, 24, 3462.
9. Margolin, A.L.; Fitzpatrick, P.A.; Dubin, P.L.; Klibanov, A.M. *Journal of the American Chemical Society* **1991**, 113(12), 4693.
10. Wallace, J.S.; Morrow, C.J. *Journal of Polymer Science: Part A: Polymer Chemistry* **1989**, 27, 2553.
11. Gross, R.A.; Xu, J.; Svirkin, Y.Y.; Henderson, L.A.; Kaplan, D.L.; Swift, G. *Polym. Matl. Sci. and Eng.* **1996**, 74, 67.
12. Chaudhary, A.K.; Kline, B.J.; Beckman, E.J.; Russell, A.J. *Polymer Preprints* **1997**, 38(2), 396.
13. Klibanov, A.M., *Chemtech* **1986**, 354.
14. Klibanov, A.M., *TIBS* **1989**, 14, 141.
15. Zaks, A.; Klibanov, A.M. *Proc. Natl. Acad. Sci. USA, Biochemistry* **1985**, 82, 3192.
16. Staas, W.H.; Spurlock, L.A. *Journal of The Chemical Society,Perkin Transactions I* **1975**, 4(17), 1675.
17. Wilkinson, S.L. *Chemical & Engineering News*, August 4, 1997, p 35.

Chapter 14

Biocatalytic Synthesis of Novel Functional Polycarbonates

Kirpal S. Bisht and Talal F. Al-Azemi

Department of Chemistry, University of South Florida, 4202 East Fowler Avenue, Tampa, FL 33620

Lipases AYS (*Candida rugosa*), AS (*Aspergillus niger*), AK (*Pseudomonas fluorescens*), PS-30 (*Pseudomonas cepacia*), PPL (porcine pancreas), Novozym-435 (*Candida antarctica*), and IM (*Mucor miehei*) were screened for their ability to catalyze ring-opening polymerization of novel functional monomer 5-Methyl-5-benzyloxycarbonyl-1,3-dioxan-2-one (MBC) in bulk at 80°C, namely. Lipase AK gave highest monomer conversion (97 %) and molecular weight (M_n = 6100); therefore, lipase AK was selected to perform the co-polymerization of MBC at various monomers feed ratios with trimethylene carbonate (TMC). Although MBC reacted more rapidly than TMC, the copolymers have a random repeat unit distribution. A good agreement with the Bernoullian model and the Fox equation supported the random repeat unit distribution. Debenzylation of poly(MBC) and its copolymers by catalytic hydrogenation led to the corresponding linear polycarbonates with pendent carboxyl groups. Presence of pendent carboxyl groups is anticipated to enhance the biodegradability of the polycarbonate and facilitate a variety of potential biomedical applications, e.g., as polymeric drug carriers in time controlled drug delivery systems.

Polycarbonates are polyesters derived from the reaction of carbonic acid or its derivatives with dihydroxy compounds. The most important commercial polycarbonate is that based on 2,2'-bis(4-hydroxyphenyl)propane (bisphenol A), i.e., poly(carbonyldioxy-1,4-phenyleneisopropylodene-1,4-phenylene) which is usually referred to as polycarbonate or PC. Bisphenol A polycarbonate has excellent resistance to acids and oxidants and a T_g = 150 °C. These properties allow this resin to fill many applications, especially in view of its attractive mechanical properties. The industrial process for the synthesis of PC involves a phosgene reaction in a stirred interfacial polymerization. The bisphenol A is usually dissolved in aqueous alkali to form the phenolate salt and then the organic solvent is added, followed by the phosgene. Phase transfer catalysts, such as quaternary ammonium or sulfonium salts, catalyze the reaction. More than 400 million pounds of PC are produced annually in the United States.

As early as the 1930s, the six-membered cyclic carbonate 1,3-dioxan-2-one or trimethylene carbonate (TMC) was thermally polymerized to polyTMC, an aliphatic polycarbonate, by Carothers et al.[1] However, until recently there was no interest in synthesis of aliphatic polycarbonates since they lack strength and durability, which are characteristics of their corresponding aromatic polycarbonates. This changed, however, with the realization that aliphatic polycarbonates, unlike their aromatic counterparts, are biodegradable. The growing interest in the synthesis of aliphatic polycarbonates and their co-polymers with cyclic esters therefore has been related of their potential application in biomedicine due their biocompatibility, low toxicity, and biodegradability.[2-4] One such aliphatic polycarbonate that has attracted much attention is poly(trimethylenecarbonate) (PTMC). The biodegradability of PTMC makes it a good candidate for use in biomedical applications as bioresorbable materials. PTMC films implanted subdermally in rats showed a weight loss of 21% and a decrease in the molecular weight of 50% over 30 weeks.[4] Similar observations have also been made with poly(ethylenecarbonate), showing almost complete degradation within three weeks when implanted in the peritoneal cavity of a rat as compared to very little hydrolytic degradation in pH 7.4 buffer.[2] These observations clearly demonstrate the susceptibility of aliphatic polycarbonates to enzymatic degradation under physiological conditions. Several other TMC-containing copolymers also have been subjects of detailed investigations. Examples from literature include trimethylenecarbonate/lactide,[5] trimethylenecarbonate/ε-caprolactone,[6] and trimethylenecarbonate/β-butyrolactone.[7]

With increasing interest in macromolecules that possess unique properties or uses, synthesis of functional polymers has been target of many directed research activities.[8,9] The properties of such materials are often dictated by the presence of chemical functional groups that are dissimilar to those of the backbone chains. Chemical heterogeneity on the polymer chain may lead to enhanced reactivity, phase separation or supramolecular assemblies. The ability of the functional polymers to form supramolecular structures is a further incentive, especially when the association or dissociation of the self-assemblies can be triggered by chemical or physical stimuli. Introduction of various functionalities into polymers, therefore, has been an active area of research for a variety of purposes including the design of drug delivery systems, immobilization of catalysts, and improvement of surface hydrophilicity, etc.[5-7,10,11] However, very few reports exist in the literature describing the introduction of functional groups into aliphatic polycarbonates.[12-14] This can be largely attributed to the difficulty in synthesis of suitable functional monomers that can be efficiently polymerized. Functional polycarbonates having pendent hydroxyl or carboxyl groups are expected to show enhanced degradability and thus may serve as attractive bioresorbable materials. For example, it has been reported that di-hydroxy aliphatic polycarbonates undergo fast *in vitro* hydrolytic chain cleavage.[12] In fact, when incubated in pH 7.4 buffer at 37 °C poly(bis-hydroxymethyl carbonate), with pendent hydroxymethylene groups, rapidly auto-degrades to pentaerythritol and carbon dioxide. It is presumed that degradation occurs via an intramolecular nucleophilic attack of the hydroxymethylene group on the carbonate link.[12]

Many 5- and 6-membered cyclic carbonate monomers have been polymerized using a wide variety of initiators and catalysts according to anionic, cationic, and coordination mechanisms.[15] While 5-membered cyclic carbonates are accompanied with loss of carbon dioxide upon polymerization resulting in ether linkages in the polymer chain, 6-membered cyclic carbonates have been polymerized without decarboxylation using anionic initiators and coordination catalysts.[15] The polymerization of 6-membered ring carbonate using cationic initiators, however, is accompanied by decarboxylation. More recently, a new synthetic method for polymer assembly has been developed using enzymes as catalysts. Polyester synthesis using isolated enzymes as catalysts have received much attention and, so far, biodegradable aliphatic polyesters have been synthesized from various monomer combinations.[16] As to enzymatic synthesis of polycarbonates, only a few reports exist in the literature describing polymerization catalyzed by lipase catalysts.[17-25] Research by others and in our laboratories have further advanced the enzymatic approach for preparations of various polycarbonates. The following paragraphs present a review of polycarbonate synthesis catalyzed by isolated enzymes.

Polymerization of dicarbonate monomers and diols by AA-BB type condensation was catalyzed by a number of different lipases (Scheme 1).[23-24] Polymers of weight average molecular weights of up to 8,000 were synthesized by polymerization between dicarbonates and diols.[23] In condensation polymerizations, however, removal of the leaving group (water, alcohol or acetaldehyde) is necessary to shift the equilibrium in favor of the polymerization. On the other hand, no leaving group is produced in a ring-opening polymerization reaction and this generally translates into higher molecular weights and narrower molecular weight distributions.

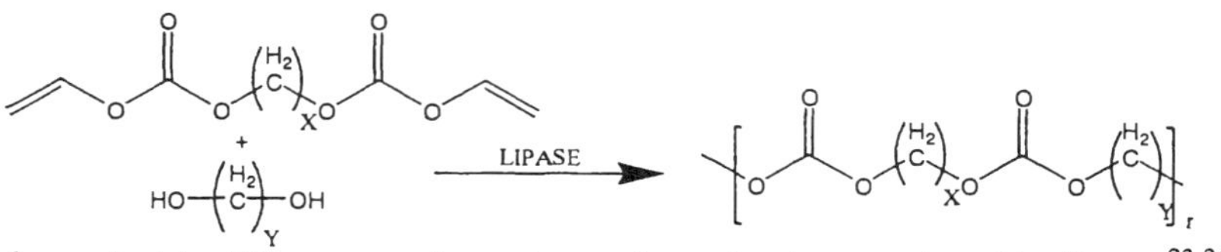

Scheme 1. AA –BB type condensation polymerization catalyzed by lipase.[23-24]

The ability of the enzymes, especially lipases, to accept cyclic carbonates as substrates for ring-opening polymerizations has been demonstrated in the lipase-catalyzed polymerization of TMC (Scheme 2).[17-21] Lipases Novozym-435 (*Candida antarctica*), PPL (porcine pancreas), PS-30 (*Pseudomonas cepacia*), AK (*Pseudomonas fluorescens*), CCL (*Candida cylindreacea*), MAP, and lipozyme-IM (*Mucor miehei*) have been evaluated for bulk polymerization of TMC. Interestingly, the molecular weight, yield, and molecular weight distribution were significantly affected by the origin of the lipase.[18] Novozym-435 gave PTMC of M_n =15,000 after 97% monomer conversion in 120 h, without decarboxylation during chain propagation.[18] Much higher molecular weight PTMC, M_w = 169,000 was reported by Matsumura *et. al.* using 0.25 % of lipase PPL at 100 °C after 24 h.[19] Kobayashi and coworkers[22] have reported lipase-catalyzed polymerization of the cyclic dicarbonates cyclobis(hexamethylene carbonate) and cyclobis(diethelene glycol carbonate) and found that lipase CA (from *Candida antractica*) was the most efficient biocatalyst (Scheme 3). The advantage of the lipase-catalyzed process for polymer synthesis was particularly manifested when oligoTMC chains were attached selectively to the primary hydroxyl of ethyl glucopyranoside (EGP) by lipase-catalyzed ring-opening of TMC in presence of EGP (Scheme 4).[21] The oligoTMC macromer was subsequently used for synthesis of star shaped block copolymer after polymerization with 3,6-dimethyl-1,4-dioxan-2,5-dione (lactide).[26]

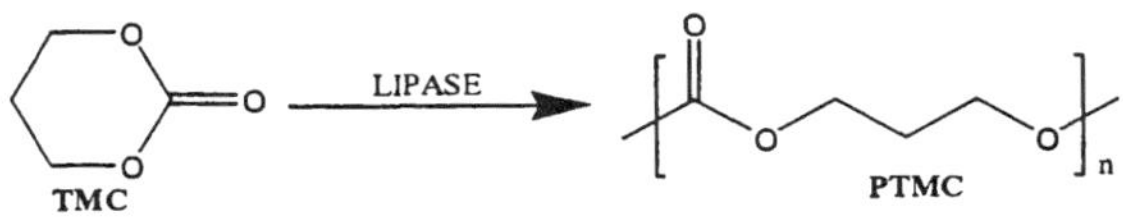

Scheme 2. Ring–opening polymerization of TMC catalyzed by lipase.[17-20]

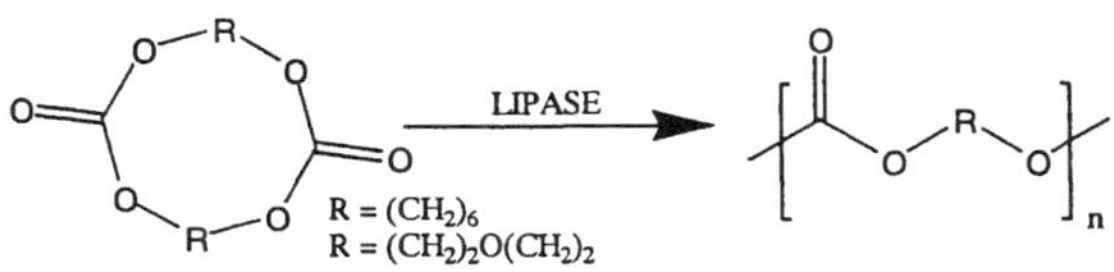

Scheme 3. Polymerization of cyclic dicarbonates catalyzed by lipase.[22]

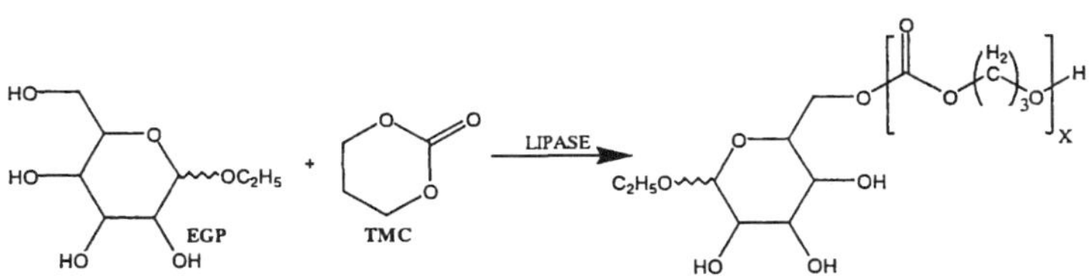

Scheme 4. Regio control in lipase-catalyzed polymerization of TMC initiated by EGP.[21]

Additionally, lipase-catalyzed copolymerization of TMC and lactide has also been investigated (Scheme 5).[25] Formation of random copolymers of M_w of up to 21,000 was reported upon copolymerization of TMC with lactide using lipase PPL at 100 °C after 168 h.[25] The TMC content in the copolymers was linearly dependent on the monomer composition in the feed. Interestingly, the lipase concentration affected the molecular weight, the highest molecular weight being obtained at PPL concentration of about 5%.[25]

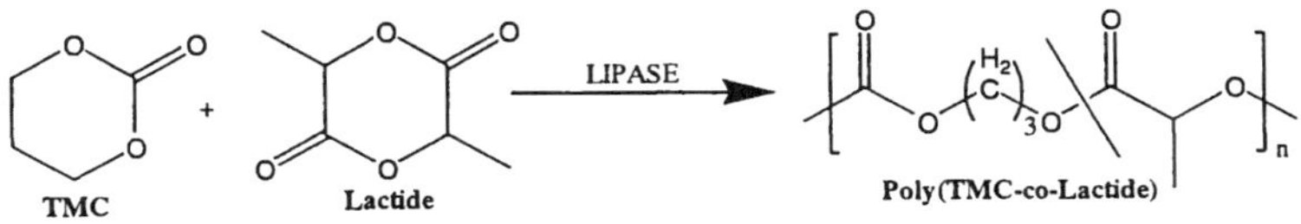

Scheme 5. Copolymerization of TMC and lactide catalyzed by lipase.[25]

A mechanism for lipase-catalyzed ring opening polymerization (ROP) of cyclic carbonates has also been proposed (Scheme 6).[18]

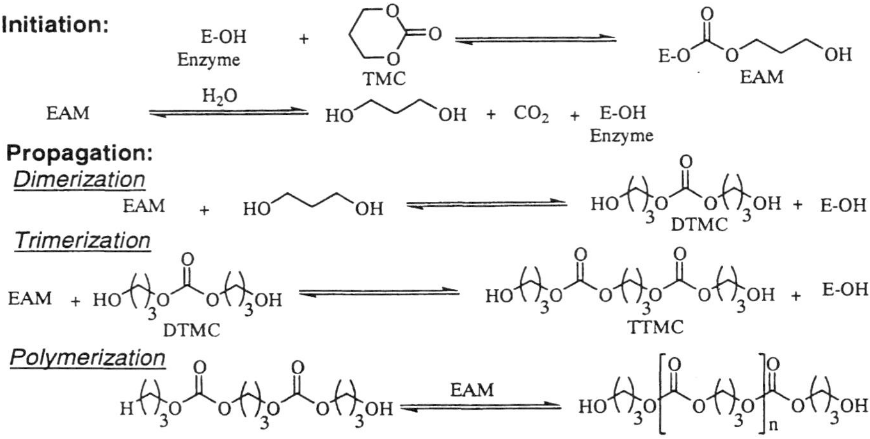

Scheme 6. Mechanism of lipase catalyzed ring-opening polymerization of trimethylene carbonate (TMC).[18]

The mechanism has been set forth based on: (a) isolation of 1,3-propanediol and dimer (DTMC) & trimer of TMC (TTMC), (b) the presence of primary hydroxyl end groups, (c) the effect of water on the molecular weight, and (d) a previous mechanism proposed for lipase-catalyzed ring-opening polymerization of lactones. In the initiation step, the formation of an enzyme-activated monomer (EAM) upon ring opening of the TMC by the lipase is proposed. In a subsequent nucleophilic attack by the initiator (water) on the EAM, rapid decarboxylation of the resulting hydroxy carbonic acid produces 1,3-propanediol. In the propagation steps, the EAM complex reacts with 1,3-propanediol to form DTMC, with DTMC to form TTMC, and so on to form the high molecular weight PTMC. The polymerization process is terminated when the monomer is consumed.

Enzymatic synthesis of polycarbonates possesses several advantages over the traditional chemical polymerization. Specifically,

- Relatively mild conditions are required for enzymatic polymerizations.
- The decarboxylation during propagation is not observed.
- Enzyme catalysts are biocompatible.
- Enzyme catalysts can be reused for many reaction cycles without significant loss of activity.
- When performed in bulk, enzymatic reactions eliminate the need for organic solvents.
- Unlike chemical catalyst/initiators that require extremely pure monomers, inert atmosphere and anhydrous conditions, enzymes can function in moderate reaction conditions.
- Organometallic initiators must be completely removed, especially for biomedical applications.

In this chapter, work carried out in our laboratories toward the synthesis of novel functional polycarbonates bearing pendant carboxylic groups utilizing enzymatic polymerization technology is summarized.[27-30] This chapter describes the homopolymerization of 5-methyl-5-benzyloxycarbonyl-1,3-dioxan-2-one (MBC), and its co-polymerization with TMC (see Scheme 7). The lipase and reaction conditions for the polymerization reaction were optimized. The relationship among monomer conversion, reaction time, and number average molecular weight (M_n) will be discussed. The spectral and thermal data for the copolymer and its correlation with its microstructure will also be analyzed.

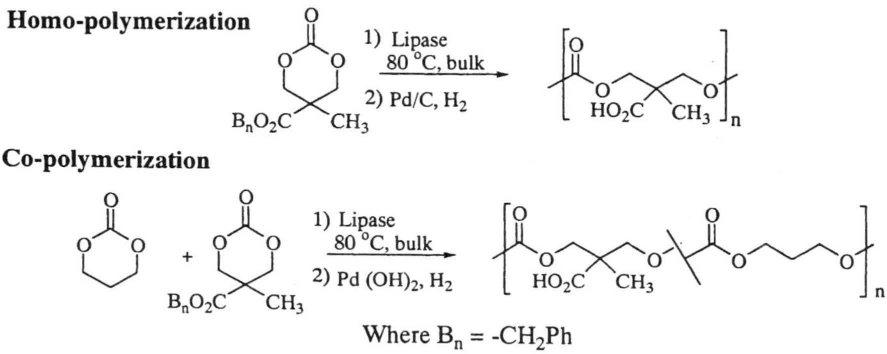

Scheme 7. Synthesis of poly(MBC) and poly(MBC-co-TMC) by lipase AK-catalyzed ring opening polymerization at 80 °C in bulk.[27,28]

Experimental

Materials: Lipases AYS, AS, AK and PS-30 were generous gifts from Amano Enzyme Co. Novozym-435, and IM lipases were kindly provided by Novo Nordisk Bioindustrial Inc. Porcine pancreatic lipase (PPL) Type II was purchased from Sigma Chemical Co. 5-Methyl-5-benzyloxycarbonyl-1,3-dioxan-2-one (MBC) was synthesized from reaction of benzyl bis–(2,2-hydroxymethyl)propanoic acid with ethyl chloroformate.[27]

NMR Spectroscopy. ^{1}H- and ^{13}C-NMR spectra were recorded on a Bruker ARX-360 spectrometer at 360 and 90 MHz, respectively. ^{1}H-NMR chemical shifts (ppm) are reported downfield from 0.00 ppm using tetramethylsilane (TMS) as an internal standard. The concentrations used were ~4% w/v in chloroform-*d* ($CDCl_3$) or DMSO-d_6. ^{13}C-NMR chemical shifts in ppm are referenced relative to the internal standard chloroform-*d* at 77.00 ppm.

Molecular weight Determination: Molecular weights were measured by gel permeation chromatography (GPC) using a Shimadzu HPLC system equipped with refractive index detector (RI), UV-Vis detector, and waters HR 4E styragel column. THF (HPLC grade) was used as eluent at a flow rate of 1.0 mL/min. The sample concentration and injection volumes were 0.5 % (w/v) and 100 μL, respectively. Molecular weights were calculated based on a calibration curve generated by narrow molecular weight distribution polystyrene standards.

Polymerization: All reactions were carried out in bulk. The monomer(s) and lipase were dried (in a drying pistol over P_2O_5, at 50 °C/0.1 mm Hg; 15 h) in separate 6 mL reaction vials. In a glove bag, maintained under a nitrogen atmosphere, the monomer(s) were transferred to a single 6 mL vial, and then the pre-weighed enzyme was added. The vials were capped with a rubber septum and placed in a constant temperature oil bath maintained at 80 °C for predetermined reaction times. Reactions were terminated by dissolving the contents of the reaction vials in chloroform and removing the enzyme (insoluble) by filtration (fritted glass filter, medium pore porosity). The filtered enzyme was washed with chloroform. The filtrates were combined, solvents were removed *in vacuo* and the crude products were analyzed by ^{1}H NMR and gel permeation chromatography (GPC). When specified, the polymer was purified by adding its chloroform solution to methanol, which precipitated the polymer.

Results and Discussion

Homopolymerization of MBC: Several different lipases, all commercially available, were investigated for the polymerization of MBC. Table I shows the data obtained after 24 and 72 h reactions carried out at 80 °C in bulk. Under the same reaction conditions, a significant variation in the monomer conversion was observed for lipases from different origins. For example, for 24 hours reactions, the lipases AS (from *Aspergillus niger*), AYS (from *Candida rugosa*) and IM (from *Mucor miehei*) showed poor conversion (12, 8 and 13 %, respectively) whereas lipases PPL, AK, and

Table I. Ring-Opening Polymerization of 5-Methyl 5-benzyloxycarbonyl-1,3–dioxan-2-one in bulk at 80°C.

Entry	*Lipase*[a]	*Time*	*Conversion (%)*[b]	*DP*[b]	M_n[c]	M_n/M_w[c]
1	AK	24	54	4.7	1500	1.076
2	AK	72	97	28.3	6100	1.647
3	PPL	24	54	4.7	1700	1.164
4	PPL	72	98	4.0	1300	1.380
5	PS-30	24	50	4.5	1450	1.079
6	NOVO-435	24	29	3.0	950	1.007
7	NOVO-435	72	86	13.4	4400	2.117
8	IM	24	13	N/A	N/A	N/A
9	AS	24	12	N/A	N/A	N/A
10	AYS	24	8	N/A	N/A	N/A
11	AK[d]	72	0	N/A	N/A	N/A

[#] Data shown is the statistical mean of duplicate experiments carried out using 2:1 monomer to enzyme ratio. [a] Enzyme (Source): AYS (*Candida rugosa*), AS (*Aspergillus niger*), AK (*Pseudomonas fluorescens*), PS-30 (*Pseudomonas cepacia*), PPL (porcine pancreas), Novozym-435 (*Candida antarctica*), and IM (*Mucor miehei*). [b] Determined by ^{1}H NMR. [c] Determined by GPC. [d] Thermally deactivated lipase. N/A – not determined due to low conversion and overlapped signals.

SOURCE: Reproduced with permission from reference 27. Copyright 1999 American Chemical Society.

PS (from porcine pancreas, *Pseudomonas fluorescens*, and *Pseudomonas cepacia*, respectively) showed considerably higher conversions (Table I).

The molecular weight of the polymer formed was also influenced by the source of the lipase. In reactions run for 72 h, at comparable conversions, much higher molecular weight polymers were obtained from lipase AK and Novozym-435 catalyzed polymerizations (6100 g/mol, entry 2 and 4400 g/mol, entry 7, Table I). This data suggested that lipase AK was an efficient catalyst for the polymerization of MBC and was further evaluated as the catalyst for the ring-opening polymerization of the MBC. Importantly, in a reaction carried out in the presence of thermally deactivated lipase AK no polymerization was observed even after 72 h and the monomer was recovered quantitatively (Table I, entry 11) suggesting that the polymerization reaction was indeed lipase catalyzed.

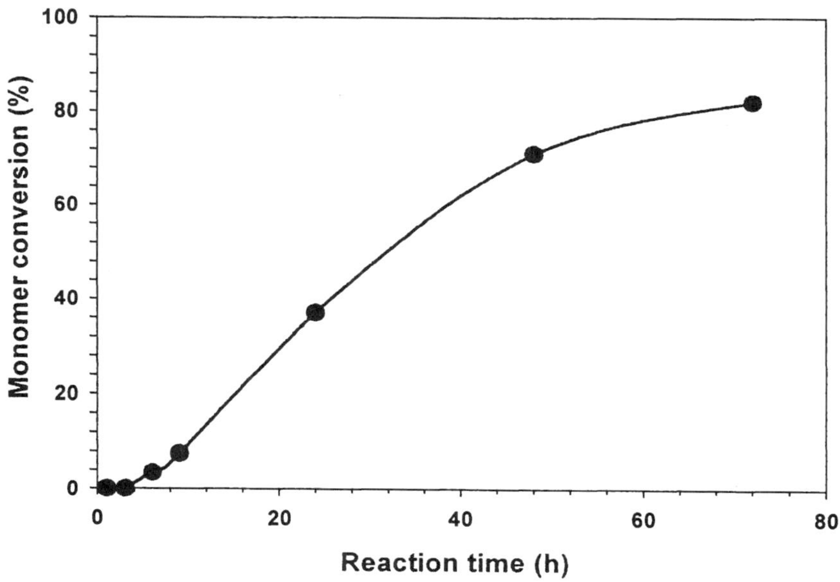

Figure 1. Percent monomer conversion (MBC) as a function of time (h) for lipase AK-catalyzed ring-opening polymerization at 80 °C in bulk.
(Reproduced with permission from reference 28. Copyright 2000 American Chemical Society.)

Relationship Between Monomer Conversion and Reaction Time: A Plot of MBC monomer conversion versus reaction time is shown in Figure 1. Interestingly, for the first three hours no monomer conversion was detected. However, beyond the first three hours the monomer conversion steadily increased to 37% by 24 hours and was 82 % after 72 hours. The lag time observed was somewhat unusual, as this has not been observed before in lipase-catalyzed polymerizations. One plausible explanation might be the high melting temperature (73 °C) of the MBC monomer. Since the reaction temperature is not much higher than the melting temperature of MBC, induction time may correlate with the melting of the MBC monomer. Higher reaction temperatures, however, were precluded in subsequent reactions due to loss of enzyme activity.

Molecular Weight to Monomer Conversion Relationship: The number average molecular weight (M_n) of PMBC increased exponentially with percent MBC conversion. Figure 2 shows number-average molecular weight M_n (g/mol) as a function of percent monomer conversion. For the initial 24 hours, when the monomer conversion was only 37 %, the molecular weight (1600 g/mol) did not show a sharp increase. However, beyond 37 % conversion the molecular weight rose sharply and was 5000 g/mol at 71 % MBC conversion and reached 6300 at 82 % (72 h) conversion. The molecular weight profile is in accordance with the chain polymerization mechanism proposed for the enzyme-catalyzed ROP.[18] The initial stagnation in molecular weight may very well indicate the initiation stage of the reaction when mostly new chains are being formed. However, as soon as the initiator (water) is consumed, at about 40 % conversion, the chain propagation dominates and is reflected in sharp molecular weight increase. Polydispersity index (M_n/M_w) registered a slight increase when monomer conversion rose from 5 to 40 % and then decreased when the monomer conversion rose to 80% (Figure 2). Importantly, the variation in the polydispersity index coincides with the molecular weight profile, thus lending support to the proposed hypothesis.

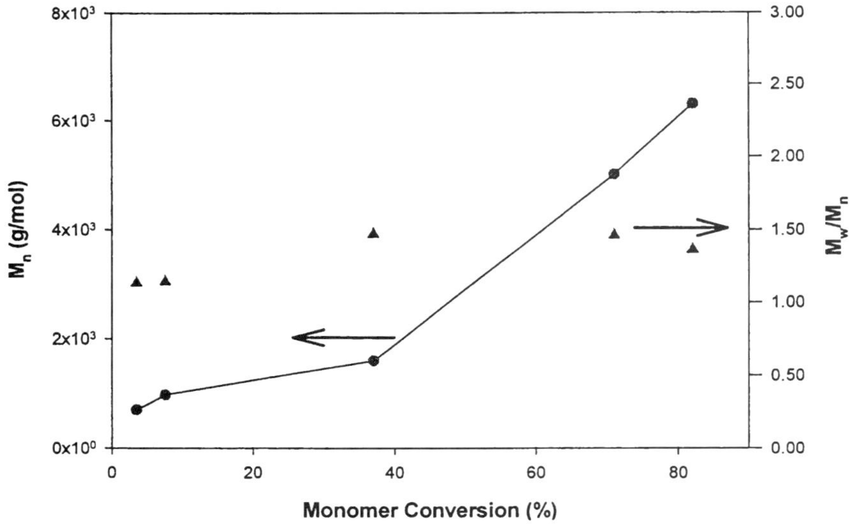

Figure 2. Number-average molecular weight and weight distribution as a function of percent MBC conversion. (Reproduced with permission from reference 28. Copyright 2000 American Chemical Society.)

Copolymerization of MBC and TMC: Aliphatic polycarbonates prepared by the lipase catalyzed ring-opening polymerization (ROP) of cyclic carbonate monomers would be of considerable interest as bioresorbable materials for medical applications. However, to meet the wide spectrum of requirements for biomedical materials, it is critical that the properties of these materials can be specifically 'tailored'. The introduction of controlled levels of functional groups along the chains of biocompatible materials is important for a number of reasons. For example, carboxylic acid side chains can be used to regulate the rate at which main chain ester linkages are hydrolyzed. We have, therefore, studied the co-polymerization of MBC with TMC to control the number of pendant groups on the polycarbonate backbone.

Co-polymers of MBC and TMC were synthesized by lipase AK catalyzed ring-opening polymerization of the two monomers (Scheme 7). Table II lists the results of lipase AK catalyzed MBC/TMC polymerizations where the monomer feed ratio was varied. The copolymerizations were carried out at 80 °C for 72 hours under a nitrogen atmosphere. High monomer conversions (> 80 %) were observed. GPC analysis of the copolymers showed a unimodal distribution and the polydispersities were in the range of 1.9 – 4.7. The polydispersity index of the copolymers was relatively higher than that observed for the homopolymers of either MBC or TMC.[18,27] Increasing the TMC content in the monomer feed did not indicate any significant change in the polymer yield. However, the molecular weight of the resultant copolymers did decrease from 9,000 to 6,500 as the TMC content was increased from 0 to 90 % (Table II). In general, the incorporation of TMC in the copolymer was less than that in the monomer feed (Table II). This result suggests that the reactivity of TMC in lipase AK catalyzed ring-opening polymerization was less than that of MBC. Interestingly, with increasing TMC content in the monomer feed, the polydispersity index (PDI) decreased to 1.9. The lower PDI might be the direct consequence of the increased fluidity in the reaction vial. TMC, a low melting monomer, may act as a solvent for the copolymer chains thus providing higher access to the growing chain ends, promoting further chain growth, and keeping the PDI low. This observation supports the previous hypothesis that the diffusion constraints in these bulk reactions prevents unhindered growth of the polymer chain.[18,31]

Table II. Lipase AK Catalyzed Ring-Opening Co-polymerizations of MBC and TMC[#]

Entry	*Monomer ratio (feed)[a] MBC:TMC*	*Polymer Yield (%)[b]*	M_n[c]	M_w/M_n[c]	*Molar Composition MBC/TMC[d]*
1	100:0	98	9000	3.2	----
2	80:20	86	8200	4.7	92:8
3	62:38	88	7600	3.9	72:28
4	50:50	85	7500	4.4	63:37
5	29:71	89	6700	3.4	35:65
6	20:80	97	6000	2.4	21:79
7	10:90	95	6500	1.9	13:87
8	0:100	70 (95)[b']	6400	2.0	-----

[#]Reactions were carried out in bulk for 72 h at 80 °C using Monomer / enzyme (w/w) = 2. [a]Monomer feed ratio in mole/mole, [b] Methanol-insoluble copolymer. [b']Value in parenthesis is for monomer conversion determined from ^{1}H NMR spectrum. [c]Determined using GPC. [d] Determined from ^{1}H NMR spectrum.

SOURCE: Reproduced with permission from reference 28. Copyright 2000 American Chemical Society.

Microstructure Analysis: The sequence distributions of the co-polymers were determined from ^{13}C NMR spectral analysis. The carbonate carbonyl resonances

(~155 ppm) were sensitive to the diad sequence. The expanded carbonyl group region (153 – 156 ppm) of the ^{13}C NMR spectrum of the poly [MBC-*co*-TMC] for three different monomer compositions is shown in Figure 3. The signal assignments at 154.2, 154.5, and 154.8 ppm were made based upon the relative signal intensities of the copolymers with different monomer compositions. At monomer composition of 35:65 (MBC:TMC, Entry 5, Table II) the signal contributions from the different diad sequences are shown in Figure 3a. Changing the composition to 92:8 (MBC:TMC, Entry 2, table II) increased the signal intensity at 154.2 ppm (*MBC-MBC*) while significantly diminishing the intensity at 154.8 ppm (*TMC-TMC*) (Figure 3c). Though similar but complementary observations were made when the copolymer composition was changed to 21:71 (MBC:TMC, Entry 6, Table II, Figure 3b). Specifically, in Figure 3b, while the signal intensity at 154.8 ppm (*TMC-TMC*) increased the signal intensity at 154.2 ppm *(MBC-MBC)* decreased. Hence, the ^{13}C-NMR signals at 154.2, 154.5, and 154.8 ppm were assigned to diad sequences *TMC-TMC*, *TMC-MBC* or *MBC-TMC*, and *MBC-MBC*, respectively. The diad fractions [MBC-MBC], [MBC-TMC], [TMC-MBC], and [TMC-TMC] were calculated from the corresponding carbonyl carbon peak area in the ^{13}C NMR spectra to analyze the repeat unit sequence distribution.

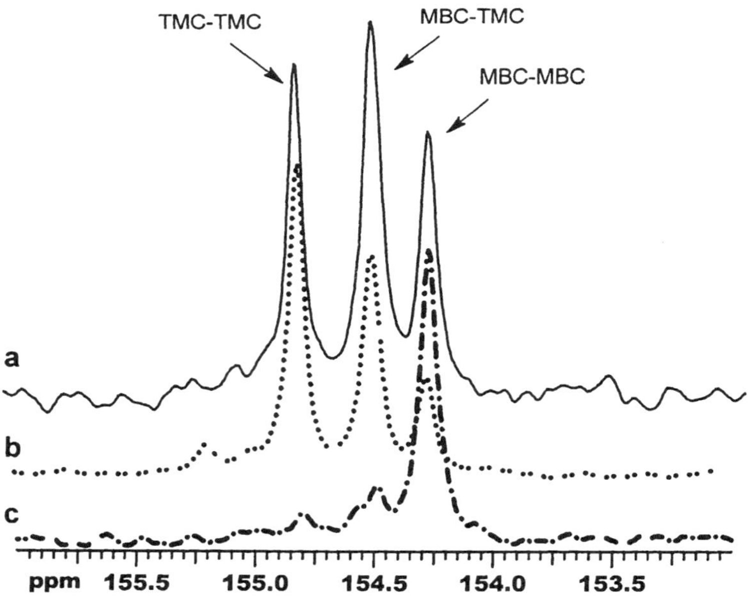

Figure 3. Expanded ^{13}C NMR carbonyl region of poly[MBC-co-TMC] for different molar composition ratio. a) 35:65 MBC:TMC (Entry 5, Table II). b) 21:79 MBC:TMC (Entry 6, Table II). c) 92:8 MBC:TMC (Entry 2, Table II).

(Reproduced with permission from reference 28. Copyright 2000 American Chemical Society.)

Assuming a Bernoullian or random statistical process for MBC/TMC co-polymerization, where F_{MBC} is the mole fraction of MBC units in the copolymer, the following equations represent the different diad sequence units in the co-polymer.

$$[MBC\text{-}MBC] = F_{MBC}^2 \quad (1)$$
$$[MBC\text{-}TMC] = [TMC\text{-}MBC] = F_{MBC}(1\text{-}F_{MBC}) \quad (2)$$
$$[TMC\text{-}TMC] = (1\text{-}F_{MBC})^2 \quad (3)$$

The experimental results obtained for copolymers having monomer composition (MBC:TMC) of 98:8, 63:37, and 87: 13 (entries 2, 4, and 7) are shown in Table (III). The diad fractions calculated from the experimental diad intensities using Bernoulli or random statistical processes were in good agreement with the experimental values. Therefore, the data suggested that the enzymatic copolymerization of MBC and TMC, catalyzed by lipase AK, resulted in the formation of a random copolymer.

Table III. Experimental and Calculated Co-monomer Diad Fractions

Entry	*Molar composition MBC : TMC*[a]	*Diad Sequence*		
		MBC-MBC [expt[a] *(calcd)*[b]*]*	*MBC-TMC + TMC-MBC [expt*[a] *(calcd)*[b]*]*	*TMC-TMC [expt*[a] *(calcd)*[b]*]*
2	92:8	0.80(0.85)	0.16(0.14)	0.04(0.01)
4	63:37	0.54(0.40)	0.36(0.46)	0.09(0.14)
7	13:87	0.06(0.02)	0.21(0.22)	0.72(0.76)

[a] Determined by measuring the peak area of the ^{13}C-NMR signals corresponding to carbonyl carbons.[b] Calculated using the equation that describe Bernoullian or random statistical copolymerization

Table IV. Thermal Properties of Poly[MBC-*co*-TMC] for Different Molar Compositions Synthesized by Lipase AK in Bulk at 80 °C.

Entry	Molar composition MBC:TMC[a]	*Tg*(°C)[b]	*Tg*(°C)[c] (Calculated)
1	100:0.0	2.52	2.52
2	87:13	-3.01	-1.99
3	72:28	0.05	-4.60
4	63:37	-2.66	-6.21
5	35:65	-6.99	-13.67
6	21:79	-14.96	-18.99
7	13:87	-19.39	-22.43
8	00:100	-29.82	-29.82

[a] Copolymer molar composition measured by ^1H NMR. [b] Measured by DSC. [c]calculated using Fox equation from the molar composition determined by ^1H NMR.

Thermal Analysis: The glass transition temperatures (T_g) of the MBC and TMC homo-polymers and the poly[MBC-*co*-TMC] of different compositions were obtained from DSC measurements. The experimental data from the DSC analyses are compiled in Table IV. No melting temperature (T_m) was observed for either the homo-polymers or the copolymers, which indicated the amorphous nature of the polymers. The glass transition temperature, T_g, of PMBC and PTMC was 2.5 and -29.8 °C, respectively. As expected, T_g of the copolymers decreased with increasing TMC in the feed ratio. In all copolymers, only one T_g was observed which increased with increasing MBC in the copolymer. This is opposed to diblock or multiblock copolymers, which would have more than one T_g value for immiscible TMC and MBC chain segments.

The glass transition temperature of many random copolymers can be calculated if the glass transition temperature of the homopolymers and the composition of the copolymers are known. The relation, below, was established by Fox;

$$1/T_g = (w_1/T_{g1}) + (w_2/T_{g2}) = w_2\,(1/T_{g2} - 1/T_{g1}) + 1/T_{g1} \qquad (4)$$

Where, w_1 and w_2 represent weight fractions; T_{g1}, and T_{g2} the glass transition temperatures of the homopolymers **1** and **2**, respectively. For random copolymers a plot of $1/T_g$ (K) Vs w_2 is linear.

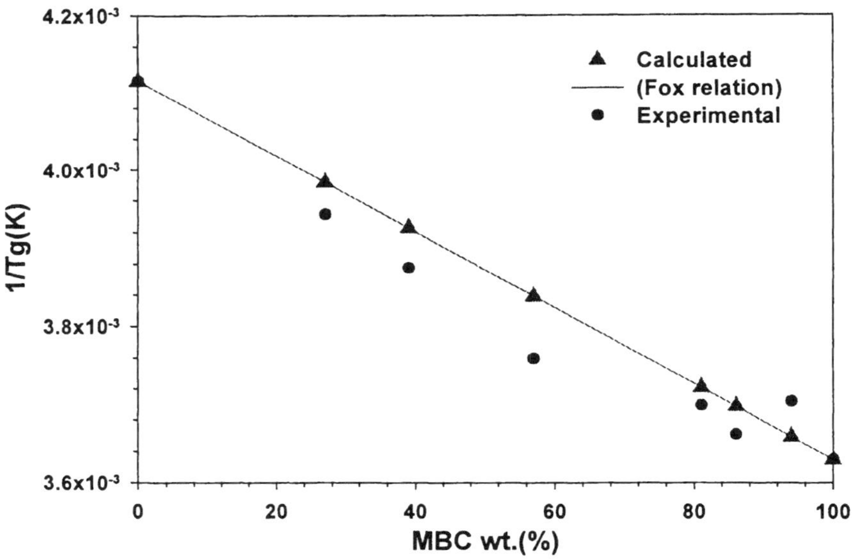

Figure 4. A graph of $1/T_g$ (K) as a function of the weight composition of MBC monomer in copolymer with different molar compositions.
(Reproduced with permission from reference 28. Copyright 2000 American Chemical Society.)

A plot of $1/T_g$ (K) Vs the MBC copolymer content (wt %) showed a linear relationship (Figure 4). That was consistent with the random distributions as opposed to diblock or multiblock copolymers, which would not give a linear plot. The experimental data exhibits a good correlation with the calculated value, however, the slight deviation observed may be due to the molecular weight effect and also possibly due to steric and polar effects. The random repeat unit distribution in the copolymers obtained fom the thermal data was in agreement with the ^{13}C NMR spectral data.

Regeneration of the Pendant Carboxylic Acid groups. The decoration of polymer chains with appropriate functional groups is required for the attachment of bioactive molecules. In addition, pendant functional groups can be used to facilitate chain cross-linking and for further functionalization as well as for use as interfacial materials. The polycarbonates, i.e., polyMBC and poly(MBC-*co*-TMC), therefore, were subjected to hydrogenolysis to remove the benzyl protecting groups. While removal of the benzyl groups in the homopolymer was easily accomplished using H_2 over Pd/C catalyst in ethyl acetate, its removal in the copolymer proved more difficult. Attempts to remove the benzyl protecting groups using H_2 over Pd/C in three different solvents, viz., ethyl acetate, acetone, and cyclohexane were unsuccessful. However, the deprotection was easily carried out in ethyl acetate using $Pd(OH)_2/C$ (20 %) as the hydrogenation catalyst. The isolated polymers, containing pendant carboxylic acid groups, were no longer soluble in chloroform and were purified by precipitation from it. In the ^{1}H-NMR spectrum, reduction in the integration value of the benzyl methylene signal (Ph-$\underline{CH_2}$) was a measure of the deprotection of the benzyl groups. The percentage of benzyl protective groups removed was calculated by comparing the peak areas of the remaining benzyl methylenes in the ^{1}H NMR spectra, before and after debenzylation. In case of PMBC, the benzyl protective groups were removed completely, whereas in case of poly(MBC-*co*-TMC) 92% of the benzyl groups were removed. A broad resonance at ~ 11 ppm was also indicative of the newly formed free carboxylic acid groups (-COO$\underline{H}$). In the ^{13}C NMR spectrum the resonances due to the aromatic carbons at ~128 ppm were absent, while a new resonance due to the free carboxylic acid (-$\underline{C}$OOH) groups appeared at ~174 ppm. The NMR spectra did not indicate any carbonate backbone cleavage and the molecular structure of the copolymer was preserved intact. GPC measurements of the polymer before and after debenzylation did not indicate loss of polymer integrity. The observed molecular weight difference was merely a consequence of the loss of the benzyl groups from the polymer.

Conclusions

The ability of various lipases to ring open a substituted cyclic carbonate has been demonstrated. The ring-opening polymerization of 5-Methyl-5-benzyloxycarbonyl-1,3-dioxan-2-one (MBC) was studied in bulk at 80 °C. Monomer conversion and the molecular weight were affected by the source of the lipase. NMR studies did not indicate evidence of decarboxylation during propagation. High monomer conversion and high molecular weight samples were prepared using lipase AK as biocatalyst. Microstructure analysis of the copolymers of MBC and TMC from ^{13}C NMR spectra indicated random repeat unit distribution. The experimental diad fractions of the copolymers were in good agreement with those calculated using Bernoullian model, which indicated a random statistical mechanism of propagation for the copolymerization. No melting transitions were detected for the MBC polymers indicative of their amorphous nature. In copolymers, the glass transition temperature (T_g) increased with increasing MBC content in the copolymers. A biocatalytic route to pendant carboxylic acid containing polycarbonates was thus demonstrated. Presence of pendent carboxyl groups is expected to enhance the biodegradability of the polycarbonate and facilitate a variety of potential biomedical applications, e.g., as polymeric drug carriers in time-controlled drug delivery systems.

Acknowledgements

This work was supported in part by the Department of Sponsored Research, USF and Moffitt Cancer Institute and Research Center-USF through their *Research & Creative Scholarship* and *ACS-IRG* grant programs, respectively.

References and Notes

1. Carothers, W. H.; Dorough, G. L.; Van Natta, F. J. *J Am Chem Soc.* **1932**,*54*,761.
2. Kawaguchi, T.; Nakano, M.; Juni, K.; Inoue, S.; Yoshida, Y. *Chem. Pharm. Bull.* **1983**, *31*, 4157. 3
3. Kojimai, T.; Nakano, M.; Juni, K.; Inoue,S.; Yoshida, Y. *Chem. Pharm. Bull.* **1984**, *32*, 2795.
4. Zhu, K. J, Hendren R. W., Jensen K.; Pitt, C. G. *Macromolecules* **1991**, *24*, 1736.
5. Kricheldorf, H. R.; Weenen-Schulz, B. *J. Polym. Sci., Part A: Polym.Chem.* **1995**, *33*, 2193.
6. Mastsuo, J.; Aoki, K.; Sanda, F.; Endo, T. *Macromolecules* **1998**, *31*, 4432.
7. Hori, Y.; Yamaguchi. *Macromolecules* **1995**, *28*, 406
8. Arshady, R. *Desk Reference of Functional Polymers: Syntheses and Applications.* ACS: Washington DC, 1997.
9. Patil, A. O.; Schulz, D. N.; Novak, B. M. *Functional Polymers: Modern Synthetic Methods and Novel Structures.* ACS: Washington DC, 1998.
10. Wang, H.; Dong, K.; Gu, Z. *J. Polym. Sci., Part A: Polym.Chem.* **1998**, *36*, 1301.
11. Acemoglu, M.; Bantle, S.; Mindt, T.; Nimmerfall, F. *Macromolecules* **1995**, *28*, 3030.
12. Vandenberg, E. J.; Tian D. *Macromolecules* **1999**, *32*, 3613.
13. Chen, X.; McCarthy S. P.; Gross, R. A. *Macromolecules* **1997**, *30*, 3470.
14. Chen X.; Gross R. A. *Macromolecules* **1999**, *32*, 3891.
15. Rokicki, G. *Prog. Polym. Sci.* **2000**, *25*, 259
16. Gross, R. A.; Kaplan, D. L.; Swift, G. *Enzyme in Polymer Synthesis.* ACS: Washington DC, 1998.
17. Kabayashi, S.; Kikuchi, H.; Uyama, H. *Macromol. Rapid Commun.* **1997**, *18*, 575.
18. Bisht, K. S.; Svirkin, Y. Y.; Henderson, L. A.; Gross R. A.; Kaplan, D. L., Swift, G. *Macromolecules* **1997**, *30*, 7735.
19. Matsumura, S.; Tsukada, K.; Toshima, K. *Macromolecules* **1997**, *30*, 3122.
20. Deng, F.; Gross, R. A. *Int. J. Biol. Macromol.* **1999**, *25*, 153.
21. Bisht, K. S.;Deng, F.; Gross R. A.; Kaplan, D. L., Swift, G. *J. Am. Chem. Soc.* **1998**, *120*, 1363.
22. Namekawa, S.; Uyama, H.; Kobayashi, S.; Kricheldorf, H. R. *Macromol. Chem. Phys.* **2000**, *201*, 261.

23. Rodney, R. L.; Satgno, J.L.; Beckman, E.J.; Russell, A.J. *Biotechnol. Bioeng.* **1999**, *62*, 259.
24. Matsumura, S.; Harai, S.; Toshima, K. *Macromol. Chem. Phys.* **2000**, *201*, 1632.
25. Matsumura, S.; Tsukada, K.; Toshima, K. *Int. J. Biol. Macromol.* **1999**, *25*, 161.
26. Deng, F.; Bisht, K. S.; Gross R. A.; Kaplan, D. L., *Macromolecules* **1999**, *32*, 5159.
27. Al-Azemi, T. F.; Bisht, K. S. *Macromolecules* **1999**, *32*, 6536.
28. Al-Azemi, T. F.; Harmon, J.P.; Bisht, K. S. *Biomacromolecules* **2000**, *1*, 493.
29. Bisht, K. S.; Al-Azemi, T. F. *Polym. Prepr.* (Am. Chem. Soc. Div. Poly. Chem.) **2000**, *41(2)*, 1865.
30. Al-Azemi, T. F.; Bisht, K. S. *PMSE. Prepr.* (Am. Chem. Soc. Div. Polym. Mat: Sci. & Eng.) **2001**, *84*, 1045.
31. Bisht, S. K.; Henderson, L. A.; Gross, A R; Kaplan L. D.; Swift, G. *Macromolecules* **1997**, *30*, 2705.

Chapter 15

Lipase-Catalyzed Polytransesterification Reactions

Ajay Kumar, Bhanu Kalra, and Richard A. Gross

NSF I/UCRC for Biocatalysis and Bioprocessing of Macromolecules, Department of Chemistry and Chemical Engineering, Polytechnic University, 6 Metrotech Center, Brooklyn, NY 11201

The capability of *Candida antartica* lipase B (Novozyme-435, immobilized) to catalyze transesterification reactions between: *i*) ε-caprolactone (CL)/pentadecalactone (PDL), *ii*) polycaprolactone (PCL)/polypentadecalactone (PPDL) and *iii*) PCL/trimethylene carbonate (TMC) are described. The reactions were normally conducted in toluene. However, solventless transesterification reactions between polyesters were also studied. Comparisons were made between PDL/TMC copolymerizations that were conducted using selected organometallic catalysts and lipases. Lipases catalyze intra-chain cleavage to form enzyme-activated-chain segments. These enzyme-activated moieties react with terminal hydroxyl units of other chains. This mechanism is consistent with the large decrease in reaction rate that resulted from acetylation of chain-end hydroxyl groups. Lipase-catalyzed polytransesterifications appear to be broadly applicable to many different systems, they occur under mild conditions, and offer new opportunities for selectivity during polytransesterifications.

Introduction

Transesterification reactions between polymer chains can be used to create block copolymers from mixtures of homopolymers or to specifically alter chain composition and repeat unit sequence distribution. These reactions offers a simple route that can improve material physico-mechanical and biological properties.[1-3] Catalysis of transesterification reactions between polymer chains has proved difficult by traditional chemical methods. Generally, the catalysts used do not sufficiently reduce the activation energy so that high-reaction temperatures are needed. Such reaction temperatures are often accompanied by high energy consumption and undesirable chain decomposition including molecular weight decrease and generation of colored substances.[1-3] In addition, some of the heavy metals (yttrium and europium) often used for transesterification reactions may be undesired components in products upon disposal.

Thus, there is a need for transesterification catalysts that allow reactions to be conducted at lower temperatures, without heavy metals or the formation of toxic residues. Another desirable feature of transesterification catalysts would be an ability to provide some type of selectivity during the exchange reactions. Catalytic proteins or enzymes can provide: *(i)* efficient substrate activation mechanisms, *ii)* high enantio- and regioselectivity, *iii)* catalyst recyclability, *iv)* activity in solventless reactions, and *v)* reaction components that are environmentally safe.[4-9] In other words, the use of enzymes as transesterification catalysts may lead to decreased reaction temperatures relative to conventional chemical catalysts and even selectivity over chain compositions that undergo interchange reactions.

A fundamentally different type of lipase-catalyzed transacylation reaction requires the cleavage of esters within chains containing multiple esters.[5c-d] If these reactions occur, enzyme-activated chain segments may be formed that can be transferred to the terminal hydroxyl group of another chain end. The occurrence of such reactions catalyzed by an immobilized preparation of *Candida antartica* lipase B (Novozyme-435) is the focus of this chapter.

(1) Copolymerization of PDL and CL monomers:

Novozyme-435 catalyzed CL/PDL copolymerizations were performed in toluene at 70 °C in solutions containing a ratio of toluene to monomer of 2:1 vol/wt. The monomer feed ratios and reaction times for the copolymerizations studied are listed in Table-1. Copolymerization of PDL and CL with a 1:1 molar feed ratio were carried out for reaction times from 1 min to 6 h (Table-1). Increase in the reaction time from 1 to 45 min results in a steady increase in the product yield and M_n (19 to 88% and 8 410 to 19 300 g/mol, respectively). Further increase in the reaction time from 45 min to 6 h results in small increases in the yield and M_n from 88 to 93%, and from 19 300 to 22 300 g/mol, respectively. Also, increase in the reaction time from 1 to 45 min gave a steady increase from 31 to 50 mol% in product CL content. This result is a consequence of the relatively faster rate that PDL is converted into the copolymer.

The four possible diad arrangements (CL*-CL, CL*-PDL, PDL*-PDL and PDL*-CL) for the poly(PDL-co-CL) copolymer is represented in Scheme-1. The microstructure of the PDL/CL copolymers was analyzed by ^{13}C-NMR spectroscopy. Calculations of diad fraction values were based on a series of equations (see *Reference 10*) that assume a Bernoulli or random statistical copolymerization of the two monomers.

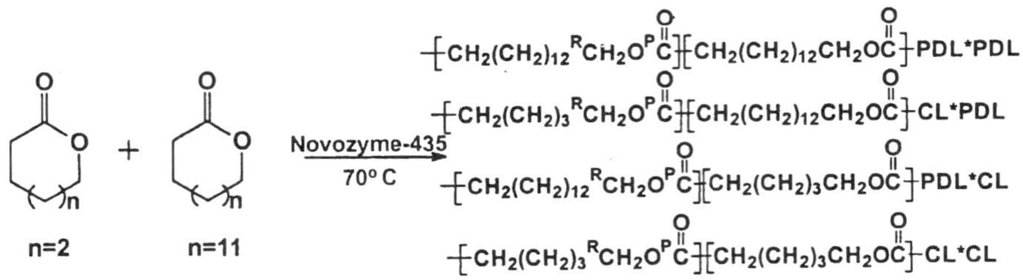

Scheme-1 The four possible diad arrangements of PDL and CL repeat units along copolymer chains.

Poly(CL-*co*-PDL), formed by copolymerization of PDL/CL (1:1 mol/mol) for times from 1 min to 6 h, gave observed and calculated diad fractions that were in good agreement. The formation of random copolymers can be explained by that, when a CL or PDL repeat unit is at the growing chain terminus, either of the incoming monomers can add with a probability approaching equality to form a random copolymer. However, based on the increase from 31 to 50 mol% in the copolymer CL content for reaction times from 1 to 45 min (1:1 monomer feed ratio), PDL must be converted more rapidly then CL into the copolymer. The reactivity ratios were then determined based on the Fineman-Ross method.[11] The reactive rate of PDL polymerization (r_1=1.742) was found to be 13 times larger then for CL polymerization (r_2=0.135).[5f] Despite the large difference in the reactivity ratios for PDL and CL copolymerization, the products isolated had repeat unit sequence distributions that approximated that of random copolymers. Furthermore, GPC traces of the copolymer series from the 1:1 monomer feed ratio all had distributions that were unimodal.

In an another experiment, PDL was polymerized as above for 3 h. This reaction time should be sufficient to form the homopolymer in high yield. Subsequently, CL was added to the reaction vessel to simulate an initial monomer feed ratio of 1:1. The reaction was maintained at 70 °C with agitation for 21 h and then was worked up as described in the general procedure. The product, poly(CL-co-47 mol% PDL), had an M_n of 18 300 g/mol and polydispersity index of 1.9. Comparison of the observed and calculated diad fraction values shows that they are in excellent agreement (entrée 6, Table-1). Thus, the addition of CL to a preformed PPDL gave a product that was random. This finding, in combination with the formation of random copolymers even though PDL reacts much more rapidly than CL, shows convincingly that in addition to catalyzing chain propagation, Novozyme-435 also promotes rapid transesterification reactions between chains.

(2) Reactions between preformed polyester chains.

Novozyme-435 catalyzed transacylation reactions between PCL and PPDL were performed at 70-75 °C for varying reaction times (15 min to 30 h, Table-2).[5c] Reaction of PCL and PPDL, M_n 9.2 × 10^3 and 4.3 × 10^3 g/mol, respectively, mixed for 1 h without enzyme or with deactivated enzyme at 70-75 °C, did not undergo transesterification (Product 1, Table-2). This conclusion was made based on no change in molecular weight and no detected CL*-PDL diads. For the reaction after 15 min with enzyme (Product 2), the ^{13}C-NMR spectrum also showed no CL*-PDL diads. With an increase in the reaction time to 30 min and 1 h (*entries* 3-4), the M_n of the corresponding products increased (5.2 × 10^3, 8.3 × 10^3 g/mol, PDI 4.03 and 1.99, respectively) and CL*-PDL/PDL*-CL diads were observed. The GPC profiles for Products 2, 3 to 4 (Table-2) changed from bimodal to unimodal and the polydispersity index decreased from 4.03 to 1.99. The average sequence lengths (μ_{CL}/μ_{PDL}, *see Table-2 footnote 2*) for Products 3-4 decreased from 18/23 to 2/2 with an increase in the reaction time from 30 min to 1 h. Consistent with the above, the calculated B-values (*for calculation see Table-2 footnote 3*) for Products 3-4 (Table-2) are 0.1 and 1.04, respectively. Thus, the copolymer formed after 30 min is best described as multiblock whereas the copolymer formed after 1 h shows good agreement with random copolymerization statistics.

The use of higher molecular weight polyesters for transacylation reactions may cause increased diffusion limitations. In addition, instead of the fluid melt that is observed at 70-75°C for mixtures of PCL (M_n 9.2 × 10^3)/PPDL (M_n 4.3 × 10^3g/mol), mixtures of higher M_n PCL (M_n 44 × 10^3)/PPDL (M_n 40 × 10^3g/mol) have not-melted PPDL as one of the substrates. Thus, to circumvent these difficulties, studies of Novozyme-435 catalyzed transacylation reactions were conducted in toluene solution (2:1, toluene: polyester vol/wt). Comparison of Products 4-5 showed that, for PCL (M_n 9.2 × 10^3) and PPDL (M_n 4.3 × 10^3), the transacylation in toluene solution proceeded within 1 h to a lower extent than in bulk giving B-values of 0.65 and 1.04, respectively. This may be a consequence of the decreased concentration of the reactants in toluene solution. Nevertheless, the M_n and PDI values of reaction 4-5 are similar.

The potential of carrying out transacylation reactions with polyester substrates of increased molecular weight was explored. (*entries* 6-9, Table 2). In *entries* 6 and 7, the M_n value of PCL/PPDL was 44 × 10^3/9.8 × 10^3 and 9.2 × 10^3/40.0 × 10^3 g/mol, respectively. By increasing the M_n of either the PCL or the PPDL component in the reaction mixture to ≥ 40 × 10^3g/mol, the reactions described by *entries* 6 and 7 proceeded similarly. After 1 h, the resulting

Table-1 Novozyme-435 catalyzed PDL and CL copolymerizations

#	PDL:CL Feed	Time	% Yield	CL/PDL Obs. Mol %	CL*-CL Obs(cal)[a]	CL*-PDL Obs(cal)[a]	PDL*-CL Obs(cal)[a]	PDL*-PDL[a] Obs(cal)	*Mn*[b] g/mol	*Mw*/*Mn*[b]
1	1:1	1min	19	31/69	0.11(0.10)	0.20(0.21)	0.19(0.21)	0.50(0.47)	8410	2.50
2	1:1	5min	34	40/60	0.19(0.16)	0.21(0.24)	0.23(0.24)	0.37(0.36)	14600	2.37
3	1:1	15min	53	50/50	0.26(0.25)	0.24(0.25)	0.22(0.25)	0.28(0.25)	17800	2.38
4	1:1	45min	88	50/50	0.28(0.25)	0.22(0.25)	0.26(0.25)	0.24(0.25)	19300	2.28
5	1:1	6h	93	52/48	0.26(0.26)	0.26(0.25)	0.24(0.25)	0.24(0.24)	22300	1.97
6	1:1	24h[c]	87	53/47	0.31(0.28)	0.23(0.25)	0.21(0.25)	0.26(0.22)	18300	1.97

Copolymerization of PDL and CL catalyzed by Novozyme-435 (1/10 w.r.t. monomer) in toluene (2.68 times wt/wt of monomers) at 70 °C. [a]Calculated diad sequences were based on equations mentioned in reference 21. [b]determined by GPC in $CHCl_3$. [c]CL added after three hours of polymerization of PDL.

Table-2 Novozyme-435 catalyzed transesterification reaction of Poly(ω-pentadecalactone) with Poly (ε-caprolactone) at 70-75 °C in bulk/toluene.

#	β	Obs $[PCL]_o$/ $[PPDL]_o$	^aM_n X10^{-3} (PCL/ PPDL)	Time	% Yield	Diad CL*-CL Obs(cal)	CL*-PDL Obs(cal)	Sequence PDL*-CL Obs(cal)	PDL*-PDL Obs(cal)	M_n X10^{-3}	*μ_{CL} /*μ_{PDL}	B	PDI
1*	bulk	56/ 44	9.2/ 4.3	1 h	66	.56	-	-	0.44	4.7			2.39
2	bulk	56/ 44	9.2/ 4.3	15 min	73	.56	-	-	0.44	3.3			3.13
3	bulk	55/ 45	9.2/ 4.3	30 min	70	.52 (.30)	.03 (.25)	.02 (.25)	.44 (.20)	5.2	18/23	.10	4.03
4	bulk	50/ 50	9.2/ 4.3	1 h	70	.25 (.25)	.25 (.25)	.27 (.25)	.23 (.25)	8.3	02/02	1.04	1.99
5	Tol	50/ 50	9.2/ 4.3	1 h	63	.27 (.22)	.20 (.25)	.21 (.25)	.32 (.27)	8.7	03/03	.65	1.92
6	Tol	53/47	44/9.8	1h	87	.45(.28)	.08(.25)	.07(.25)	.39(.28)	26.0	06/06	.30	1.89
7	Tol	36/64	9.2/40	1 h	75	.29(.13)	.06(.33)	.06(.23)	.58(.41)	20.6	06/11	.33	2.42
8	Tol	38/ 62	44/ 40	1 h	86	.36 (.14)	.02 (.24)	.03 (.24)	.62 (.38)	18.2	19/21	.10	1.92
9	Tol	50/ 50	44/ 40	30 h	83	.26 (.25)	.25 (.25)	.23 (.25)	.27 (.25)	31.2	02/02	.95	1.87
10	bulk	49/51	3.3/4.1	1 h	87	.47(.24)	.02(.25)	.06(.25)	.45(.26)	3.5	25/09	.16	2.09

1. Reaction 1 was a control, minus immobilized enzyme in the reaction mixture. 2. $\mu_{CL} = f_{CL^*\text{-}PDL} + f_{CL^*\text{-}CL} / f_{CL^*\text{-}PDL}$. $\mu_{PDL} = f_{PDL^*\text{-}CL} + f_{PDL^*\text{-}PDL} / f_{PDL^*\text{-}CL}$, *where μ is the average sequence length and f = integral of the corresponding diad signal in the inverse gated carbon spectrum. 3.* $B = (f_{PDL^*\text{-}CL}/f_{PDL^*\text{-}PDL} + f_{PDL^*\text{-}CL})/2F_{PDL} + (f_{CL^*\text{-}PDL}/f_{CL^*\text{-}CL} + f_{CL^*\text{-}PDL})/2F_{CL}$. *Where f= integral of the corresponding diad signal and F is the mol fraction of the observed ω-pentadecalactone(PDL) or ε-caprolactone(CL) units.*

copolymers had B-values of about 0.30. In addition, the GPC of Products 6 and 7 showed that they had unimodal molecular weight distributions and M_n values of 26.0 × 10^3 and 20.6 × 10^3 g/mol, respectively. Therefore, in the presence of one component of molecular weight ≥ 40.0 × 10^3 g/mol, Novozyme-435 retained good activity for PCL/PPDL transacylation. Comparison of *entries* 6 and 7 to 5 in Table 2 shows that, an increase in the molecular weight of one of the polyester substrates decreased the extent of transacylation that occurred within the 1-h reaction time.

Next, the extent of transacylation when both polyester components in the reaction mixture had M_n values ≥ 40.0 × 10^3 g/mol was studied. Hence, *entries* 8 and 9 describe Novozyme-435 catalyzed transacylation reactions in toluene between PCL (M_n 44.0 × 10^3, PDI 1.65) and PPDL (M_n 40.0 × 10^3, PDI 1.71). Comparison of the *entries* 6 and 7 to 8 shows the large decrease in the extent of transacylation that occurred when both instead of one of the polyester substrates had M_n values ≥ 40.0 × 10^3g/mol. In other words, the absence of at least one of the two polyester substrates with M_n < 10.0 × 10^3 g/mol resulted in a large decrease in the rate of Novozyme-435 catalyzed polytransacylation reactions. Nevertheless, Product 8 is a multiblock copolymer with μ_{CL}/μ_{PDL} values of 19/21 (Table-2). When the reaction described by *entry* 8 was repeated but with an increase in the reaction time (*entry* 9), a product closely approximating a random copolyester resulted. Product 9 had M_n and B values of 31.2 × 10^3 and 0.95, respectively. Thus, although the reaction rate was substantially decreased by increasing the substrate polyester molecular weights, ultimately, Novozyme-435 reshuffled the repeat unit sequences of the substrates with M_n ≥ 40.0 × 10^3g/mol by a series of transacylation reactions to provide a high molecular weight random copolymer.

The lipase-catalysis of polytranesterification reactions between pre-formed chains is believed to occur as is shown in *Scheme-2*. At the lipase-box of Novozyme-435, the breaking of an intra-chain ester group is catalyzed to give an enzyme-activated-chain-segment (EACS). Subsequently, a terminal hydroxyl group of another chain, that is associated with the enzyme, reacts with the EACS to give an ester group. At the onset of these reactions, diblock and multiblock copolymers are formed. However, as the reaction progresses at extended reaction times, the resulting block copolymers can further react as above to ultimately give copolymers that are random. Undoubtedly, the rate of transacylation reactions between polyesters will be a function of the main chain structure. At present, our laboratory is actively investigating how the kinetics of lipase-catalyzed transacylations is effected by many factors including chain composition.

This work showed that polyester chains with lower molecular weight averages more rapidly undergo Novozyme-435 catalyzed transacylation reactions. This is consistent with the above reaction mechanism since chains with

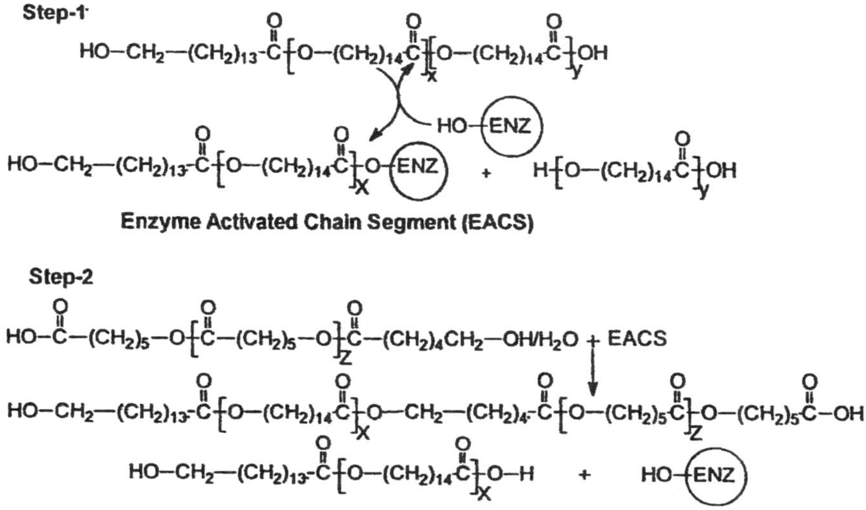

Scheme-2 Mechanism of transesterification of polyester chains

lower molar mass will have a higher concentration of hydroxyl terminal groups per unit weight of polymer. It should then follow that acylation of the chain-end hydroxyl groups of these polyesters will reduce the kinetics of transacylation reactions. To confirm this hypothesis, PCL (3.3×10^3g/mol PDI 1.66) and PPDL (4.1×10^3 g/mol PDI 2.36) were completely acetylated using acetic anhydride and a catalytic amount of sulphuric acid. These hydroxy-terminal acetylated polyesters were then subjected to Novozyme-435 in bulk at 70-75 °C for 1 h. Analysis of the product by ^{13}C-NMR showed a large decrease in the extent of transacylation that occurred. Thus, the product formed (M_n 3.5×10^3 g/mol, PDI 2.09) had PDL*PDL, PDL*CL, CL*PDL, and CL*CL diad fractions of 0.45, 0.06, 0.02, and 0.47, respectively. The fact that a low level of transesterification still occurred may be due to a low concentration of non-acetylated chain-end hydroxyl groups or to low-level lipase-catalyzed chain hydrolysis that generates new chain-end hydroxyl units.

***(3)* Transesterification of polyester co-polycarbonate**

Aliphatic polycarbonates and their copolymers with polyesters are of interest for use in bioresorbable suture filament, artificial skin, prostheses, bone fixation plates, ligature clamps and galenic formulations.[12] Poly(ω-pentadecalactone), poly(PDL), is a semi-crystalline polymer with a melting point of about 95 °C,[13] while poly(trimethylene carbonate), poly(TMC), is normally amorphous.[12] An ability to synthesize copolymers of PDL and TMC with control of the copolymer repeat unit sequence distribution would allow a full interrogation of PDL/TMC copolymer properties.

This portion of the Chapter is concerned with whether lipase-catalysis provides unique attributes for lactone/carbonate copolymerizations. PDL/TMC copolymerizations were performed in monophasic organic media (toluene) using Novozyme-435 and other lipases.[5d] A schematic of possible copolymer triads and the corresponding linkages between units is shown in Scheme-3. The assignments of the peaks in the ^{1}H-NMR spectrum of poly(PDL-*co*-50mol% TMC) is shown in Figure-1. These assignments were based on previous ^{1}H-NMR studies of the respective homopolymers.[14] Since a triplet at δ 3.45 was not observed in the ^{1}H-NMR spectrum, we conclude that decarboxylation during propagation did not occur.[14] The integral ratio of the signals at 2.28 ppm (m, $COCH_2$, P) and 2.02 ppm (m, $COCH_2CH_2$, T) was used to determine the repeat unit composition of the copolymer. The observed diad sequences (*see Table-3*) were calculated based on the OCH_2 signals (4.28-4.04ppm) in the ^{1}H NMR spectra. Specifically, based on our previous work on respective homopolymers,[14] the integrals of signals at 4.24 (m, 4H, OCH_2), 4.04 (t, 2H, OCH_2) were used to calculate the T*-T and P*-P diads, respectively. The integral of the signals between 4.08-4.18 (m, 4H, OCH_2) was assigned to the T*-P/P*-T codiads. The diad, triad, and tetrad assignments for the ^{13}C-NMR spectrum of poly(PDL-co-50%TMC) were made by comparison of the signals to those in spectra recorded of the corresponding homopolymers [poly(PDL)[5f], poly(TMC)][14] and copolymers of different repeat unit composition.[5c] The carbonyl carbon signals of PDL and TMC repeat units for poly(PDL-co-50mol%TMC) were observed in the spectral regions of 173.7-174.3 ppm and 154.9-155.5 ppm, respectively. The signals at δ 68.4-68.0 and 60.5-60.9 can only appear when we have PP*T, TT*P and PT*T, TP*P types of triads and the increased intensity of these signals supports random copolymerization.[5c]

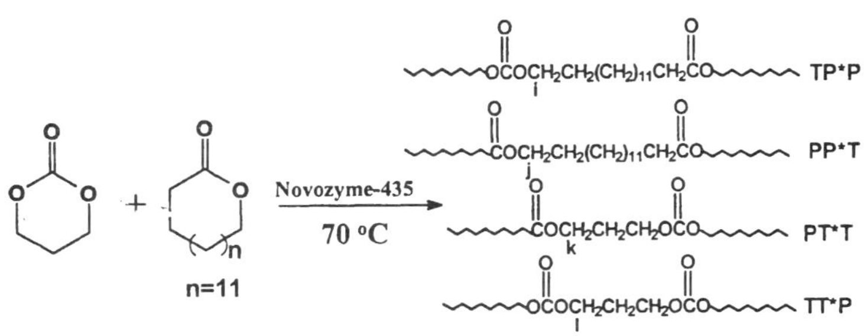

Scheme-3 Schematic representation of triads from Poly(PDL-co-TMC).

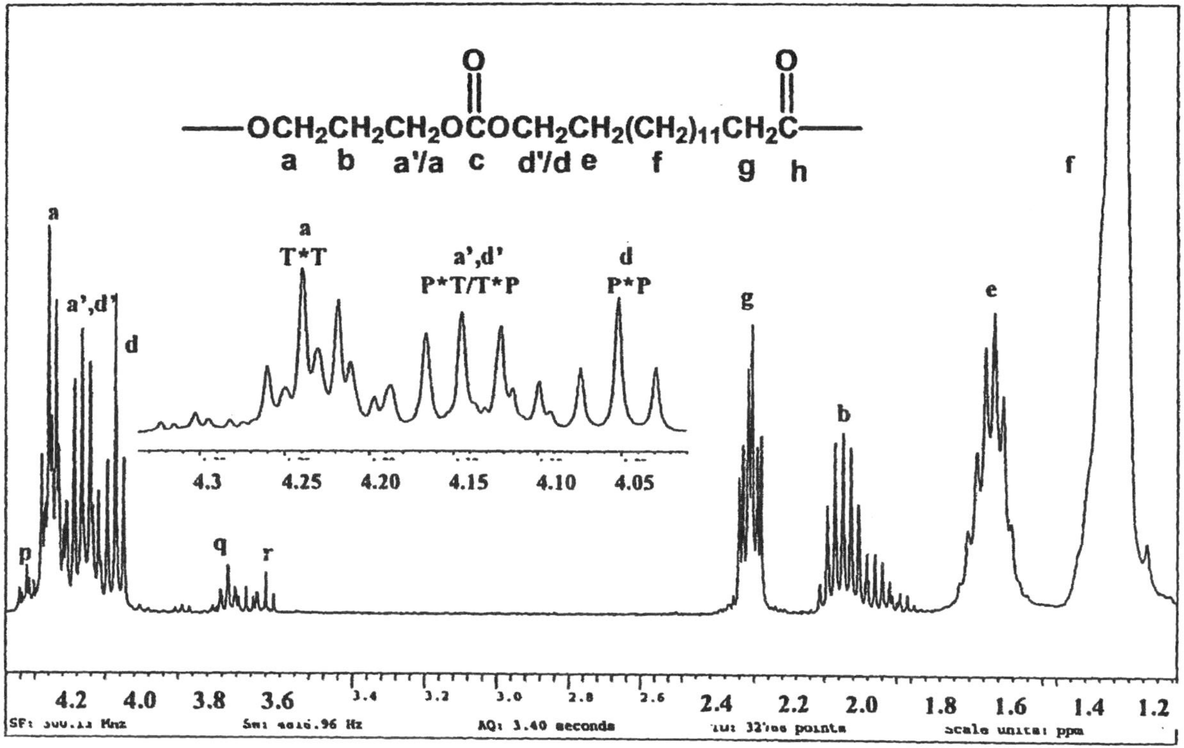

p=$HOCH_2CH_2CH_2O$, q= $HOCH_2CH_2CH_2O$, r= $HOCH_2CH_2(CH_2)_{11}$

Figure-1 ^{1}H-NMR spectrum of poly(PDL-*co*-50mol% TMC)

The ability of the lipases from *Aspergillus niger (AK), Candida antarctica, Candida rugosa(AY), Pseudomonas cepacia (PSC)*, and porcine pancreas (PPL) to catalyze PDL/TMC copolymerization was assessed. Novozyme-435 (immobilized *Candida antarctica* lipase B) was found to be the best of those studied for the formation of copolymers with high molar mass.[5c] Further studies of Novozyme-435 catalyzed copolymerization of TMC and PDL at 70 °C in toluene (2:1, toluene: monomers) were carried out with varying reaction times and monomer feed ratios (*see Table-3*). At F_{PDL}/F_{TMC} 1:1, the isolated yield and M_n of the copolymer increased from 5 to 83% and 5.4×10^3 to 18.8×10^3 from 5 min to 24 h, respectively (reactions 1-6,Table-3).By one hour, the isolated yield of the copolymer was 62 % and the copolymer composition was 80/20 PDL/TMC. Thus, PDL was consumed more rapidly than TMC. By extending the reaction time to 24 h, the copolymer contained 50 mol% of both TMC and PDL. Analysis of the copolymer microstructure showed the product to be random. This may be explained based on transesterification reactions between TMC and the PDL-rich copolymer chains.

For polymerizations carried out with a PDL/TMC feed ratio of 1:1, the average sequence lengths of PDL and TMC units (L_{PDL} and L_{TMC}, respectively) at reaction times of 5 min and 1 h, were 15/1 (L_{PDL}/L_{TMC}) and 13/3, respectively (*Table-3*). Thus, at these relatively short reaction times, copolymers were formed that consist of PDL blocks inturrupted by one or two TMC units. Increase in the reaction time to 3 and 24 h resulted in decreased block sizes (L_{PDL}/L_{TMC} 6/2 and 2/2, respectively). This is further illustrated by that the observed and calculated diad fractions were close in value after a 24 h reaction time (reaction 6). These results are explained by the catalysis of transesterification or transacylation reactions by Lipase B from *Candida antartica*.

The ability of Novozyme-435 to catalyze transacylation reactions for chains of mixed ester-carbonate linkages was further evaluated. TMC was added to preformed poly(PDL) chains (M_n 12×10^3, M_w/M_n 2.0) under reaction conditions that were identical to those used for PDL/TMC copolymerizations. The molar ratio of TMC to the repeat units of preformed poly(PDL) was 1 to 1. The product formed after 48 h was poly(PDL-co-53mol% TMC), in 90% isolated yield, with M_n 5.2×10^3 g/mol and M_w/M_n 1.57. The observed and calculated (assuming a random distribution) diad fractions were P*-P 0.21, T*-T 0.27, P*-T/T*P 0.54 and P*-P 0.23, T*-T 0.27, P*-T/T*P 0.50, respectively. At present, we do not know whether this result is due to Novozyme-435 catalyzed transacylation reactions between: *i*) poly(PDL) chains that have terminal TMC units, *ii*) poly(PDL) chains with terminal TMC chain segments, *iii*) the formation of oligomeric TMC chains that undergo transacylation reactions with poly(PDL), or *iv*) a combination of the reactions described in *i-iii*. (*see Scheme-4)*

The mechanism for transacylation reactions for chains of mixed ester-carbonate linkages is believed to occur similarly to that was described above for PPDL and PCL transacylations. The only potential difference is that, Lipase B from

Table-3 Copolymerization of PDL and TMC by Novozyme-435 at 70 °C

	F_{PDL}/F_{TMC}	Time	% Yield	[b] [PDL]/[TMC]	P*-P Obs (cal)	P*-T Obs (cal)	T*-P Obs (cal)	T*-T Obs (cal)	[c] L_{PDL}/L_{TMC}	[d] M_n g/mol	[d] M_w/M_n
1	1:1	5min	5	92/08	.85 (.85)	.06 (.07)	.06 (.07)	.03 (.01)	15/1	5390	2.69
2	1:1	15min	18	88/12	.84 (.77)	.06 (.11)	.06 (.11)	.04 (.02)	15/2	5570	2.44
3	1:1	30min	44	87/13	.81 (.75)	.07 (.11)	.07 (.11)	.05 (.07)	12/2	7100	2.31
4	1:1	1h	62	80/20	.84 (.64)	.06 (.16)	.06 (.16)	.04 (.04)	13/3	7310	2.39
5	1:1	3h	69	75/25	.61(.56)	.13 (.19)	.13 (.19)	.13 (.06)	06/02	13000	2.07
6	1:1	24h	83	50/50	.25 (.25)	.23 (.25)	.23 (.25)	.29(.25)	02/02	18800	1.65

designates the reaction conditions/reagents and the product formed. [b] *Observed PDL/TMC fraction by* ^1H *NMR.* [c] $L_{TMC} = f_{TMC*PDL} + f_{TMC*TMC} / f_{TMC*PDL}$, $L_{PDL} = f_{PDL*TMC} + f_{PDL*PDL} / f_{PDL*TMC}$, *where L is the average sequence length and f = integral of the corresponding diad signal in the proton spectrum* [d] M_n *and* M_w/M_n *of the samples were determined by GPC in* $CHCl_3$

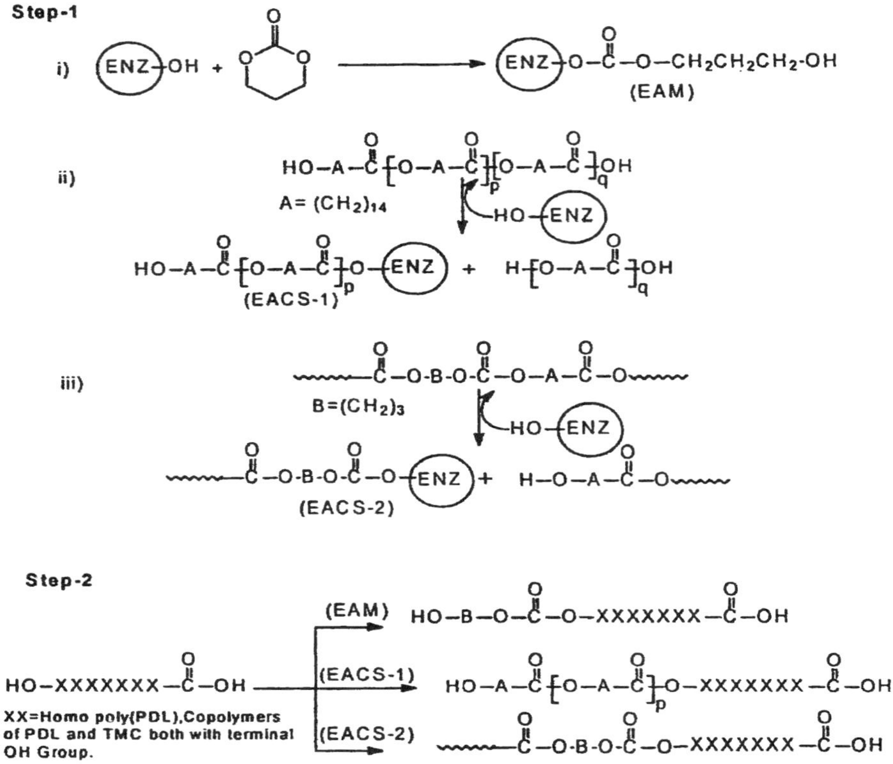

Scheme-4 Mechanism for transacylation of PDL and TMC. In this Scheme, A is $(CH_2)_{14}$ and B is $(CH_2)_3$.

Candida antartica might also cleave carbonate linkages within chains that can be transferred to the terminal hydroxyl group of another chain (*see Scheme-4*). The relative frequency at which carbonate and ester linkages are cleaved during Novozyme-435 catalyzed acylation is currently under study and will be the subject of a separate paper.

Summary of Results

Novozyme-435 catalyzed copolymerization of CL and PDL at 70 °C (toluene to PDL 2:1 vol/wt) occurred rapidly. Experimental determination of the reactivity ratios showed that the rate of PDL polymerization (r_1=1.742) was 13 times greater then for CL polymerization (r_2=0.135). Despite the large difference in reactivity of these monomers, random copolymers were formed.

These results were attributed to an active pathway by which Novozyme-435 catalyzes polytransesterification reactions. Subsequently, it was discovered that Novozyme-435 catalyzes transacylation reactions between pre-formed aliphatic polyester chains with M_n values in excess of 40.0 × 10^3g/mol. The rate of these lipase-catalyzed transacylation reactions is a function of the polyester chain length and the availability of chain-end hydroxyl groups. As the chain length increased, or the hydroxyl chain-end concentration decreased, the rate of transacylation reactions decreased. The higher reactivity of PDL relative to TMC, and the ability of Novozyme-435 to catalyze the conversion of poly(PDL-co-TMC) from multiblock to random copolymers, led us to conclude that Novozyme-435 actively promotes transesterification or transacylation reactions for chains that consist of mixed ester/carbonate linkages. The ability to manipulate the block lengths in these copolymers is expected to allow fine-tuning of the physical and biological properties of these copolymers. Furthermore, the potential to apply lipase-catalysis for a range of low-temperature transesterification reactions opens up a number of new opportunities in macromolecular synthesis. Additional work is underway to extend these findings to other systems, to better understand the factors that promote and disfavor the transesterification pathway, and to better understand the mechanism of these **reactions**.

References

1. Park,S. S.; Chae, S. H.; Im, S. S. *J. Polym.Sci. A, Polym. Chem.* **1998**, *36*, 147-156.
2. Lee, S. C.; Yoon, K. H.; Park, H.; Kim, H. C.; Son, T. W. *Polymer* **1997**, *38*, 4831-4835.
3. Gogolewski, S. *Polymer*, **1977**, *18*, 63-68.
4. (a) Tramper, J.; Vander Plas, H. C.; Linko, P. In *Biocatalysis in Organic Synthesis*: Elsevier Science Publishers, Amsterdam, **1985**.; (b) Drauz, K.; Waldmann, H. *Enzyme Catalysis in Organic Synthesis*:VCH, Germany, **1995**.; (c) Kline, B. J.; Beckman, E. J.; Russel, A. J. *J. Am. Chem. Soc.* **1998**, *120*, 9475.; (d) Santaniello, E.; Ferraboschi, P.; Grisenti, P.; Manzocchi, A. *Chem. Rev.* **1992**, 1071. (e) Klibanov, A.M. *Acc. Chem. Res.* **1990**, *23*, 114. (f) Klibanov, A. M. *CHEMTECH* **1986**, *16*, 354.; (g) Morrow, C. J.; Wallace, J. S. *US Patent No. 5*, 147791. Issued: Sept. 15, **1992,** 993.
5. (a) Kumar, A.; Kalra, B.; Gross, R. A. *Chemical Reviews* **2001**, 101, 2097.; (b) Kumar, A.; Kalra, B.; Gross, R. A. *Applied microbiology and Biotechnology* **2001** *55*, 655.; (c) Kumar, A.; Gross, R. A. *J. Am. Chem. Soc.* **2000**, *122*, 11767.; (d) Kumar, A.; Garg, K.; Gross, R. A. *Macromolecules* **2001**, *34*, 3527.; (e) Kumar, A.; Gross, R. A. *J. Org. Chem.* **2000**, *65*, 7800.; (f) Kumar, A.; Kalra, B.; Dekhterman and Gross, R. A. *Macromolecules* **2000**, *33*, 6303.; (g) Kumar, A.; Gross, R. A. *Biomacromolecules* **2000**, *1*, 133-138.; (h) Gross, R.

A.; Kaplan, D. L.; Swift, G. (Eds.). In *Enzymes in Polymer Synthesis.* ACS symposium series 684: American Chemical Society, Washington, DC, **1998**.; (i) Deng, F.; Bisht, K. S.; Gross, R. A.; Kaplan, D. L. *Macromolecules* **1999**, *32*, 5159.; (j) Deng, F.; Gross, R. A. *I.J.B.M.* **1999**, *25*, 153.; (k) Bisht, K. S.; Deng, F.; Gross, R. A.; Kaplan, D. L.; Swift. G. *J. Am. Chem. Soc.* **1998**, *120*, 1363.

6. (a) Kobayashi, S.; Uyama, H.; Namekawa, S.; Hayakawa. H. *Macromolecules* **1998**, *31*, 5655.; (b) Kobayashi, S.; Kiyosada, T.; Shoda, S. *J. Am. Chem. Soc.* **1996**, *118*, 13113.; (c) Kobayashi, S.; Wen, X.; Shoda, S. *Macromolecules* **1996**, *29*, 2698.; (d) Uyama, H.; Kikuchi, H.; Takeya, K.; Kobayashi, S. *Acta Polym.* **1996**, *47*, 357.;

7. (a) Chaudhary, A. K.; Lopez, J.; Beckman, E. J.; Russell, A. J. *Biotechnol. Prog.* **1997**, *13*, 318.; (b) Reetz, M. T.; Zonta, A.; Simpelkamp, J. *Biotechnol. Bioeng.* **1996**, *49*, 527.

8. Gutman, A. L.; Bravdo. T. *J. Org. Chem.* **1989**, *54*, 4263.

9. Dordick, J. S.; Martin, B.; Linhardt, R. J. *U.S. Patent No.* 5,474,915. Issued **Dec. 12. 1995**.

10. Calculations of diad fractions were used to determine what the relative diad distribution would be assuming Bernoulli random statistics, where P is the probability of finding the same monomer units next to each other. For example, the fraction of diads where PDL units neighbor PDL units is given as PDL*-PDL=P^2_{PDL}, PDL next to CL or CL next to PDL is given as PDL*-CL=CL*-PDL=$2P_{PDL}(1\text{-}P_{PDL})$, similarly CL next to CL is given as CL*-CL=$(1\text{-}P_{PDL})^2$.

11. In the Fineman-Ross method, $F/f \times (f\text{-}1) = r_1(F^2/f)\text{-}r_2$ where F = [PDL]/[CL] and f = d[PDL]/d[CL], y = F/f x (f-1) and x =(F^2/f)

12. (a) Zhu, K.J.; Hendren, R.W.; Pitt, C.G.*Macromolecules*. **1991**, *24*, 17362. (b) Schindler, A.; Jeffcaat, R.; Kimmel, G.L.; Pitt, C.G.; Wall, M.E.; Zweidinger, R. in *contemporary Topics in Polymer Sciences,* Vol.2, Pearce, E.M. and Schaefgen, J.R. Eds. Plenum, New York, 1997, p251.

13. Focarete, M. L.; Scandola, M.; Kumar, A.; Gross, R. A. *J of Polymer Science, Part B: Polymer Physics.* **2001**, *39(15)*, 1721-1729.

14. Bisht, K. S.; Svirkin, Y. Y.; Henderson, L. A.; Gross, R. A.; Kaplan, D. L.; Swift. G. *Macromolecules* **1997**, *30*, 7735.

Polysaccharides

Chapter 16

Synthesis and Modification of Carbohydrates through Biotechnology

Peter R. Andreana, Wenhua Xie, and Peng George Wang*

Department of Chemistry, Wayne State University, Detroit, MI 48202

Chemoenzymatic approaches to the modification of polysaccharides plays an important role in discovering novel polymeric structures with varying properties while establishing a non-toxic rapport with the environment. This review will explore the utilization of thermophilic glycosidases, and thermophilic lipases in the ring-opening polymerization/ modification of hydroxy-containing compounds, as well as introduce LiCl as a non-toxic polymer catalyst.

US demand for natural polymers is projected to approach $2.8 billion in 2003 (*1*). The fastest growth will take place in fermentation products; rapid short term growth will come from the use of these polymers as food ingredients, while in the longer term, strong growth will be fueled by increasing demand for feedstocks in the production of degradable plastics. In addition, protein-based polymers should experience robust growth as baby boomers age and new applications emerge for collagen. While wheat gluten will remain the largest single product by volume, cellulose derivatives (including microcrystalline cellulose) are expected to remain the largest product class in dollar terms. Cellulose ethers constitute a fairly mature market, however, so novel food applications and the advent of new products (e.g., cationic cellulose ethers) should buoy growth somewhat. In addition to fermentation products and protein-based polymers, marine polymers should also fare well. Despite a shake-out in the alginate industry, these products will be among the fastest growing natural polymers. Moreover, significant increases in chitin and chitosan production capacities will allow domestic producers to meet rising US demand. Natural phenolics and polyphenols will exert a moderating effect on the natural polymers market as a whole, given that this segment is dominated by mature products such as tannins and vanillin. Nevertheless, improvements in polyterpene chemistry should serve to spur demand. New technologies have changed the area of natural polymers over the last two decades, and this trend should continue. In order for those products, with the largest market potential (e.g., lactic acid and wheat polymers) to succeed, further advances must be made in order to lower production costs to the point where these biodegradable polymers can compete with petrochemical-derived polymers on the basis of price. The best opportunities for natural polymers will continue to be in food and beverages, led by wheat gluten and xanthan gum. The trend toward health-consciousness will continue to be a boon to the food and beverages sector, as exudate, vegetable and fermentation gums gain appeal by virtue of their abilities to replace fat, add texture and provide dietary fiber. The fastest growing non-food markets are expected to be packaging and textiles. The packaging market will be the first beneficiary of the rapid gains made by polylactic acid (PLA) and other carbohydrate-based polymers. Textiles will make more significant gains over the longer term, as biodegradable polymers such at polylactic acid and polytrimethylene terephthalate (PTT) are converted to fibers for use in apparel manufacture.

The temptation to improve upon nature has always been great and has rarely been resisted. When scientists link the special properties of these substances (physical properties such as tensile strength and flexibility) to the sizes and functionalities of their molecules the next logical step involved chemical modifications of naturally occurring polymers.

Modification of Carbohydrates

Carbohydrates can be modified in order to widen their range of potential application. In this way it is possible to produce a wide range of carbohydrate ingredients with specific functional properties. Naturally, the structure - function relationship plays an important role in this. The application of enzymes to industrial productions has developed rapidly since 1960, particularly in the food and beverage industry. Today biotechnology and the use of enzymes has integrated with traditional chemical industry for the production of drugs and vitamins. The substrate and product specificity of enzymes is used for the production of compounds, which only with great difficulty, can be synthesized by conventional organic chemistry. Therefore, the synthesis of biomass-based carbohydrates with specific functional properties for use in drugs, personal care and food products is receiving increasing attention. The functional properties are controlled by the degree of polymerization and the type, position and degree of substitution. These parameters are essential to industrial production and application of carbohydrates.

Thermophilic microorganisms produce enzymes with unique characteristics such as high temperature and chemical and pH stability. The enzymes can be applied as biocatalysts in already existing industrial processes to replace the presently used, often polluting, chemical reagents. Thermophilic enzymes are of further interest for production of food additives, such as oligosaccharides.

The synthetic potential of sugar-modifying enzymes such as glycosidases has been investigated for the production of biologically active oligosaccharides. Novel micro-organisms have been isolated and characterized and suitable enzymes identified by screening procedures. Structure and function of the enzymes have been studied with respect to substrate specificity, stability, and product yield. Process improvement as well as the influence of the substrate and of the medium composition have been investigated to obtain novel products and improved yields in synthetic processes.

Thermophilic Glycosidase Library

Enzymes evolving from thermophilic and hyperthermophilic organisms have attracted considerable attention due to their potential as biocatalysts with unprecedented properties for industrial applications. The glycosidase CLONEZYME™ library, which currently contains 10 unique thermostable glycosidases, has been developed by Diversa (San Diego, CA) through cloning and automated high through-put screening systems (*2*). Each enzyme displays a variety of activities ranging from galactosidase, glucosidase to fucosidase. As shown in Scheme 1, a CLONEZYME™ library was screened for the synthesis of *N*-acetyllactosamine and lactosamine, whose core structure exists in many glycoproteins and glycolipids (*3,4*).

All thermophilic enzymes from the library were screened as potential catalysts for the glycosylation reaction. Table I lists the percent hydrolysis and yields for the varying thermophiles.

Modification of Hydroxyethylcellulose by Transgalactosylation with β-Galactosidases

In recent years, enzymatic transglycosylation catalyzed by glycosidases has been the focus of considerable interest (*5-9*). These enzymatic syntheses demonstrate transglycosylation ability toward a wide array of acceptors. By utilizing the thermophilic CLONEZYME™ glycosidase library as well as glycosidases from three mesophilic sources (*A .oryzae, B. circulans, and E. coli*), it was demonstrated that hydroxyethylcellulose (HEC) could be transgalactosylated using lactose as the donor Scheme 2 (*10*). HEC has been

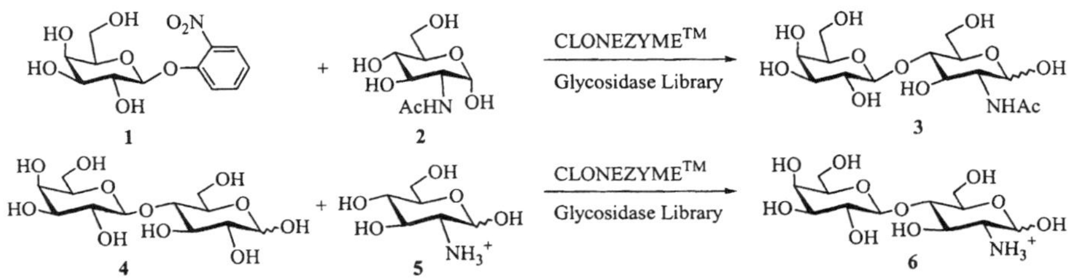

Scheme 1. CLONEZYME™ glycosidase library in the synthesis of N-acetyllactosamine and lactosamine.

Table I. Enzymatic Synthesis with Recombinant Thermophilic Glycosidase CLONEZYME™ Library and Conventional Mesophilic β-Galactosidase

Enzyme	*Hydrolysis(%)*[a]	*Yield(%)*
Gly001-02	92	-
Gly001-06	9	46
Gly001-07	8	45
Gly001-08	86	-
Gly001-09	18	61

[a] Hydrolysis was determined according to the free galactose content measured enzymatically by the Hans-Otto Beutler method using galactose dehydrogenase.

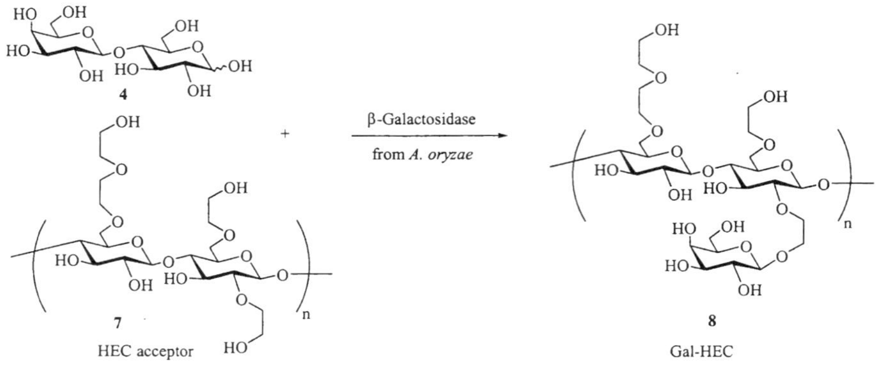

Scheme 2. Transgalactosylation between lactose and HEC.

widely used as rheology modifiers, thickening agents, protective colloids, and a variety of other applications (*11*).

Since HEC is a statistical polymer, the orientations of the hydroxyl groups are randomly distributed along the polymer backbone. Treatment with the enzymes would result in random transgalactosylation. ^{13}C NMR spectroscopy is a conventional means to determine the loci of newly formed glycosidic bonds (*12*), as well as the degree of substitution on HEC. From initial studies directed at the elucidation of the parent HEC and its galactosylated forms, it was concluded that they were indistinguishable. Thus in order to quantify substituted galactose on the HEC polymer, a highly sensitive enzymatic assay (*13*) was employed. The galactose oxidase-catalyzed reaction was highly specific for and reportedly more efficient with polymers containing terminal D-galactose as illustrated in Scheme 3.

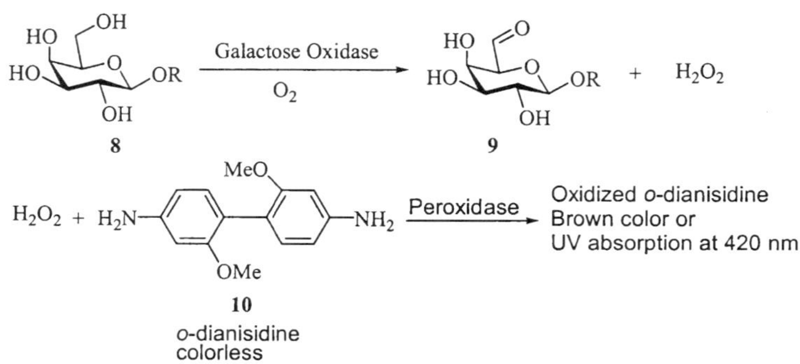

Scheme 3. Enzymatic assay via the peroxidase-chromogen test.

The peroxide-chromogen test was instrumental in the determination of the degree of substitution (DS) which is defined in terms of each galactosyl unit. The results indicated that β-galactosidase from neither *B. circulans* nor *E. coli* generated galactosylated HEC. In the case of *A. oryzae*, the DS of trans-galactosylation determined was far less than 0.015 from the peroxide-chromogen test. In order to reduce the hydrolysis effect, higher concentrations of HEC were used along with an appropriate organic co-solvent such as acetonitrile. However, none of the three enzymes turned out to be efficient. On the contrary, reduced activity of the enzyme from *A. oryzae* was observed when 50% organic solvent was used.

Effects of different ratios and concentrations of donor and acceptors were studied to search for a better degree of substitution for the transgalactosylation reaction. The optimal results for concentrations and ratios of donor and acceptor used in combination with different enzymes are summarized in Table II.

Table II. Transgalactosylation of HEC with Galactosidases

Enzyme	*[HEC] mM*	*[Lactose] M*	*DS*[a]
Aspergillus oryzar	*0.08*	1.0	0.03
	0.04	1.0	0.03
CLONZYME™	*0.80*	2.4	0.02
Gly001-09	*0.08*	1.0	0.01

[a] DS values were obtained in comparison to a control measurement with HEC only.

Galactosidase-catalyzed transglycosylation can be employed to galactosylate water-soluble, cellulose-based polymers. The most efficient β-galactosidase was found to be the one from *A. oryzae*, and the reactions resulted in increased yields of Gal-HEC without organic co-solvent. It is important to note that depolymerization of polysaccharides should not be neglected in the course of the transglycosylation reactions.

Ring-Opening Polymerization of β-Butyrolactone by Thermophilic Lipases

Enzyme-catalyzed ring-opening polymerization of lactones has recently received much attention as a new approach for the synthesis of polyesters, especially poly(hydroxyalkanoate)s (PHAs), (*14-18*) a family of biodegradable polymers which are of great interest for biomedical applications. Researchers including Kobayashi, (*19-21*) and Gross (*22,23*) have shown that a number of lipases catalyze the ring-opening polymerization of ∈-caprolactone affording polymers with molecular weights (M_n) from 1600 up to 7700 g/mol. The unique stability of thermophilic enzymes over a range of pH, temperature, and solvent conditions permits the exploration of reaction conditions without significant loss in enzyme activity (*24-26*). An example of ring-opening polymerization is shown in Scheme 4 where racemic β-butyrolactone was converted to optically active (*R*-enriched poly(3-hydroxybutyrate) PHB (27).

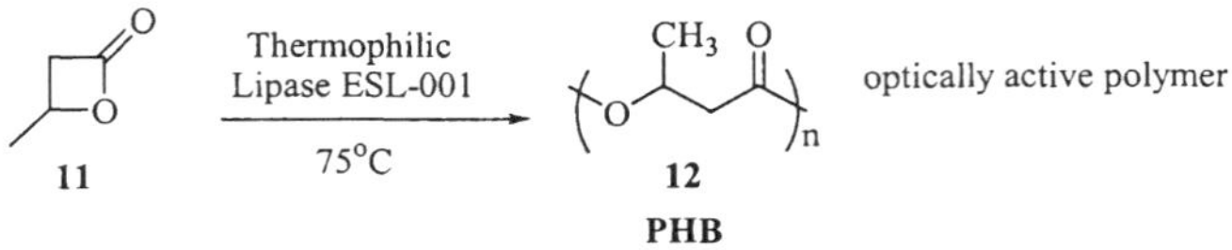

Scheme 4. Thermophilic Lipase ESL-001 catalyzes ring-opening polymerization.

Reactions for the ring-opening polymerization were performed in either bulk or in isooctane with substantial yields in both situations. Due to the unique thermal stability of the enzyme, the polymerization could be carried out at temperatures ranging from 60 to 80°C, which greatly reduced the reaction time. It is important to note that a trace amount of water in the system acted as an initiator.

Most of the polymerization results under different conditions are summarized in Table III. It was found that polymerizations performed in bulk afforded higher molecular weights than those performed in isooctane. Also, higher monomer conversions were achieved in bulk. When performed in bulk at 60°C with prolonged reaction time, the polymerization gave a polymer with a relatively lower molecular weight.

Table III. Lipase ESL-001-02 Catalyzed Ring-Opening Polymerization of β-Butyrolactone[a].

Entry	*Temp (°C)*	*Solvent*	*Time (h)*	*Yield (%)*	M_w[b]	M_w/M_n[b]	*ee (%)*
1	*80*	*isooctane*	*66*	*51*	*2800*	*2.5*	*37*
2	*80*	*bulk*	*66*	*62*	*3900*	*2.8*	*27*
3	*70*	*isooctane*	*94*	*58*	*2900*	*2.9*	*25*
4	*70*	*bulk*	*94*	*65*	*3800*	*2.9*	*20*
5	*60*	*bulk*	*140*	*62*	*2300*	*2.7*	*23*

[a] The polymerization was carried out in bulk at 75°C for 108 h, using 0.720 g of β-butyrolactone and 8.4 mg of the enzyme.

[b] Determined by GPC.

The PHBs obtained from ESL-001-02 catalyzed ring-opening polymerization were found to be optically active. They showed the same sign of optical rotation with that of natural PHB, which contains only (*R*)-hydroxybutyrate (HB) repeating units. This suggested that the polymers from the ring-opening polymerization were enriched with R-configuration repeating units. To evaluate the stereoselectivity, the resulting polymers were degraded by controlled methanolysis as illustrated in Scheme 5.

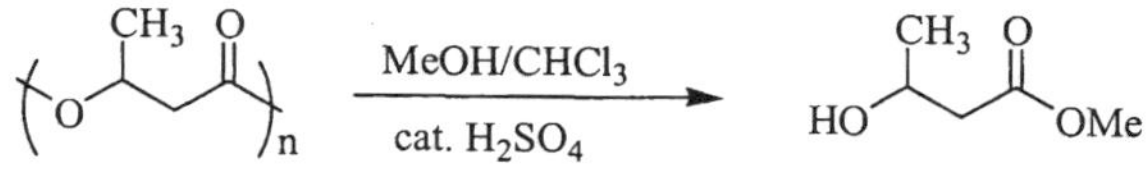

Scheme 5. Polymer degradation by controlled methanolysis in the determination of enatiomeric excess.

According to the mechanism proposed for lipase catalyzed polymerization of lactones, the polymerization was initiated by the reaction of enzyme with the lactone to from an acyl-enzyme intermediate. The intermediate then reacted with either a nucleophile to accomplish the initiation or with the hydroxy group of a growing polymer chain to continue the propagation. The formation of *R*-enriched polymers can be attributed to the rate difference between the reaction of the lipase with (*R*)- or (*S*)-β-butyrolactone and/or the rate difference between the reaction of the acyl-enzyme intermediate with *R*- or *S*-configuration chain ends.

Six other lipases from the thermophlic enzyme library CLONEZYME™ ESL-001 were also examined. For comparison, the polymerizations were all conducted in bulk at 75°C for 108 h. The results are presented in Table IV.

Table IV. Ring-Opening Polymerization of β-Butyrolactone Catalyzed by Thermophilic Lipase Library ESL-001[a].

Enzyme	*Yield (%)*	M_w[b]	M_w/M_n[b]	$[\alpha]^{25}_{365}$	*ee (%)*
ESL-001-01	70	2700	2.4	+1.7	12
ESL-001-02	63	2400	2.3	+5.1	31
ESL-001-03	52	1200	2.2	+3.8	21
ESL-001-04	67	2600	2.4	+1.5	11
ESL-001-05	78	2100	2.7	+2.1	15
ESL-001-06	83	2700	3.0	+1.0	6
ESL-001-07	41	900	2.4	+5.6	40

[a] The polymerization was carried out in bulk at 75°C for 108 h, using 0.720 g of β-butyrolactone and 8.4 mg of the enzyme.

[b] Determined by GPC.

As shown in Table IV, ESL-001-02 turned out to be the best candidate catalyst from the viewpoint of both molecular weight and stereoselectivity. The thermophilic lipase-catalyzed stereoselective ring-opening polymerization of racemic β-butyrolactone presents a unique approach to the synthesis of optically active PHB enriched *R*- repeating units.

Polycaprolactone-Modified HEC by Lipase-Catalyzed Ring-Opening Polymerization

Polyesters and polysaccharides represent two traditional and most useful hydrophobic and hydrophilic polymers, respectively. Polysaccharides such as cellulose, in its various forms, constitute about half of all of the polymers consumed industrially in the world. Polyesters such as poly(∊-caprolactone) and poly(lactic acids) play important roles in the design of biodegradable polymers containing polyesters and polysaccharides that impart novel physical properties (*28*) to materials which can be used in injection molding operations for shaped articles, biodegradable detergents, compatibilizers for polymer blends, and water-repellent materials or oil absorbents.

It has been shown that HEC, as an organic solvent-insoluble polysaccharide derivative, when deposited as a film, can be grafted by ∊-caprolactone in bulk using lipases derived from porcine pancreatic lipase (PPL) and from a thermophilic lipase from the CLONEZYME™ ESL-001 library (29) which was shown above to be effective for the formation of optically active (*R*-enriched poly(3-hydroxybutyrate) Scheme 6.

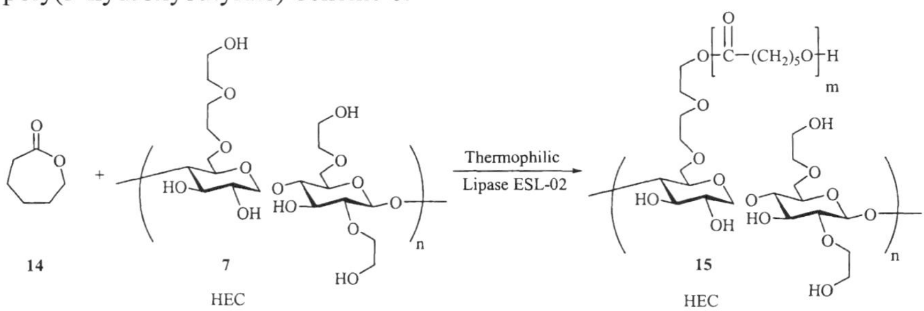

Scheme 6. Polycaprolactone-modified HEC via lipase-catalyzed ring-opening polymerization.

FT-IR spectroscopy was used to identify whether enzymatic grafting of HEC occurred. A comparison of the spectrum of PPL lipase-treated HEC (3days) to that of parent HEC at the same reaction condition without lipase showed a new peak in the region around 1730 cm^{-1} corresponding to a C=O group of the ester in the treated film. The degree of substitution (DS) designated the average number of hydroxyl groups on the anhydroglucose ring that have been treated with ∊-caprolactone. The DS value was determined by taking the integration of all of the nonexchangable protons of each anhydroglucose unit (7 H for a pyranose ring and 10 H for ethylene oxide units with degree of substitution of 2.5 in parent HEC) and comparing it to the two α-H in ∊-caprolactone.

For HEC modified in the presence of PPL at 60°C for 3 days, 74 monomeric ∊-caprolactone units were grafted on the HEC. That indicated ca. 0.18 monomers were associated with each anhydroglucose unit. The low substitution was due to the small amount of accessible hydroxyl groups on the surface of the films exposed to enzymes. Graft copolymerization of ∊-caprolactone was screened against six enzymes in a thermophilic CLONEZYME™ lipase library and the results are illustrated in Table V.

Table V. Degree of Substitution (DS) and Degree of Polymerization (DP) Resulting From Lipase-Catalyzed ∊-Caprolactone HEC Backbone.

	ESL-01	*ESL-02*	*ESL-03*	*ESL-04*	*ESL-05*	*ESL-06*	*PPL*
DS	*0.17*	*0.32*	*0.21*	*0.23*	*0.10*	*0.16*	*0.18*
DP	*4*	*10*	*8*	*15*	*10*	*3*	*n.d.*

This work illustrates the potential of lipase-catalyzed ring-opening graft copolymerization to graft hydrophobic polyesters onto hydrophilic cellulose-based polymers.

Lithium Chloride as Catalyst for the Ring-Opening Polymerization of Lactide in the Presence of Hydroxyl-Containing Compounds

There has been an increasing interest in the development of simple efficient catalytic systems for the ring-opening polymerization of lactide, which provides a convenient route to homo- and copolymers of polylactide (PLA), a family of important biomaterials widely used in clinical and biomedical research (*30,31*). A number of catalysts have been used to effect the ring-opening polymerization of lactide including stannous octoate (*32-34*), alkali-metal hydrides (*35-38*), aluminum and zinc compounds (*39-45*), as well as other miscellaneous reagents such as $BuSnCl_3$ (*46*), tin-substituted silica molecular sieves (*47*), and Grignard reagents. Trivalent lanthanide alkoxides were also reported to effect the polymerization of lactide (*48-52*). The main issue surrounding the use of those various catalysts is two fold: 1) the removal of the catalyst or the purification of the compound is inefficient and 2) the catalysts, such as stannous octoate, invoke cytotoxicity. In order to develop an efficient catalyst as well as one that would serve biomedical applications, a number of catalysts were screened and LiCl proved to be very economical and virtually represented a nontoxic reagent which could catalyze the ring-opening polymerization of lactide in the presence of hydroxyl-containing compounds effectively (*53*). A number of amphiphilic PLA copolymers were prepared by utilizing poly(ethyleneglycol) (PEG), hydroxyethylcellulose (HEC), and hydroxypropylcellulose (HPC) as macroinitiators. Scheme 7 illustrates the ring opening-polymerization of PLA applied to both ethyleneglycol (EG) and methyl-α-D-glucopyranoside.

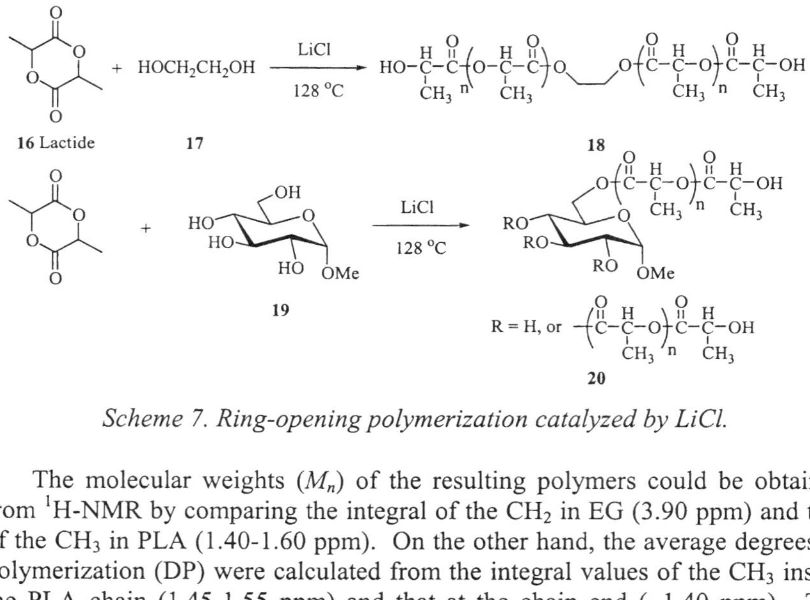

Scheme 7. Ring-opening polymerization catalyzed by LiCl.

The molecular weights (M_n) of the resulting polymers could be obtained from ^{1}H-NMR by comparing the integral of the CH_2 in EG (3.90 ppm) and that of the CH_3 in PLA (1.40-1.60 ppm). On the other hand, the average degrees of polymerization (DP) were calculated from the integral values of the CH_3 inside the PLA chain (1.45-1.55 ppm) and that at the chain end (~1.40 ppm). The average degrees of substitution (DS) were found to be very close to 2, indicating that both hydroxy groups in EG were involved in the initiation step. With a $[M]_o/[I]_o$ ratio of 50:1 ($[M]_o$, the concentration of the monomer; $[I]_o$, the concentration of EG molecules), the polymerization carried out at 128°C for 48 h afforded a polymer with a molecular weight of 6060. Assuming DS=2, the DP for the PLA chains in this polymer were estimated to be 21.

In the case of methyl-α-D-glucopyranoside the average DS, estimated similarly from the corresponding molecular weights (based on the integral of the

OCH_3 in the glucoside unit and that of the CH_3 in PLA) and average DP (by end-group analysis) were in the range of 2.7-3.2. This suggested that not all of the four hydroxyl groups in MGlc were equally involved in the ring-opening polymerization. This was not surprising because the four hydroxyl groups actually were of different reactivity. It was difficult, however, to estimate the proportions of the different types of the hydroxyl groups involved, either by ^{1}H- or ^{13}C-NMR. The results along with the theoretical M_n values calculated from the $[M]_o/[I]_o$ ratios are summarized in Table VI.

Table VI. LiCl Catalyzed Ring-Opening Polymerization of Lactide[a].

Entry	*Initiator*	$[M]_o/[I]_o$[b]	M_n (calcd)[c]	M_n (NMR)	M_n (GPC)	M_w/M_n (GPC)
1	EG	10 : 1	1502	960	1100	2.2
2	EG	20 : 1	2942	1940	2100	2.4
3	EG	50 : 1	7262	6060	6500	2.7
4	MGlc	12 : 1	1922	1090	2300	2.3
5	MGlc	20 : 1	3074	2050	3600	2.6
6	MGlc	24 : 1	3650	2870	4200	2.5
7	MGlc	30 : 1	4514	3560	4900	2.8

a Polymerization reactions were carried out at 128°C for 10 h unless otherwise indicated.

b $[M]_o$, the concentration of the monomer; $[I]_o$, the concentration of the initiator molecules.

c Calculated from $M_n = 144[M]_o/[I]_o + M.W.$ (initiator).

Macroinitiators such as PEG, HEC, and HPC were also used in the LiCl-catalyzed polymerization of lactide as shown on Scheme 8.

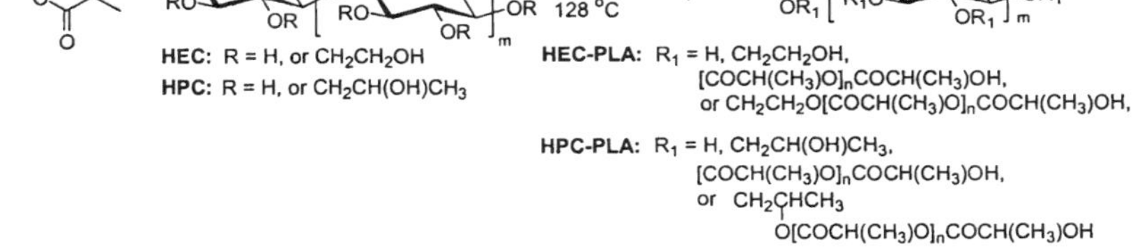

Scheme 8. Macroinitiators PEG, HEC, and HPC in the ring-opening polymerization catalyzed by LiCl.

A PLA-PEG-PLA polymer with a molecular weight of 3100 was obtained when PEG (M_n~600) was employed with a [lactide]/[PEG] ratio of 24 : 1. In the cases of HEC and HPC, *N,N*-dimethylacetamide (DMAC) was utilized as solvent to make the reaction system homogeneous. Novel amphiphilic polymers, HEC-PLA and HPC-PLA were afforded after the ring-opening reaction. When the [lactide]/[glucose] ratio was 9 : 1, the amounts of lactide grafted onto every glucose unit in the backbone of HEC and HPC were estimated, respectively from ^{1}H-NMR to be approximately 6.3 (from the integral of CH_2CH_2 in HEC and that of CH_3 in PLA) and 5.1 (from the integral of CH_3 in HPC and that of CH_3 in PLA). Compared with the parent HEC and HPC, these polymers were no longer water-soluble, but could be readily dissolved in chloroform as a result of the grafting.

Thermophilic enzymes have been shown to facilitate the modification of hydroxy-containing compounds through a transgalactosylation reaction with D-lactose as the donor. They have also been used in the ring-opening polymerization of β-butyrolactone and in the modification of HEC by grafting ϵ-caprolactone through similar means. They are able to induce a stereoselective polymerization as illustrated in the β-butyrolactone example, prove to be stable under increased temperatures, and in the cases with glycosidases do not require a UDP-cofactor which decreases the cost significantly. LiCl was discovered to be an important nontoxic polymer catalyst with the potential to replace conventional cytotoxic ones such as stannous octoate. Under high temperature conditions, LiCl proved successful in the ring-opening polymerization of lactide (3,6-dimethyl-1,4-dioxane-4,6-dione) for the modification of hydroxy containing compounds such as HEC, HPC and PEG.

References:

1. Internet address: www.findarticles.com/m0FVP/2_256/55158395/p1/article. jhtml.
2. Internet address: http://www.biocat.com.
3. Li, J.; Wang, P. G. *Tetrahedron Lett.* **1997**, *38*, 7967-7970.
4. Fang, J. W.; Xie, W. H.; Li, J.; Wang, P. G. *Tetrahedron Lett.* **1998**, *39*, 919-922.
5. Nilsson, K. G. I. In *Modern Methods in Carbohydrate Synthesis;* Khan, S.H., O'Neill, R.A., Eds.; Harwood Academic Publishers: The Netherlands, 1996; pp 518-547.
6. Gijsen, H. J. M.; Qiao, L.; Fitz, W.; Wong, C. H. *Chem. Rev.* **1996**, *96*, 443-473.
7. Fernandez-Mayoralas, A. *Top. Curr. Chem.* **1997**, *186*, 1-20.
8. Murata, T.; Usui, T. *Biosci. Biotech. Biochem.* **1997**, *61*, 1059-1066.
9. Crout, D. H.; Vic, G. *Curr. Opin. Chem. Biol.* **1998**, 98-111

10. Li, J.; Cheng, H. N.; Nickol, R. G.; Wang, P. G. *Carb. Res.* **1999**, *316*, 133-137.
11. Bahulekar, R.; Tokiwa, T.; Kano, J.; Matsumura, T.; Kojima, I.; Kodama, M. *Biotechnol. Technol.* **1998**, *12*, 721-724.
12. Usui, T.; Yamaoka, N.; Matuda, K.; Tuzimura, K.; Sugiyama, H.; Seto, S. *J. Chem. Soc., Perkin Trans. 1* **1973**, 2425-2432.
13. Scaman, C. H.; Lipari, F.; Herscovics, A. **Glycobiology 1996**, *6*, 265-270.
14. Doi, Y. *Microbial Polyesters*; VCH Publishers Inc.: New York, NY 1990.
15. Reusch, R. N. *Proc. Soc. Exp. Bio. Med.* **1989**, *191*, 377-381.
16. Reusch, R. N.; Sadoff, H.L. *Proc. Natl. Acad. Sci. U.S.A.* **1988**, *85*, 4176-4180.
17. Muller, H. M.; Seebach, D. *Angew. Chem., Int. Ed, Engl.* **1993**, *32*, 477-502.
18. Yim, K. S.; Lee, S. Y.; Chang, H. N. *Biotechnol. Bioeng.* **1996**, *49*, 495-503.
19. Uyama, H.; Kurioka, H.; Sugihara, J.; Kobayashi, S. *Bull. Chem. Soc. Jpn.* **1996**, *69*, 189-193.
20. Uyama, H.; Takeya, K.; Hoshi, N.; Kobayashi, S. *Macromolecules* **1995**, *28*, 7046-7050.
21. Uyama, H.; Kikuchi, I. I.; Takeya, K.; Kobayashi, S. *Acta Polym.* **1996**, *47*, 357-360.
22. Henderson, L. A.; Svirkin, Y. Y.; Gross, R. A.; Kaplan, D.; Swift, G. *Macromolecules* **1996**, *29*, 7759-7766.
23. Endo, M.; Aida, T.; Inoue, S. *Macromolecules* **1987**, *20*, 2982-2988.
24. Robertson, D. E.; Mathur, E. J.; Swanson, R. V.; Marrs, B. L.; Short, J. M. *SIM News* **1996**, *46(1)*, 3.
25. Govaraham, C. P.; Margolin, A. L. *Chem. Ind.* **1995**, *17*, 689-693.
26. Brennan, M. B. *Chem. Eng. News* October 4, 1996, p 31.
27. Xie, W. H.; Li, J.; Chen, D. P.; Wang, P. G. *Macromolecules* **1997**, *30*, 6997-6998.
28. Li, S.; Vert, M. In *Degradable Polymers. Principles & Apllications;* Scott, G., Gilead, D., Ed.; Chapman & Hall: London, 1995 and references herein.
29. Li, J.; Xie, W. H.; Cheng, H. N.; Nickol, R. G.; Wang, P. G. *Macromolecules* **1999**, *32*, 2789-2792.
30. Langer, R. *Science*, **1990,** *249*, 1527-1533.
31. Barrera, D. A.; Zylstra, E.; Lansbury, P. T.; Langer, R. *J. Am. Chem. Soc.*, **1993**, *115*, 11010-11011.
32. Sawhney, A.; Pathak, C. P.; Hubbell, J. A. *Macromolecules* **1993**, *26*, 581-587.
33. Zhang, X.; MacDonald, D. A.; Goosen, M. F. A.; McAuley, K. B. *J. Polym. Sci. Part A: Polym. Chem.* **1994**, *32*, 2965-2970.
34. Han, D. K.; Hubbell, J. A. *Macromolecules* **1996**, *29*, 5233-5235.
35. Jedlinski, Z.; Walach, W. *Mokromol. Chem.* **1991**, *192*, 2051-2057.

36. Li, S. M.; Rashkov, I.; Espartero, J. L.; Manolova, N.; Vert, M. *Macromolecules* **1996**, *29*, 57-62.
37. Rashkov, I.; Manolova, N.; Li, S. M.; Espartero, J. L.; Vert, M. *Macromolecules* **1996**, *29*, 50-56.
38. Li, S.; Anjard, S.; Rashvok, I.; Vert, M. *Polymer* **1998**, *39*, 5421-5430.
39. Dubois, P.; Jacobs, C.; Jerome, R.; Teyssie, P. *Macromolecules* **1991,** *24*, 2266-2270.
40. Xu, J.; McCarthy, S. P.; Gross, R. A. *Macromolecules* **1996,** *29*, 4565-4571.
41. Trollsas, M.; Atthoff, B.; Claesson, H.; Hedrick, J. L. *Polym. Prepr.* **1998,** *39(2)*, 232-233.
42. Schwach, G.; Coudane, J.; Engel, R.; Vert, M. *Polym. Bull.* **1996,** *37*, 771-776.
43. Schwach, G.; Coudane, J.; Engel, R.; Vert, M. *Polym. Int.*, **1998,** *46*, 177-182.
44. Kricheldorf, H. R.; Damrau, D. O.; *Macromol. Chem. Phys.*, **1998,** *199*, 1747-1752.
45. Kricheldorf, H. R.; Berl, M.; Scharnagl, N. *Macromolecules* **1988,** *21*, 286-293.
46. Kricheldorf, H. R.; Mahler, A. *Polymer* **1996,** *37*, 4383-4388.
47. AbdelFattah, T. M.; Tarek M.; Pinnavaia, T. J. *Chem. Commun.* **1996**, 665-666.
48. Kricheldorf, H. R.; Lee, S.-R. *Polymer* **1995,** *36*, 2995-3003.
49. Stevels, W. M.; Ankone, M. J. K.; Dijkstra, P. J.; Feijen, J. *Macromolecules* **1996,** *29*, 3332-3333.
50. Stevels, W. M.; Ankone, M. J. K.; Dijkstra, P. J.; Feijen, J. *Macromol. Chem. Phys.* **1995,** *196*, 1153-1161.
51. McLain, S. J.; Ford, T. M.; Drysdale, N. E. *Polym. Prepr.* **1992,** *33*, 463-464.
52. Le Borgne, A.; Pluta, C.; Spassky, N. *Macromol. Rapid. Commun.* **1994,** *15*, 955-960.
53. Xie, W. H.; Chen, D.-P.; Fan, X. H.; Li, J.; Wang, P. G. *J. Polymer Sci.* **1999**, 3486-3491.

Chapter 17

Enzyme-Catalyzed Reactions of Polysaccharides

H. N. Cheng and Qu-Ming Gu

Hercules Incorporated Research Center, 500 Hercules Road, Wilmington, DE 19808–1599

Enzyme-catalyzed reactions have been used to modify polysaccharides in the industrial context in order to improve the end-use properties. Typical reactions include molecular weight reduction, addition of charged or polar functional groups, hydrophobic modification, derivatization to form reactive functionalities, and synthesis of reactive oligomers. Many of these reactions are reviewed in this paper, using primarily hydrolases and oxidoreductases.

Introduction

Polysaccharides are important articles of commerce. They are used variously as thickeners, gelling agents, stabilizers, interfacial agents, flocculants and encapsulants, in applications such as foods, coatings, construction, paper, pharmaceuticals, and personal care *(1)*. For some applications, polysaccharides need chemical modifications in order to improve their end-use properties. Indeed, many chemical reactions have been devised to modify the polysaccharides *(2)*. A substantial understanding has been obtained of the structure-property correlations for these modified polysaccharides.

Whereas chemical means have been successful in modifying polysaccharides, enzymatic reactions may be useful to complement and to supplement these means. Since polysaccharides are natural materials, many enzymes are available in nature that can carry out biotransformations *(3)*. In addition, enzymes are often specific (regio-, enantio-, or chemospecific); thus many enzyme-catalyzed reactions can provide products with well-defined or stereospecific structures *(4)*. Another benefit is the mild conditions under which most enzymatic reactions can be done, often leading to products with less color or odor, and decreased occurrence of by-products.

In this paper, several useful enzyme-catalyzed reactions of polysaccharides are reviewed. For convenience, these reactions are grouped into five categories: 1) molecular weight reduction, 2) addition of charge, 3) addition of polar group, 4) hydrophobic modification, and 5) formation of reactive oligomers. In order to limit the scope of this paper, only derivatives of cellulose and guar are covered, with an emphasis on the work done at Hercules Incorporated.

Results and Discussion

Molecular Weight Reduction

In many applications, it has been found necessary to lower the molecular weights of polysaccharides. In commercial processes, low molecular weights are often achieved via two chemical means, namely, hydrolysis with mineral acids and oxidation with hydrogen peroxide *(5)*. However, these methods sometimes produce colored products or undesirable by-products, e.g., oxidized species and/or fragments with very low molecular weights. Interestingly, the use of enzymes can often obviate these difficulties.

An example of a low-molecular-weight polysaccharide is guar. It is known from the literature that in the presence of a mannanase, guar can be degraded in molecular weight *(6)*. Low-molecular-weight guar has been reported to be useful as a dietary fiber and as a medically efficacious substance *(7)*. In a combinatorial experiment *(8)*, the degradation of guar was reported in the presence of four enzymes (lipase, hemicellulase, pectinase, and protease). As expected, hemicellulase showed a large decrease in molecular weight. The combination of hemicellulase and protease gave the lowest molecular weights achieved. This result is perhaps not surprising in view of the fact that guar contains up to 7% proteins. Thus, the protein in guar is likely to contribute to the observed molecular weight of guar.

The use of enzymes to degrade cellulosic derivatives is well known *(9)*. A recent example was reported by Sau *(10)*. Through the use of the enzyme cellulase, a number of cellulosic derivatives were degraded to lower molecular weights. The resulting materials are reported to be biostable; i.e., their resistance towards biological degradation has been increased.

Addition of a Polar Substituent

The nature of a substituent on cellulose has a large effect on the properties of the cellulosic derivative. Thus, hydroxyethylcellulose (HEC) is soluble in water at all temperatures, but methylcellulose starts to gel at about 45°C, and hydroxypropylcellulose precipitates from water at 40-45°C. An example of the effect of polar groups on solubility is shown for the following acetylation reaction.

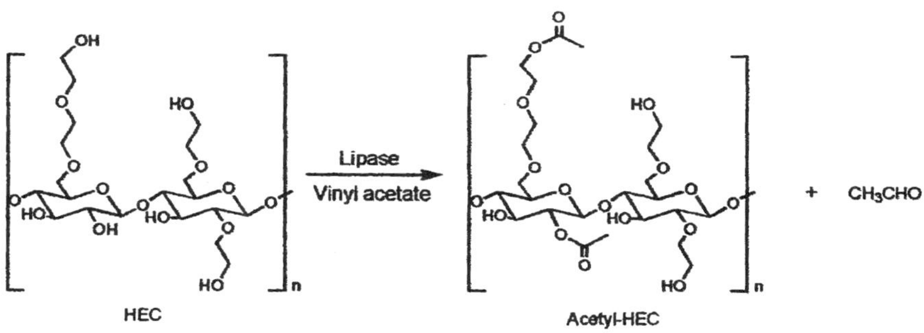

Figure 1. Lipase-catalyzed synthesis of acetylated HEC

The reaction is very facile and readily takes place in the presence of a lipase. HEC is water-soluble, but when acetylated, the resulting polymer becomes much less soluble in water.

A good example of the manipulation of substituents is the addition or removal of galactose from polysaccharide side chains. Galactose side chains can be removed from galactomannans through α-galactosidase *(11)*. Conversely, galactose can be added through the use of lactose and β-galactosidase. This can be achieved either at the chain ends of an oligosaccharide *(12),* or at the end of ethylene oxide units on HEC *(13)*. Another way to add the galactose to a polysaccharide is to use glycidyl galactose (*vide infra*).

Addition of Charge

For some industrial applications, a charged polysaccharide can be highly desirable. For example, an anionic polymer may bind specifically to cationic materials, whereas a cationic polymer may bind to anionic substrates. The co-addition of both a polycation and a polyanion may produce a polyelectrolyte complex (either stoichiometric or non-stoichiometric) which can be used as gelling agents, encapsulants, and flocculants *(14)*.

An example of the addition of a cationic charge to a polysaccharide is shown for the following reactions with carboxymethylcellulose (CMC), where X = $(CH_2)_n$, and $n \geq 2$.
As reported earlier *(15)*, the reaction with CMC (where R=H) could proceed at 5-20% yield for subtilisin Carlsberg, papain, and protease from *Aspergillus saitoi* in N,N-dimethylformamide (DMF) solvent. The reaction with CMC ester (R=methyl or ethyl) was, however, more facile, and higher yields were achieved.

An example of the addition of an anionic group is shown for the following acylation reaction. Lipase AK (*Pseudomonas sp.*, from Amano) was found to have excellent activity for this reaction, giving maleated guar.
A related reaction can be carried out using succinic anhydride. Earlier, guar had been shown to react with succinic anhydride to give succinated guar in the presence of a lipase *(8)*. The reaction is also possible with HEC, to give the corresponding succinated HEC.

Hydrophobic Modification

Hydrophobic modification is a common method to synthesize surfactants or rheology modifiers. For rheology modifiers, the hydrophobes associate with one another in aqueous solutions at low concentrations, and thereby increase the viscosity of the solution at low shear frequencies *(16)*. A number of hydrophobically modified water-soluble polymers are commercially available, e.g., HMHEC *(17)*, HEUR *(18)*, and HASE *(19)*.

An enzyme can also be used as a catalyst to put a hydrophobic unit onto a cellulose derivative. For example, a stearic ester can be grafted onto HEC through the use of vinyl stearate *(20)*. Since vinyl stearate is expensive, a more commercially viable route is to use methyl stearate. The latter reaction also works, except that the reaction rate is slower.

Formation of Reactive Oligomers

Enzymatic reactions can be used under mild reaction conditions to generate reactive functionalities from saccharides and polysaccharides. An example is the guar aldehyde where only an enzymatic process (through galactose oxidase) can produce the reactive aldehyde functionality in the C6 position of galactose

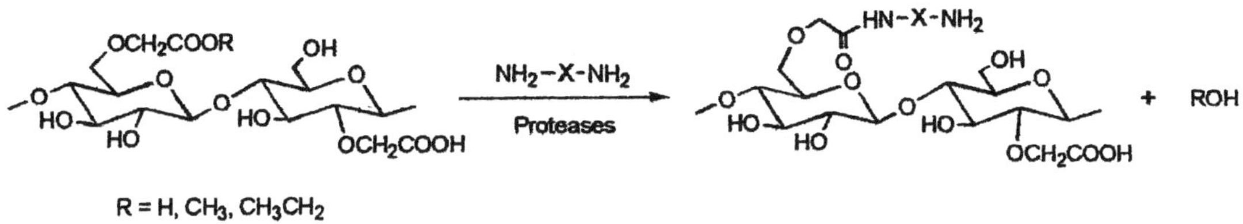

Figure 2. Aminolysis of CMC and CMC esters

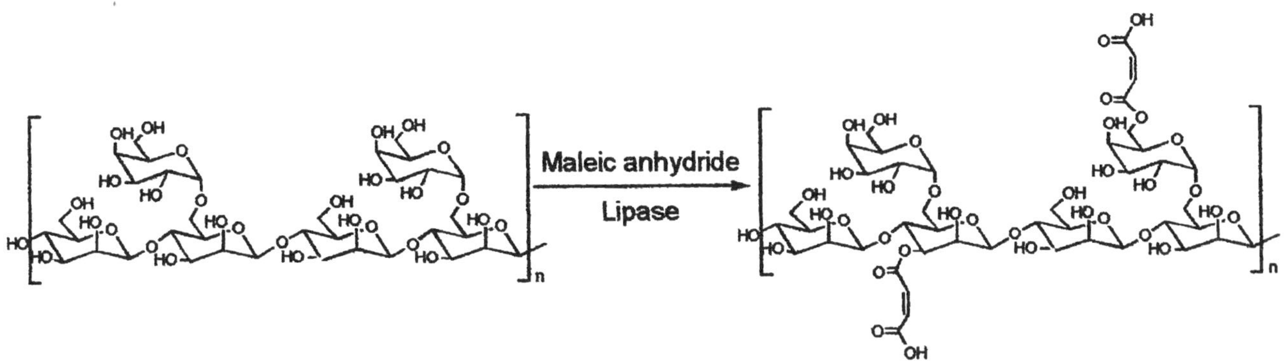

Figure 3. Synthesis of maleated guar

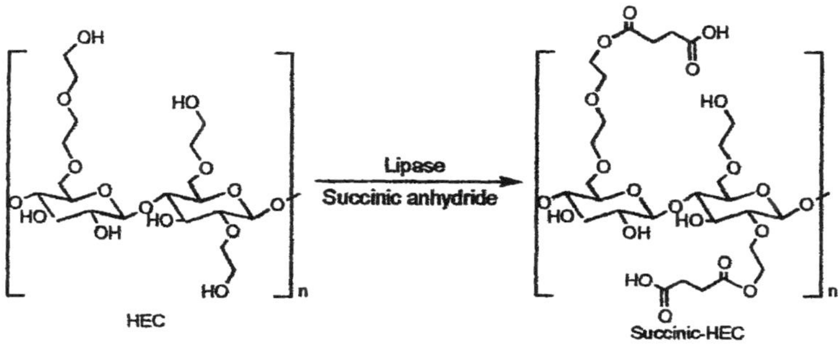

Figure 4. Synthesis of succinated HEC

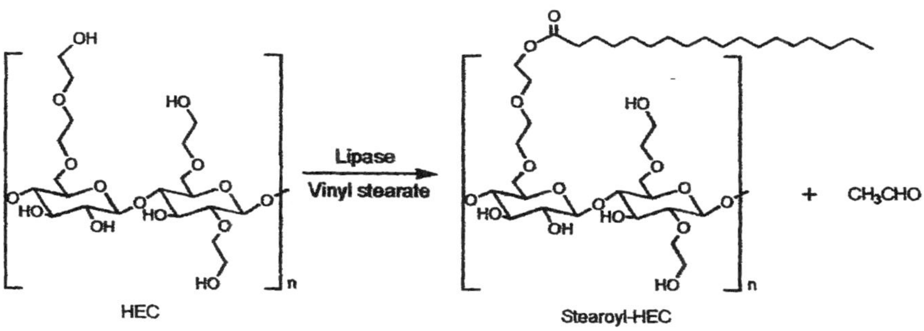

Figure 5. Hydrophobic modification of HEC

(21,22). A more general reaction is the use of vinyl acrylate and a lipase to add an acrylate functionality onto HEC. This HEC-acrylate can then be used for further reactions *(20)*.

A reactive compound can also be made from a monosaccharide. Thus, one can use lactose and β-galactosidase to form glycidyl galactose *(8)*.
The glycidyl galactose can then be used to react (via epoxide ring-opening reaction) with an amine or an alcohol. This is one way to add a galactose moiety to a polysaccharide.

An interesting structure can be made via the oxidation of N-allylamine to form the corresponding epoxide in organic solvents in the presence of an enzyme from *Aspergillus saitoi*. Under basic pH conditions, the compound automatically oligomerizes to a trimer product.

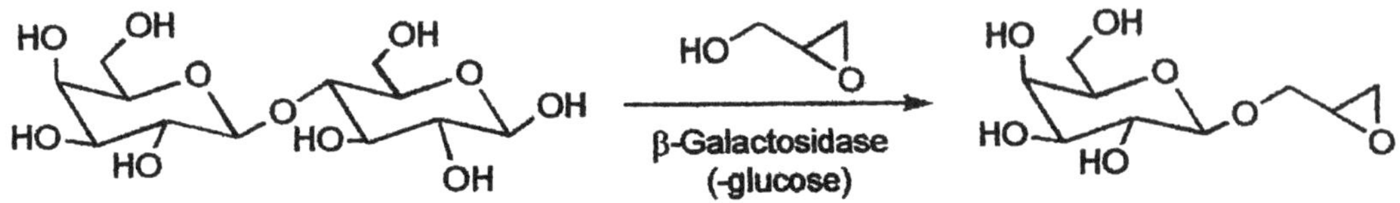

Figure 6. Synthesis of glycidyl galactose using β-galactosidase

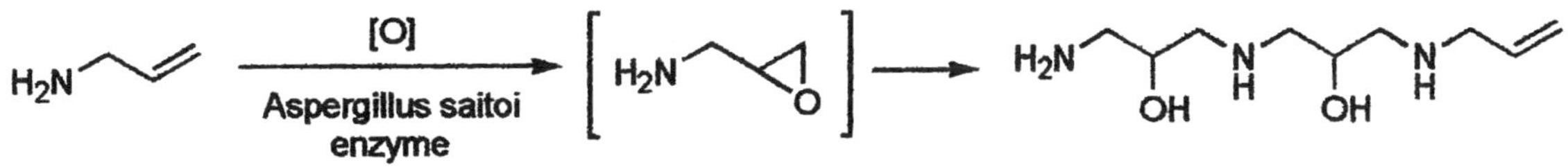

Figure 7. Oligomerization of allylamine using an epoxidase

Under neutral pH conditions in the presence of water, epoxidation of allylamine leads to the enantiomerically pure 3-amino-1,2-propanediol, which is a useful building block for many analogs of phosphatidic acids.

Comments

The polysaccharide modification reactions delineated above may be summarized in Table 1. Both hydrolase and oxidoreductase enzymes have been used in the presence of organic solvents, water, or mixed solvents.

In order to achieve optimal yield, each reaction needs to be separately studied, and the reaction parameters systematically varied to produce the best results. The key parameters are the enzyme used, the solvent medium, reactant and enzyme concentrations, and temperature (in that order). Enzyme screening is a major part of this optimization process. Recent developments in directed evolution and gene shuffling will be of great help in producing improved enzymes *(23)*.

Table 1. Summary of enzyme-catalyzed reactions cited in this work

Polymer	Reaction Type	Reactant	Enzyme	Solvent
HEC	MW control	-	cellulase	water
	charge	succinic anhydride	lipase	DMAc
	hydrophobe	vinyl stearate	lipase	DMAc
	polar group	vinyl acetate	lipase	DMAc
	reactive functionality	vinyl acrylate	lipase	DMAc or DMF
CMC	charge	diamine	lipase or protease	DMF/water
Guar	MW control	-	mannanase	water
	reactive functionality	oxygen	galactose oxidase	water

EXPERIMENTAL

Materials

Vinyl stearate, vinyl acetate, succinic anhydride, N-allylamine, maleic anhydride, N,N-dimethylacetamide (DMAc) and t-butyl methyl ether were obtained from Aldrich. Lipases from *Pseudomonas fluorescens* (Lipase AK) and from *Pseudomonas cepacia* (Lipase P) were obtained from Amano Enzyme USA. Lipase from *Candida antarctica* (Novozym® 435) came from Novozymes A/S. Protease from *Aspergillus saitoi* was purchased from Sigma Co. Neutral guar, cationic guar, hydroxyethylcellulose (HEC), and carboxymethylcellulose (CMC) used for the study are the products of Hercules Incorporated.

Brookfield Viscosity Measurements

The Brookfield viscosities of aqueous solutions of the modified polysaccharides were measured at 1% concentration at pH 6.5 and room temperature. A LV type Brookfield viscometer was used with the spindle speed set at 30 rpm.

Preparation of Succinated HEC

A sample of HEC (Natrosol® Pharm-250MR, Hercules Incorporated, 5 g) was suspended in 50 ml of DMAc. To this solution 1 g of succinic anhydride and 0.2 g of Lipase P (*Pseudomonas cepacia*) were added. The mixture was warmed up to 50°C, which turned into a slurry in about 10 minutes. The resulting slurry was incubated at 50°C for 16 hours and then treated with acetone to precipitate the product. After washing with acetone and air-drying, 4.9 g of HEC acetate was obtained as a white solid. IR: 1750 cm-1. The control sample (without enzyme) gave also a band at 1750 cm-1 with about 30% of the intensity.

Preparation of Maleated Guar

Flake guar (Hercules Incorporated, 1 g) and maleic anhydride (0.5 g) were mixed in 10ml t-butyl methyl ether. Lipase P (*Pseudomonas cepacia* , 0.2 g) was added and the mixture was stirred for 48 hrs at 50°C. After the mixture was cooled down, isopropanol (20 ml) was added, and the solid residue washed with isopropanol three times and dried under vacuum to give 0.95 g of the maleated guar. IR spectral analysis of the product showed the presence of a strong ester band at 1755 cm^{-1}, whereas the spectrum of the control sample (without enzyme) gave a much smaller band at 1755 cm^{-1}.

Preparation of Acetylated HEC

A sample of HEC (Natrosol® Pharm-250MR, Hercules Incorporated, 5 g) was suspended in 50 ml of DMAc, followed by the addition of 1 g of vinyl acetate and 0.5 g of Lipase P (*Pseudomonas cepacia*). After thorough mixing in 10 minutes, the resulting mixture gradually turned into a slurry. The mixture was subsequently incubated at 50°C for 24 hours. The slurry was treated with isopropanol (IPA) to precipitate the product. After washing with IPA and air-drying, 4.5 g of HEC acetate was obtained as a white solid. IR: 1745 cm-1. The control sample (without enzyme) gave also a band at 1745 cm-1 but with only 10% of the intensity. The acetylated HEC was visibly less soluble in water, giving a cloudy solution at 1%.

Hydrophobic Modification of HEC

An HEC sample (Natrosol® Pharm-250MR, Hercules Incorporated, 4 g) was suspended in 20 ml of DMAc, followed by the addition of 1 g of vinyl stearate and 0.5 g of Lipase P. After thorough mixing in 5-10 minutes, the resulting mixture turned into a slurry that was subsequently incubated at 50°C for 48 hours. The yellowish slurry was treated with acetone/IPA (1:1) to give precipitates. After washing with IPA and air-drying, 3.8 g of the hydrophobically modified HEC was obtained as a white solid. IR: 1750 cm^{-1}. The DS (degree of substitution) was estimated to be 0.1, based on IR and comparing with the carbonyl intensities of known mixtures of methyl palmitate/cationic guar. The Brookfield viscosity of the product was significantly higher than the starting materials, indicating that the hydrophobe has been grafted onto the HEC. (It is known that the Brookfield viscosity of a hydrophobically modified polysaccharide varies with the degree of substitution of the hydrophobe.)

Oligomerization of N-Allylamine

N-Allylamine (0.4 ml) was dissolved in a mixture of 4 ml of toluene and 30 μl water, and an *Aspergillus saitoi* protease containing epoxidase activity (Sigma, 20 mg) was added. The mixture was stirred at 37°C for 16 hours when the solution became cloudy and the viscosity increased. TLC indicated a major new spot formed. The material was purified by flash chromatography (silica gel, dichloromethane/methanol, 8:2) to give a yellowish liquid product. The yield was 0.2 g. The trimer structure was confirmed as the one shown in Figure 7. ^1H-NMR (D_2O, 300MHz): 5.80-5.65 ppm (m, 1H, -CH=), 4.95-4.80 ppm (d, 2H, CH2=), 3.85-3.70 ppm (m, 2H, N-CH_2-C=C), 3.65-3.00 ppm (m, 10H); ^{13}C (D_2O, 75.5MHz) 135.7 ppm (-CH=), 116.9 ppm (CH_2=), 72.8 ppm (C-OH), 72.5 ppm (C-OH). MS (m/z): 204 (M^++1).

Acknowledgements. The authors wish to acknowledge the technical assistance of S. Mital, G. F. Tozer, and A. J. Walton, and helpful discussions with colleagues at Hercules Incorporated. This is Hercules Incorporated Contribution Number 2388.

References

1. For example, (a) *Industrial Gums*, 3rd Ed; Whistler, R. L.; BeMiller, J. N., Eds.; Academic Press: San Diego, CA, 1993. (b) *Food Polysaccharides and their Applications*; Stephen, A. M., Ed.; M. Dekker; New York, NY, 1995. (c) *Application of Polymers in Foods*; Cheng, H. N.; Cote, G. L.; Baianu, I. C., Eds.; (Macromolecular Symposia 140); Wiley-VCH; Weinheim, Germany, 1999.
2. For example: (a) *Cellulose Derivatives. Modification, Characterization, and Nanostructures*; Heinze, T. J.; Glasser, W. G., Eds.; (ACS Symp. Ser. 688); American Chemical Society, Washington, DC, 1998. (b) *Industrial Polysaccharides*; Yalpani, M., Ed.; Elsevier, Amsterdam, NL, 1987.
3. For example, (a) *Enzymes in Industry*; Gerhartz, W., Ed.; VCH; Weinheim, Germany, 1990, Chapter 4. (b) *Industrial Enzymology*; Godfrey, T.; West, S., Eds.; Macmillan; London, UK, 1996. (c) *Source Book of Enzymes*; White, J. S.; White, D. C.; CRC Press, Boca Raton, FL, 1997.
4. For example, (a) *Biocatalysis for Fine Chemicals Synthesis*; Roberts, S. M., Ed.; Wiley, Chichester, UK, 1999. (b) *Enzyme Catalysis in Organic Synthesis*; Drauz, K.; Waldemann, H., Eds.; VCH, Weinheim, Germany, 1995. (c) *Biotransformations in Organic Chemistry*, 3rd Ed.; Faber, K.; Springer, Berlin, Germany, 1997.
5. For example: (a) Vink, H. *Makromol. Chem.* **1963**, *67*, 105. (b) *Comprehensive Cellulose Chemistry*; Klemm, D.; Philipp, B.; Heinze, T.; Heinze, U.; Wagenknecht, W. ; Wiley-VCH, Weinheim, Germany, 1998. (c) *Cellulose and Cellulose Derivatives*; Ott, E.; Spurlin, H. M.; Grafflin, M.W., Eds.; Interscience, New York, NY, 1954.
6. For example: (a) Tayal, A.; Kelly, R. M.; Khan, S. A. *Macromolecules* **1999**, *32*, 294. (b) Cheng, Y.; Prud'homme, R. K. *ACS Polym. Prepr.* **2000**, *41*(2), 1868.
7. For example: (a) Greenberg, N. A.; Sellman, D. *Cereal Foods World* **1998**, *43*, 703. (b) Alam, N. H.; Meier, R. *J. Pediatr. Gastroenterol. Nutr.* **2000**, *31*, 503. (c) Ishihara, N.; Chu, D.-C.; Akachi, S.; Juneja, L. R. *Poult. Sci.* **2000**, *79*, 689.
8. Cheng, H. N.; Q.-M. Gu in *Glycochemistry: Principles, Synthesis, and Applications*; Wang, P. G.; Bertozzi, C. R., Eds.; M. Dekker, New York NY, 2001, pp. 567-579.
9. For example, Saake, B.; Horner, S.; Puls, J *ACS Symp. Ser.* **1998**, *688*, 201.
10. (a) Sau, A. C. *U.S. Patent* 5,879,440, March 8, 1999. (b) Sau, A. C. *U.S. Patent* 5,989,329, November 23, 1999.

11. (a) MacCleary, B. V.; Bulpin, P. V. *U. S. Patent* 5,234,825, August 10, 1993. (b) Bulpin, P. V.; Gidley, M. J.; Jeffcoat, R.; Underwood, D. R. *Carbohydrate Polymers* **1990**, *12*, 155.
12. Gu, Q.-M. *J. Environ. Polym. Degrad.* **1999**, *7*, 1.
13. Li, J.; Cheng, H. N.; Nickol, R. G.; Wang, P. G. *Carbohydrate Res.* **1999**, *316*, 133.
14. For example, Tsuchida, E.; Kokufuta, E.; Dubin, P. L. *J. Macromol. Sci., Pure Appl. Chem.* **1994**, *A31*, 1.
15. Cheng, H. N.; Gu. Q.-M. *ACS Polym Prepr.* **2000**, *41*(2), 1873.
16. For example, *Associative Polymers in Aqueous Media*; Glass, J. E., Ed.; (ACS Symp. Ser. 765); American Chemical Society: Washington, DC, 2000.
17. For example: (a) Landoll, L. M. *J. Polym. Sci., Polym. Chem. Ed.* **1982**, *20*, 443. (b) Gelman, R. A.; Barth, H. G. *Adv. Chem. Ser.* **1986**, *213*, 101.
18. For example: (a) Emmons, W. D.; Stevens, T. E. *U.S. Patent* 4,079,028, March 14, 1978. (b) Glass, J. E. *Adv. Chem. Ser.* **1986**, *213*, 391.
19. For example: (a) Shay, G.D.; Eldridge, E.; Kail, E. *U.S. Patent* 4,514,552, 1985. (b) Shay, G. D.; Kravitz, F.K.; Brizgys, P.V.; Kersten, M.A. *U.S. Patent* 4,801,671, 1989. (c) Shay, G.D.; Kravitz, F. K.; Brizgys, P.V. *ACS Symp. Ser.* **1991**, *462*, 121.
20. Gu, Q.-M. *ACS Polym Prepr.* **2000**, *41*(2), 1834.
21. For example: (a) Yalpani, M.; Hall, L. D. *J. Polym. Sci., Polym. Chem. Ed.* **1982**, *20*, 3339. (b) Frollini, E.; Reed, W. F.; Milas, M.; Rinaudo, M. *Carbohydrate Polymers* **1995**, *27*, 129. (c) Chiu, C.-W.; Jeffcoat, R.; Henley, M.; Peek, L. *U.S. Patent* 5,554,745, September 10, 1996.
22. (a) Brady, R. L.; Leibfried, R. L. *U.S Patent* 6,022,717, February 8, 2000. (b) Brady, R. L.; Leibfried, R. L.; Nguyen, T.T. *U.S Patent* 6,124,124, September 26, 2000. (c) Brady, R. L.; Leibfried, R. L.; Nguyen, T. T. *U.S Patent* 6,179,962, January 30, 2001.
23. For example: (a) Lin, Z.; Thorsen, T.; Arnold, F. H. *Biotech. Prog.* **1999**, 15, 467, and references therein. (b) Stemmer, W. P. C. *Proc. Nat. Acad. Sci. USA*, **1994**, *91*, 10747. (c) Reetz, M. T.; Jaeger, K.-E. *Topics Current Chem.* **1999**, *200*, 31.

Chapter 18

Modification of Cellulose Solids by Enzyme-Catalyzed Transesterification with Vinyl Esters in Anhydrous Organic Solvents

Jiangbing Xie and You-Lo Hsieh

Fiber and Polymer Science, University of California at Davis, Davis, CA 95616

Enzyme-catalyzed transesterification of several cellulose solids in organic media has been investigated. Protease and lipase enzymes were made soluble in organic media through ion-paired enzyme-surfactant complexes. Of the enzymes studied, Subtilisin Carsberg was found to be most catalytically active in the transesterification of cellulose in anhydrous pyridine. Following transesterfiction with vinyl propionate and vinyl acrylate, the presence of ester carbonyl groups on acylated cellulose were confirmed by FTIR and the modified cellulose surfaces became hydrophobic. From reactions on specifically substituted cellulose, the enzyme-catalyzed transesterification was confirmed to regioselectively target the primary hydroxyl group of cellulose.

Introduction

Enzymes are natural catalysts in biological systems. They have found many industrial applications including food processing, detergent additives, and textile finishing (*1*). Despite the conventional belief that enzymes could only function in an aqueous environment, the pioneering work of Klibanov et. al. (*2-3*) demonstrated that solid enzymes dispersed in anhydrous organic solvents also exhibited high catalytic activity. A variety of enzyme-catalyzed reactions in organic solvents have since been reported. These findings offer opportunities to use enzyme to catalyze reactions, such as esterification and transesterification, that are impossible to carry out in aqueous medium due to thermodynamics and/or kinetics reasons.

The advantages of enzyme catalysis in organic syntheses are many:

(i) Enzymes promote reactions that are difficult or impossible to emulate using traditional synthetic sequence, sometimes resulting in shortcuts.
(ii) Enzymes exhibit enantioselectivity and regioselectivity in certain reactions, preferentially reacting with specific stereoisomer and/or site, offering easy control of chemical structure.
(iii) Enzyme-catalyzed reactions are usually performed under mild conditions with regard to temperature, pH, and pressure, improving energy efficiency.
(iv) Enzymes are natural catalysts without detrimental properties to either human or the natural environment.
(v) The products from enzymatic reactions are biodegradable, the most attractive aspect of these reactions from an environmental perspective.
(vi) Enzymes can be recycled for repeated use, providing an advantage of resource conservation.

Enzyme-catalyzed transesterifications have been described as follows (*4*). It begins with binding of the ester donor, or acylating agent, to the active site of the enzyme to form an enzyme-ester complex which then transforms into the acyl-enzyme intermediate with the concomitant release of the alcohol product (Scheme 1). The acyl-enzyme intermediate then combines with nucleophiles, such as compounds containing hydroxyl groups, to form secondary binary complex, which finally free the enzyme and form the transesterified product.

Many lipases and proteases have been explored to catalyze esterification or transeterification reactions in organic solvents. The choice of enzymes is limited by the reaction systems and their catalyzing activities vary a great deal. For instance, out of dozens of commercially available lipases, only very few exhibit low activities in pyridine (*5-6*). No lipase was found active in DMF and DMSO yet. However, a few proteases, such as chymotrypsin and subtilisin, have shown relatively high activities in the highly polar pyridine (*6*) and DMF (*5*) in catalyzing reactions of soluble substrates, such as sugars.

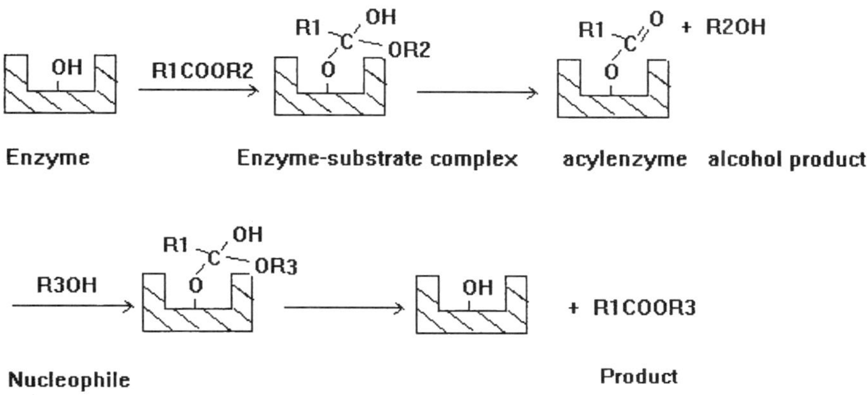

***Scheme 1**: Mechanism of enzyme-catalyzed transesterification*

Various organic solvents, from non-polar to polar, have been documented to support enzymatic transesterifications. Generally speaking, the more hydrophobic or less polar the reaction medium, the more catalytically active the enzyme (*2-3, 7*). This is attributed to the fact that hydrophilic solvents are more capable of striping bonded water from the enzymes, deteriorating their stability and activity. In addition, water, either originally contained in the polar solvents or deprived from the enzymes, causes hydrolysis of the ester product and is unfavorable to transesterification. On the other hand, the solubility of substrates may need to be considered in solvent selection. Often, highly polar solvents, such as pyridine and DMF, are necessary to dissolve the substrates.

In most cases, special esters are chosen as acylating agents to facilitate the reaction thermodynamically and/or kinetically. Activated esters, such as trihaloalkyl esters, are commonly used to increase the electrophilicity of the acylating agent (*5*). The bulky leaving groups are also desirable in reducing the rate of reverse reaction through steric hindrance. Vinyl esters are also excellent acylating agents which not only accelerate the rate of acyl transfer but also shift the reaction equilibrium forward continuously because the enol byproducts tend to tantomerize to ketones or aldehydes.

Enzyme-catalyzed transesterifications in organic solvents have been explored for the synthesis of both macromolecules and small molecules. Polyesters can be synthesized by either ring-opening self-condensation of hydroxyl ester (A-B type) (*8-10*), or condensation between diester and diol (A-A and B-B type) (*11-15*). Although most are synthesized from aliphatic monomers, polyesters with molecular weight ranging from 400 to 1,400 daltons have also been reported from aromatic monomers (*14, 16*). The reaction rates and molecular weight are highly dependent of the solvent medium and the removal of byproducts (*10-15, 17*). Some polyesters by enzyme-catalyzed transesterification have been characterized by narrower molecular weight distribution, more uniform structure, and, for chiral monomers, optically active products (*11, 18-19*).

Enzyme-catalyzed transesterification has also been studied for the acylation of small molecules which contain either mono-hydroxyl groups (alcohols) (*2-3, 19-20*) or multi-hydroxyl groups (sugars) (*5-7, 21-23*). The regioselectivity of reactions on sugars is particularly significant because the great advantage over conventional reactions involving tedious block-deblock processes.

Enzymes are insoluble in organic solvents. Most enzyme-catalyzed transesterifications, either synthesis of polyesters or acylation of small molecules, are carried out with suspended enzymes but soluble reactants in organic solvents. However, many reactants are either insoluble in organic media, or only soluble in solvents that may denature enzymes. For insoluble substrates, conventional suspension technology cannot support reactions with insoluble enzymes because of their poor interaction. Dordick et. al. (*24-26*) developed a method to solublize enzymes in hydrophobic organic solvents through the formation of enzyme-surfactant ion pairs. This approach differs from other methods in which enzymes are made soluble in organic solvents by either chemical modification or encapsulation in reversed micelles. The advantages of this approach include the extremely low water content ($<0.01\%$) around the enzyme molecules and minimal enzyme deactivation. Protease subtilisin Carsberg prepared by this approach has shown to catalyze the acylation of insoluble polysaccharides in isooctane (*27-28*). Acylation of amylose, β-cyclodextrin and hydroxyl ethyl cellulose (HEC), either casted as thin films or suspended as milled powders, can also be catalyzed by subtilisin protease in isooctane. More over, amylose is regioselectively acylated. These are the first successful attempts to modify insoluble polymers through enzyme-catalyzed transesterification in organic solvents.

The aim of our research is to investigate enzyme-catalyzed transesterifiactions on cellulose solids (Scheme 2). This exploitation is of considerable scientific and practical interest. First, enzyme-catalyzed transesterifiactions on cellulose is expected to be regioselective. Acylation of small sugars (*5-7, 21-23*) and amylose (*27*) by enzyme-catalyzed transesterification has been found to occur only on the primary hydroxyl groups. Considering the similarity between the chemical structures of cellulose and these saccharidic compounds, enzyme-catalyzed transesterification on cellulose should be regioselective. As regioselectively substituted cellulose has important applications but is difficult to produce, this approach is a valuable alternative. Secondly, the enzyme-catalyzed reactions should be limited to the solid surface due to the large dimensions of enzymes. It brings novel surface properties while not affecting the bulk properties of cellulose. Lastly, the cellulose derivatives are expected to be biodegradable through reversed enzyme-catalyzed hydrolysis in aqueous media.

Experimental

Materials. All enzymes, surfactant and buffer reagents were obtained from commercial sources and used as received. Vinyl acrylate, vinyl propionate and vinyl neodecanoate (Aldrich) were passed through inhibitor removal columns. Cellulose acetate (39.8% acetyl content, M_n=30,000 dalton) from Aldrich and triphenylmethyl chloride from Acros were used as received. All solvents involved in reactions except commercial anhydrous solvents were pre-dried using 4 Å molecular sieves for 24 hr. The cellulose materials used were rayon fabric (#208, Test Fabrics Inc.), cotton fabric (400R, Test Fabrics Inc.) and cellulose filter paper (Q2, Fisher Scientific). Fibers were taken from the cotton and rayon fabrics. Cellulose granules was milled (Laboratory Mill, Intermediate Model, Thomas-Wiley) from filter paper and passed through #60 mesh (about 0.18 mm^2 per pore). All cellulose materials were cleaned with 1 M NaOH solution at boil under N_2 for 1 hr, washed with water and dried in 90°C vacuum overnight. All cotton samples were also mercerized in 6 M NaOH at 40°C for 1 hr.

Enzyme Transfer. A previously reported method (*24-26*) was expanded to extract enzymes from aqueous solutions to organic solvents. Each enzyme was first dissolved in an aqueous buffer (1.1 mg/ml) at specific pH and ionic strength. An equal volume of isooctane containing 2 mM AOT was added and the two-phase solution was stirred at 25°C for 30 min. The mixture was settled for 30 min and then centrifuged at 2000 rpm for 5 min to enhance phase separation. The organic phase was recovered and the enzyme concentration was

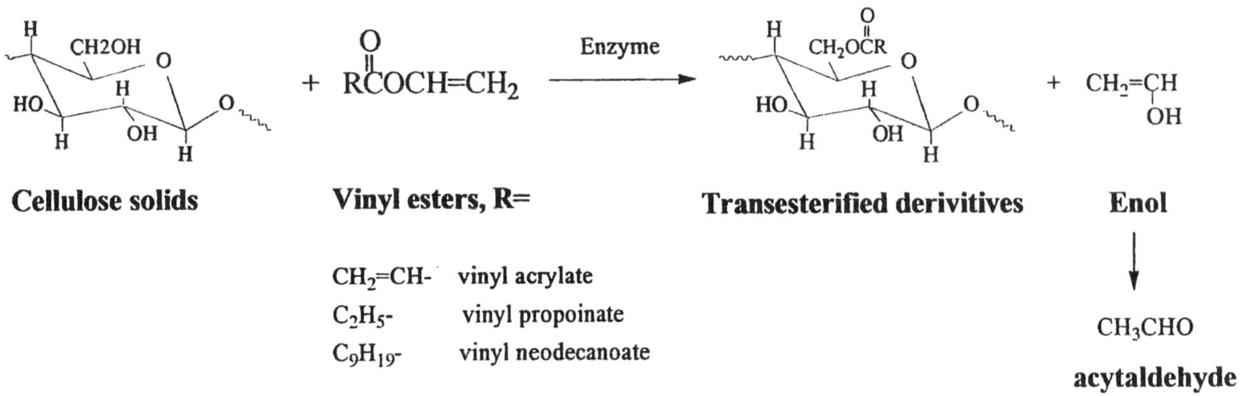

Scheme 2: *Enzyme-catalyzed transesterification on cellulose solids*

determined spectrophotometrically at 280 nm (HITACHI U-2000 Spectrophotometer). To transfer soluble enzymes from isooctane to other solvents, isooctane was first evaporated at room temperature by passing through dry N_2 gas, followed by addition of another solvent.

Enzyme-catalyzed Transesterification on Cellulose. Cellulose solids in the forms of fabrics (rayon and cotton), lose fibers (rayon and cotton), paper (cellulose filter) and milled granules (cellulose filter) were used to perform transesterification. For a typical reaction, 10-mg cellulose solid was reacted with 50-mg vinyl esters in 2-ml pyridine containing 1 mg/ml subtilisin Carsberg. The reactions were carried out at 37°C under constant shaking (60 rpm, Dubroff metabolic shaking incubator, Precision) for 5 days. A blank experiment was performed under same conditions without the enzyme. After the reaction, the sample was washed thoroughly first with water then with acetone in order to remove enzyme, surfactant and unreacted vinyl ester.

Regioselectivity of Enzyme-catalyzed Transesterification. Enzyme-catalyzed transesterifications were carried out with specifically substituted cellulose derivatives. 6-O-triphenylmethyl-cellulose and 2,3-O-methyl-cellulose were prepared following a previous described procedure (*29*). Briefly, cellulose was regenerated by deacylation of cellulose acetate in a 15% aq. ammonium hydroxide solution. Triphenylmethyl chloride was added to the regenerated cellulose to produce 6-O-triphenylmethyl-cellulose ("6-O-trityl-cellulose"). The reaction was confirmed to occur predominately on the C6 hydroxyl group, giving the degree of substitution around 1. Methylation of 6-O-trityl-cellulose with methyl iodide produced 2,3-O-methyl-6-O-trityl-cellulose which was then detritylated in HCl gas to produce 2,3-O-methyl-cellulose. The enzyme-catalyzed transesterifications on regenerated cellulose, 6-O-trityl-cellulose and 2,3-O-methyl-cellulose follow the same procedure as before.

Characterization. Cellulose solids were characterized by Infrared measurements (Nicolet FTIR spectrometer 560 with a Spectra Tech microscope attachment). Three modes were used to collect the FTIR spectra. Granular samples were made into KBr pellets for FTIR measurement. Spectra of single fibers were collected using the transmission mode on the microscope. ATR spectra of filter paper, fabric, and single fibers were collected through the microscope using a slide-on crystal. All spectra were recorded at 4 cm^{-1} resolution. The spectra from KBr were collected from 32 scans, and other spectra were collected from 200 scans. Water wetting behaviors of the cellulose samples were described by water contact angle (WCA) calculated from the wetting force measured on a tensiometer (*30*).

Results and Discussion

Enzyme Transfer. Transferring of enzymes from aqueous buffers to isooctane was conduced using several proteases and lipases. The transfer efficiency was found to be less than 10% except for subtilisin Carsberg and α-chymotrypsin whose transferring efficiencies reached 60% and 85% (Table 1). Both the ionic strength and pH of the aqueous buffers were found to strongly influence the transferring efficiency. In the case of transferring subtilisin Carsberg from 8.5mM HEPES containing 6mM KCl, the efficiency was improved by increasing ionic strength of the buffer with added $CaCl_2$. On the other hand, too high the ionic strength could lower the extraction efficiency, an effect reported by others as well (*24*). This is due to the combined effects of reduced partitioning of AOT into aqueous buffer and the weakened interaction between enzymes and AOT.

The enzyme-AOT complexes are found to be soluble in several common organic solvents, irrespective of their polarities. We have confirmed that enzyme-AOT complexes in isootane can be transferred into polar solvents, such as THF, pyridine, DMF, and vinyl esters (vinyl acrylate, vinyl neodecanoate and vinyl propionate). The ability to transfer enzyme-AOT complexes between solvents gives great promise for broadening the use of enzymes in organic solvents where either direct extraction or phase separation between aqueous and organic media does not work.

Enzyme-catalyzed Transesterification. Cellulose was first transesterified with enzymes in isooctane. Although both the acylating agents and enzymes are

Table 1. Enzymes transferring from aqueous buffers to isooctane

Enzyme	Aqueous buffer	Transferring efficiency
subtilisin Carsberg	8.5 mM HEPES, pH=7.8, 6mM KCl, 10 mM CaCl2	12%
subtilisin Carsberg	10 mM bis-tris propane, pH=7.8, 2mM CaCl2	50%
a-chymotrypsin	8.5 mM HEPES, pH=7.8, 6mM KCl, 10 mM CaCl2	85%
a-chymotrypsin	10 mM KAc, pH=5.0, 10 mM KCl, 6mM CaCl2	50%

soluble in isooctane, the FTIR spectrum of the transesterified cellulose shows no evidence of ester structure. This is likely a result of poor interaction between the hydrophobic solvent and the hydrophilic substrate.

To improve the solvent-solid interaction, a more polar solvent, pyridine, was used. Several cellulose solids, including cotton (fabric and fibers), rayon (fabric and fibers) and cellulose filter (paper and granules), were transesterified with vinyl acrylate, vinyl neodecanoate and vinyl propionate in pyridine. With subtilisin Carsberg, the reaction products from vinyl acrylate exhibit FTIR peaks at 1715-1731 cm^{-1}, a clear evidence of the ester carbonyl groups (Figure 1b-d). The peak intensities and noises in FTIR spectra vary by the detection methods, i.e., the KBr method gives the strongest signals, the microscope with the ATR give the weakest with the highest noise level. Depending on the acylating agents, the cellulose derivatives exhibited ester C=O peaks between 1700 and1740 cm^{-1}. The ability of pyridine to swell cellulose appears to be important to support the reaction.

Transesterification with vinyl acrylate and vinyl propionate was confirmed to occur on all forms of cellulose solids studied and the ester carbonyl peaks from FTIR were similar among all transesterified products. A blank reaction under the same conditions but with no enzymes produced no evidence of the ester, indicating the enzyme was indeed the catalyst for the reaction. The cellulose transesterified with vinyl neodecanoate showed no evidence of carbonyl.

The water wetting contact angles of the cellulose derivatives acylated by vinyl acrylate and vinyl propionate are much higher than those of the untreated cellulose substrates (Table 2). The lowering in water wettability of the cellulose solids acylated by vinyl acrylate and vinyl propionate is expected as it is consistent with the replacement of the cellulose hydroxyl groups with hydrophobic groups. The derivatives from reaction with vinyl neodecanoate remain hydrophilic with unchanged water contact angles. This is consistent with the lack of FTIR evidence of the transesterified cellulose with vinyl neodecanoate. Although vinyl decanoate was shown to acylate solid amylose, β-cyclodextrin and hydroxy ethyl cellulose (HEC) through enzyme-catalyzed transesterification (*27-28*), vinyl neodecanoate appear to be unable to acylate cellulose. The reason could be the steric hindrance of the bulky alkyl chain in vinyl neodecanoate.

When α-chymotrypsin was employed in the transesterification of cellulose solids in either isooctane or pyridine, no evidence of ester in the products was found. This observation suggests that α-chymotrypsin cannot catalyze transesterification on cellulose in these organic media.

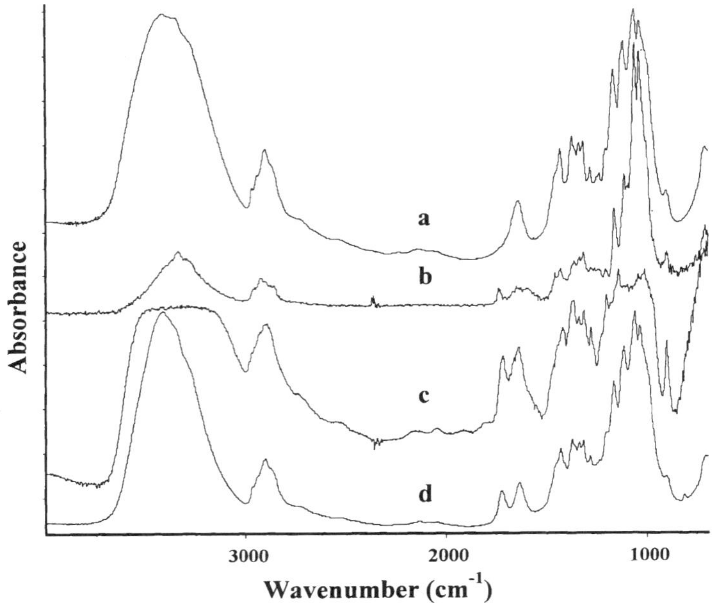

***Figure 1**: FTIR spectra of (a) cellulose (filter granular) by KBr and transesterified cellulose with vinyl acrylate by varying collection modes: (b) cotton fibers by microscope-ATR; (c) cotton fiber by microscope transmission; (d) filter paper granular by KBr.*

Table 2: Water contact angles of cellulose and transesterified cellulose

Cellulose source	Sample	Water contact angle (degree)
cotton	control	36
	cellulose acrylate	58
	cellulose neodecanoate	34
filter paper	control	36
	cellulose acrylate	51
	cellulose propoinate	57
	cellulose neodecanoate	37

Regioselectivity of Enzyme-catalyzed Transesterification. Regioselectivity of subtilisin Carsberg-catalyzed transesterification of cellulose with vinyl acrylate was studied by performing this reaction on cellulose with hydroxyl groups selectively blocked. The cellulose derivatives used were 6-O-trityl-cellulose whose primary –OH blocked and 2,3-O-methyl-cellulose whose secondary -OH groups were blocked. Following transesterification, cellulose regenerated from cellulose acetate and 2,3-O-methyl-cellulose exhibit strong carbonyl peaks at 1730 cm^{-1} (Figure 2a and 2b). However, the carbonyl peak is missing from the transesterification product of 6-O-trityl-cellulose (Figure 2c), which indicates that subtilisin Carsberg cannot catalyze transesterification on the secondary hydroxyl groups. The ester carbonyl band of 2,3-O-methyl-cellulose acrylate (Figure 2b) gives clear evidence that subtilisin Carsberg-catalyzed transesterification does occur on the primary hydroxyls.

The presence of carbonyl ester in reaction product of 2,3-O-methyl-cellulose and its absence in transesterified 6-O-trityl-cellulose confirm that subtilisin Carsberg-catalyzed transesterification site specifically targets the primary hydroxyl group of the cellulose soids. Although regioselectivity has been reported on enzyme catalyzed transesterification on soluble saccharidic compounds, such as small sugars and amylose (*5, 7, 22, 27*), our report here is the first evidence of this regioselective reactions on cellulose solids.

It should also be noted that the transesterified 2,3-O-methyl-cellulose acrylate shows a much stronger ester peak than the cellulose acrylate (Figure 2a and 1d). Since both have the same amount of primary hydroxyl groups, this difference in the extent of reaction may be ascribed to their different solubility between the two substrates in pyridine. Cellulose is not soluble in pyridine. Therefore, the enzyme-catalyzed transesterification on cellulose is heterogeneous. 2,3-O-methyl-cellulose, on the other hand, is partial soluble in pyridine under the same condition. The better accessibility of 2,3-O-methyl-cellulose to the acylating agents and the catalyst explains the enhanced reaction.

Conclusion

Ever since the discovery of enzymes activity in organic solvents, enzyme-catalyzed transesterification has been reported for the acylation of small molecules containing hydroxyl groups, i.e., alcohols and sugars. The regioselectivity of the reactions on sugars is particularly significant because of the great advantage over conventional reactions involving tedious block-deblock processes. Enzyme-catalyzed transesterification of insoluble polysaccharides in the forms of casted films or powders has been demonstrated in isooctane.

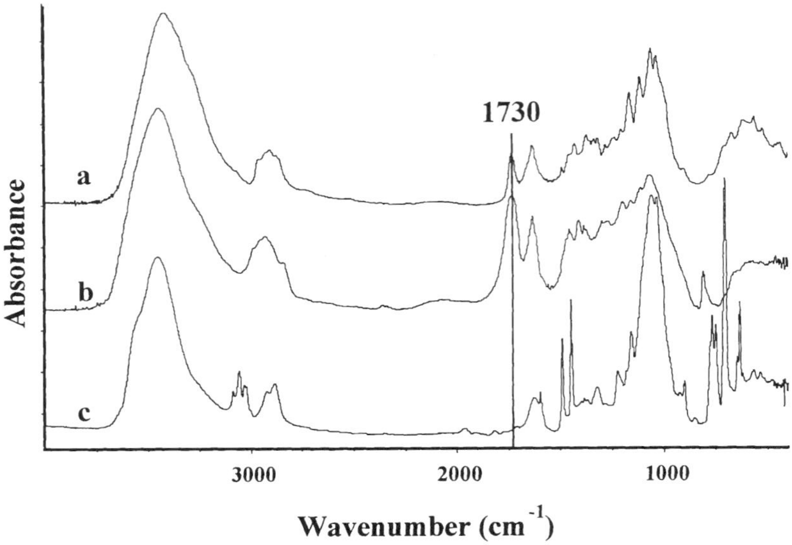

***Figure 2**: FTIR-KBr spectra of enzyme-catalyzed transesterification on (a) cellulose and regio-specifically substituted (b) 2,3-O-methyl-cellulose and (c) 6-O-trityl cellulose*

This study confirms that transesterification reactions of cellulose in various solid forms of fabric, paper and fibrous is possible with subtilisin Carsberg as the catalyst in organic media. Extraction of enzymes from aqueous buffers to isooctane was accomplished by forming enzyme-surfactant ion-paired complexes. The transferring can be 80% and 65% efficient for chymotrypsin and subtilisin Carsberg, respectively. These enzymes can be further transferred from isooctane to other organic solvents such as pyridine, DMF and vinyl esters. Subtilisin Carsberg catalyzed transesterifications with vinyl acrylate and vinyl propionate has been confirmed on several cellulose solids including cotton (fabric and fibers), rayon (fabric and fibers), and cellulose filter (paper and milled granules). These reactions were effective in pyridine, but not in isooctane. Vinyl neodecanoate was found not to be an effective acylating agents for cellulose. The regioselectivity of subtilisin Carsberg-catalyzed transesterification of cellulose has been verified. Although regioselectivity has been reported on enzyme-catalyzed transesterification on other saccharidic compounds, this study provides first evidence of the regioselective reactions on cellulose solids.

References

1. Kaplan, D. L.; Dordick, J. Gross, R. A.; Swift, G. In *Enzymes in Polymer Synthesis*; Gross, R; Kaplan, D. L.; Swift, G., Eds; Americal Chemical Society: Washington DC, 1998; Chap. 1, p2-16.
2. Zaks, A; Klibranov A. M. *Proc. Natl. Acad. Sci. USA* **1985**, 82, 3192-3196.
3. Zaks, A; Klibranov A. M. *J. Bio. Chem.* **1988**, 263, 3194-3201.
4. Chaudhary, A. K.; Beckman, E. J.; Russell, A. J. In *Enzymes in Polymer Synthesis*; Gross, R; Kaplan, D. L.; Swift, G., Eds; Americal Chemical Society: Washington DC, 1998; Chap. 2, p18-57.
5. Rova, S.; Chopineau, J.; Kieboom, A.P.G.; Klibranov A. M. *J. Am. Chem. Soc.* **1988**, 110, 584-589.
6. Patil, D. R.; Rethwisch, D. G.; Dordick J. S. *Biotech. Bioengi.* 37, 639-646, 1991.
7. Rich, J. O.; Bedell, B. A.; Dordick J. S. *Biotech. Bioengi.* **1995**, 45, 426-434.
8. Gutman, A. L.; Oren, D.; Boltanski, A.; Bravdo, T. *Tetra. Lett.* **1987**, 28, 5367-5368.
9. Bisht, K. S.; Deng, F.; Gross. R. A.; Kaplan, D. L.; Swift, G. *J. Am. Chem. Soc.* **1998**, 120, 1363-1367.
10. Knani, D.; Gutman, A.L.; Kohn, D. H. *J. Polym. Sci. Part A: Polym. Chem.* **1993**, 31, 1221-1232.
11. Binns, F.; Roberts, S. M.; Taylor, A., Williams, C. F. *J. Am Chem. Soc. Perkin Trans.* **1993**, 1, 899-904,.
12. Chaudhary, A. K.; Lopez, J.; Beckman, E. J.; Russell, A. J. *Biotech. Prog.* **1997**, 13, 318-325.
13. Linko, Y. Y.; Wang, Z. L.; Seppala J. *Enz. Microb. Tech.* **1995**, 17, 506-511.
14. Park, H. Y.; Chang, H. N.; Dordick, J. S. *Biotech. Lett.* **1995**, 17, 1085-1090.
15. Uyama, H.; Kobayashi, S. *Chem. Lett.* **1994**, 1687-1690.
16. Park, H. G.; Chang, H. N.; Dordick, J. S. *Biocatalysis* **1994**, 11, 263-271.
17. Morrow, C. J. *MRS Bulletin* **1992**, 17, 43-47.
18. Wallace, J. S.; Morrow, C. J. *J. Polym. Sci. Part A: Polym. Chem.* **1989**, 27, 2553-2567.
19. Wang, Y. -F.; Lalonde, J. J.; Momongan, M.; Bergbreiter, D. E.; Wong, C. -H. *J. Am. Chem. Soc.* **1998**, 110, 7200-7205.
20. Dabulis, K.; Klibranov A. M. *Biotech. and Bioengi.* **1993**, 41, 566-571.
21. Martin, B. D.; Ampofo, S. A.; Linhardt, R. J.; Dordick J. S. *Macromolecules* **1992**, 25, 7081-7085.

22. Therisod, M.; Klibranov A. M. *J. Am. Chem. Soc.* **1986**, 108, 5638-5640.
23. Therisod, M.; Klibranov A. M. *J. Am. Chem. Soc.* **1987**, 109, 3977-3981.
24. Paradkar, V. M.; Dordick, J. S. *Biotech. and Bioengi.* **1994**, 43, 529-540.
25. Paradkar, V. M.; Dordick, J. S. *J. Am. Chem. Soc.* **1994**, 116, 5009-5010.
26. Wangikar, P. P.; Michels, P. C.; Clark, D. C.; Dordick, J. S. *J. Am. Chem. Soc.* **1997**, 119, 70-76.
27. Bruno, F. F.; Akkara, J. A.; Ayyagari, M.; Kaplan, D. L.; Gross, R.; Swift, G.; Dordick, J. S. *Macromolecules* **1995**, 28, 8881-8883.
28. Bruno, F. F.; Dordick, J. S.; Kaplan, D. L.; Akkara, J. A. In *Enzymes in Polymer Synthesis*; Gross, R; Kaplan, D. L.; Swift, G., Eds; Americal Chemical Society: Washington DC, 1998; Chap. 9, p167-174.
29. Hsieh, Y. L.; Xie, J. *Polym. Preprint* **2001**, 42(N1), 512-513.
30. Hsieh, Y. L.; Yu, B. *Textile. Res. J.* **1992**, 62, 677-685.

Chapter 19

Grafting Renewable Chemicals to Functionalize Chitosan

C. Justin Govar[1,4], Tianhong Chen[2,3], Nai-Chi Liu[1,5], Michael T. Harris[1], and Gregory F. Payne[2,3,*]

[1]Department of Chemical Engineering, 2112 Institute for Physical Science and Technology Building, University of Maryland, College Park, MD 20742
[2]Center for Agricultural Biotechnology, 5115 Plant Sciences Building, University of Maryland Biotechnology Institute, College Park, MD 20742
[3]Chemical and Biochemical Engineering, 1000 Hilltop Circle, University of Maryland Baltimore County, Baltimore, MD 21250
[4]Current address: Battery Technology Group, Naval Sea Systems Command, 9500 MacArthur Boulevard, West Bethesda, MD 20817
[5]Current address: Industrial Science and Technology Network, Inc., CyberCenter, 2101 Pennsylvania Avenue, York, PA 17404

The need for environmentally-friendly processing has stimulated research on biocatalysis, while a growing interest in sustainability is stimulating interest in the use of renewable resources. We are currently examining an enzymatic approach to graft renewable phenols onto the biopolymer chitosan. In this approach, tyrosinase is used to convert phenols into reactive *o*-quinones which undergo non-enzymatic reactions that lead to grafting onto chitosan. As discussed, tyrosinase-catalyzed chitosan modification results in dramatic changes in functional properties.

In the early 1990s there were growing safety, health, and environmental concerns for traditional polymers and polymer processing operations. The Bhopal disaster was forcing people to re-consider the routine use of highly reactive compounds for synthesis, while the poor biodegradability of some synthetic polymers was leading to recycling programs for plastics. These concerns were also stimulating interest in exploiting natural polymers (e.g. cellulose and xanthan) and using biotechnology to "engineer" functional polymers (e.g. proteins and polyhydroxyalkanoates). There was also interest in developing cell-free biocatalytic operations for polymer synthesis and modification (*1*).

Cell-Free Biocatalysis for Polymer Processing

Typical biosynthetic approaches for polymer synthesis are difficult to transfer to cell-free systems. Often, biosynthetic processes to polymer synthesis involve templates (e.g. for protein and DNA biosynthesis) and/or activated monomers (e.g. for polysaccharide synthesis). Although it may be technically feasible to obtain cell-free synthesis when templates and activated monomers are required, these approaches are currently too cumbersome for practical application to large-volume polymer production.

An alternative biosynthetic process for polymer synthesis involves the use of an enzyme to generate a reactive intermediate that undergoes subsequent non-enzymatic reaction. In nature, reactive free radicals are intermediates in the biosynthesis of lignin polymers. Compared to other biosynthetic routes to polymers, the free radical route can be considered to be biochemically simple because few enzymes are involved, and templates and complex cofactors are not required. However, the ability of the free radicals to undergo various reactions makes control and even analysis of the product polymer difficult. For instance, lignin's three-dimensional structure is still being elucidated, while the control of lignin biosynthesis is an emerging area of study (*2,3*). There has been some study of the use of free-radical generating peroxidases for cell-free polymer synthesis and modification (*4-6*).

Developments in non-aqueous enzymology (*4*) suggested the potential for using common hydrolytic enzymes (e.g. proteases, lipases, and polysaccharide-hydrolyzing enzymes) under non-physiological conditions to generate linkages common in industrial polymers (e.g. amide, ester, and glycosidic). As evidenced by many chapters in this text, there has been substantial progress over the last decade on the use of hydrolytic enzymes in polymer synthesis and modification.

Tyrosinase for Cell-Free Polymer Processing

In our studies, we are using an enzyme that converts phenols into reactive *o*-quinone intermediates that undergo non-enzymatic grafting reactions. Enzymes that convert phenols into *o*-quinones are ubiquitous in nature and are referred to as tyrosinases or phenol oxidases. When we began our study, we reasoned that tyrosinase may offer several advantages for polymer grafting reactions. First, tyrosinases use molecular oxygen as the oxidant and these enzymes have simple co-substrate requirements. Second, tyrosinases have broad substrate ranges and react with various natural and synthetic phenols, including phenol-containing polymers (*7-9*). Third, the *o*-quinone product of the reaction is freely diffusible and can be grafted without the need for the enzyme to recognize a second substrate. We reasoned that this approach would be less prone to steric limitations, compared to approaches that required two substrates to react at the enzyme's active site (e.g. an enzyme-bound acyl intermediate is transferred to a nucleophile by lipase and protease enzymes). Fourth, tyrosinase-catalyzed and subsequent non-enzymatic reactions do not appear to have equilibrium limitations and reactions can be performed in aqueous solutions - provided the phenolic substrate is soluble. The major limitation to the use of tyrosinase for grafting is that the reactive *o*-quinone can undergo a complex reaction cascade leading to phenol (quinone) oligomerization and grafting may result in a heterogeneous array of moieties.

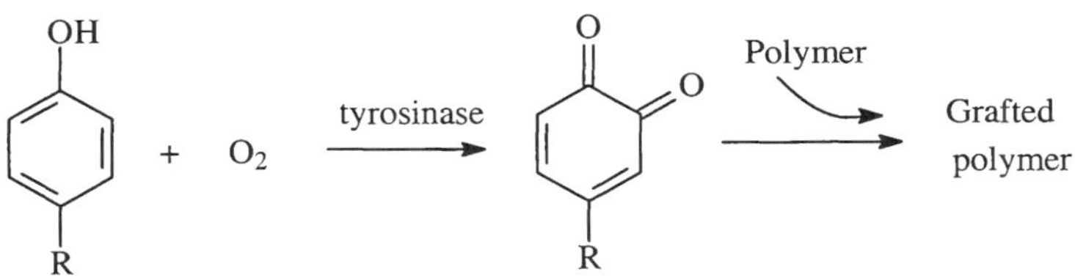

In nature there is precedence that enzymatically-generated *o*-quinones are used to confer functional properties to biological materials. For instance, enzymatically-generated *o*-quinones act to crosslink (i.e. tan) proteins in insect sclerotization (i.e. hardening of the "shell") (*10-12*). Additionally, the enzymatic oxidation of tyrosyl and/or dihydroxyphenylalanyl residues of polyphenolic adhesive proteins are believed to play a role in conferring water-resistant adhesive properties to mussel glue (*13-15*).

Renewable Phenolic Substrates for Grafting

In addition to the emphasis on developing environmentally-friendly chemistries for synthesis, there is an emerging emphasis toward sustainability. This emphasis on sustainability is stimulating interest in obtaining chemicals from renewable resources (*16,17*). Phenols are abundant in nature and there are increasing efforts to recover such compounds from renewable resources (*18*). For instance, recent studies are showing that a diverse array of functional phenols are present in foods (e.g. antioxidants) (*19*). It seems plausible that such compounds could provide a safer and renewable alternative to some synthetic phenols. Also, lignins are an abundant source of polyphenolic compounds and lignins are generated in large amounts as a byproduct of the pulp and paper industry. Although these lignins are primarily burned for energy, there is growing interest in exploiting these byproducts as a feedstock for chemical production (*20,21*). Since tyrosinases react with many natural phenols we reasoned that this enzyme could be used to graft various functional phenols onto polymers.

Chitosan, the Biopolymer for Grafting

To exploit tyrosinase for polymer grafting, the reactive *o*-quinone must be capable of reacting with the polymer. Since quinones undergo reaction with various nucleophiles, an obvious candidate biopolymer is the amine-rich polysaccharide chitosan. Chitosan is derived from chitin which is the second most abundant natural polysaccharide on earth. Chitosan's primary amino groups have pKa's of about 6.3. Below the pKa, the amino groups are protonated making chitosan a water-soluble cationic polyelectrolyte. Above the pKa, chitosan's amino groups are unprotonated, nucleophilic, and can undergo a variety of reactions with electrophiles (including enzymatically-generated *o*-quinones).

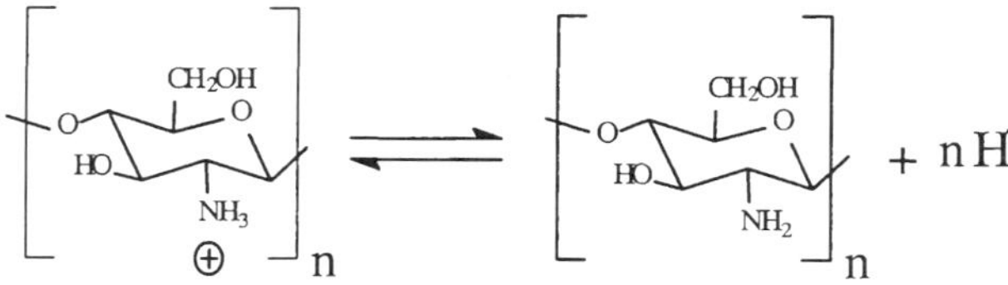

Interestingly, the pH-responsive solubility of chitosan allows this polymer to be modified under either heterogeneous or homogeneous conditions. For instance, chitosan can be cast into insoluble films at neutral pH and these films will react with *o*-quinones that are generated in solution. In contrast, chitosan can be modified under homogeneous conditions by performing the reaction at a pH of about 6 such that the chitosan remains in solution during modification.

Conferring Functionality to Chitosan by Enzymatic Modification

In initial studies we determined that tyrosinase can be exploited to graft various phenolic substrates onto chitosan (*22,23*). In later studies we focused on conferring desirable functional properties to chitosan. In these studies, phenolic substrates were selected based on "educated guesses" while considerable trial and error was required to identify appropriate phenolic substrates and reaction conditions.

Chitosan Modification to Confer Base Solubility

Our first goal in functionalization was to obtain chitosan derivatives that were soluble under alkaline conditions. We selected chlorogenic acid as our phenolic substrate. It should be noted that the name, chlorogenic acid, is a misnomer as there is no chlorine in the molecule. Rather chlorogenic acid is a natural product found in high concentrations in various foods (e.g. coffee) (*24*). The catecholic moiety of chlorogenic acid undergoes reaction with tyrosinase, and we reasoned that the quinic acid moiety offers hydrophilic and carboxylate functionality that could confer base-solubility.

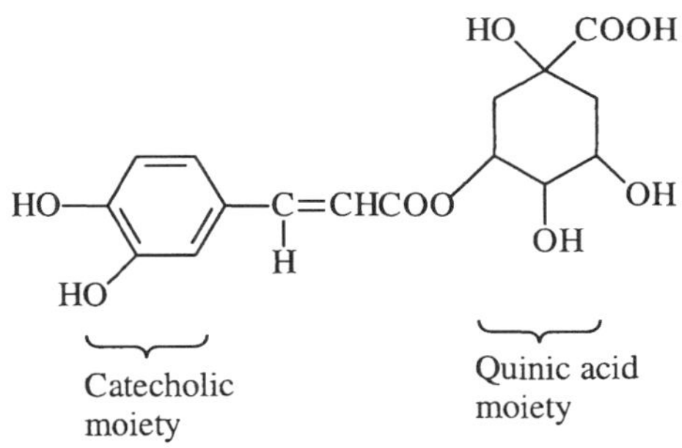

To graft chlorogenic acid onto chitosan, we performed reactions under homogeneous conditions. After recovering and washing the modified chitosan, we observed that the chlorogenic acid-modified chitosan offered unique pH-responsive behavior (*25*). Specifically, this modified chitosan was soluble under acidic and basic conditions but had a "window of insolubility" at near-neutral pHs.

Tyrosinase-catalyzed Chitosan Crosslinking

In the biological process of insect sclerotization, tyrosinases are reported to convert low molecular weight phenols into reactive *o*-quinones that undergo reactions to crosslink proteins in the insect's integument. We examined an analogous approach to determine if tyrosinase-generated *o*-quinones could be used to cross-link chitosan. For this we considered *p*-cresol as the phenolic substrate. Cresol could be obtained from renewable resources as it has been reported to be a major product of lignin pyrolysis (*26*). We reasoned that cresol could be an appropriate substrate because it's *o*-quinone may offer the di- (or) multi-functionality necessary for crosslinking. Specifically, the *o*-quinone generated from tyrosinase-catalyzed oxidation of *p*-cresol has a site for Michael-type adduct formation, two sites for Schiff-base formation, and possibly a site for reaction through the quinone methide.

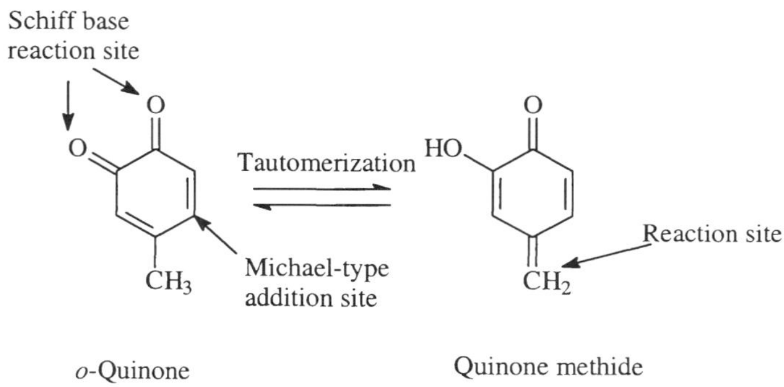

When dilute solutions of cresol, chitosan, and tyrosinase were mixed under homogeneous conditions, we visually observed that the solutions were converted into gels. Rigorous characterization using dynamic rheological measurements confirmed that gelation had occurred (*27*). This observation provides physical evidence for tyrosinase-catalyzed crosslinking. Because of the complexity of the quinone reactions we have not been able to characterize the underlying chemical reactions.

Conferring Water-resistant Adhesive Properties

The adhesion of marine animals to wet or submerged surfaces requires a polyphenolic protein that is rich in dihydroxyphenylalanine (DOPA) residues. Some of these DOPA residues are required to confer adhesive strength between the protein and the surface (*28*). Other DOPA residues are believed to be oxidized by tyrosinases and these oxidized residues undergo protein crosslinking reactions necessary to confer cohesive strength (*29*).

To examine if tyrosinase could be used to confer water-resistant adhesive properties to chitosan, we selected dihydroxyphenylethylamine (Dopamine) as the phenolic reactant. When chitosan solutions were reacted with tyrosinase and dopamine under homogeneous conditions, we observed a dramatic increase in the solution viscosity (consistent with crosslinking) and samples from these reacted chitosans were observed to offer adhesive properties with wet and submerged surfaces (*30*).

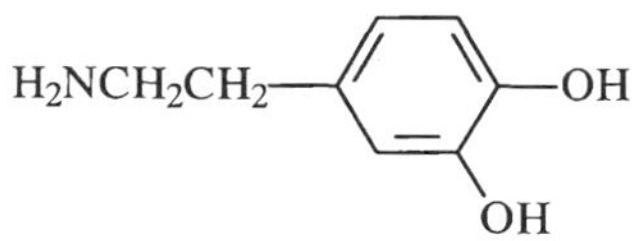

Dopamine

Grafting of Hydrophobic Phenols onto Chitosan

There is considerable recent interest in creating water soluble polymers that have a small number of hydrophobic substituents. To investigate the potential for enzymatically generating hydrophobically-modified chitosan derivatives, we initially examined the synthetic substrate 4-*n*-hexyloxyphenol. To perform these reactions under homogeneous conditions requires a solvent that can dissolve both the chitosan polymer and the phenolic substrate. Also, the solvent must not inactivate the enzyme. In our studies, we used aqueous methanolic mixtures (50 w/v %) and observed that the chitosan that had been enzymatically-modified using hexyloxyphenol behaves as an associative thickener. Specifically, Figure 1 shows that viscosity of the hexyloxyphenol-modified chitosan increases dramatically with polymer concentration compared to the unmodified chitosan control (*31*).

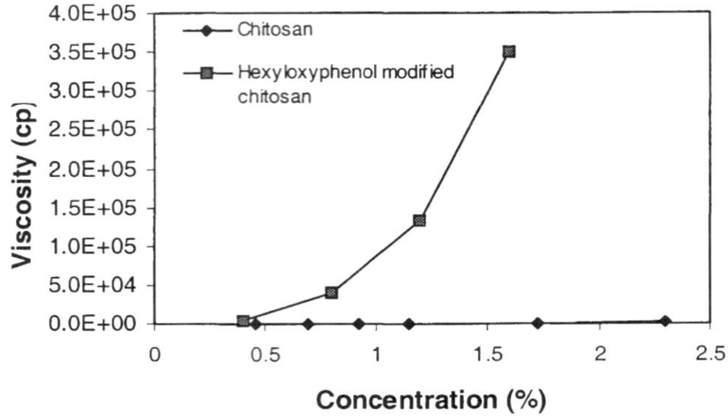

Figure 1.Solution viscosity for chitosan and hexyloxyphenol-modified chitosan. The modified chitosan was prepared under homogeneous conditions by reacting hexyloxyphenol (0.3 mM) with chitosan (20 mM repeating unit). Adapted from reference 31.

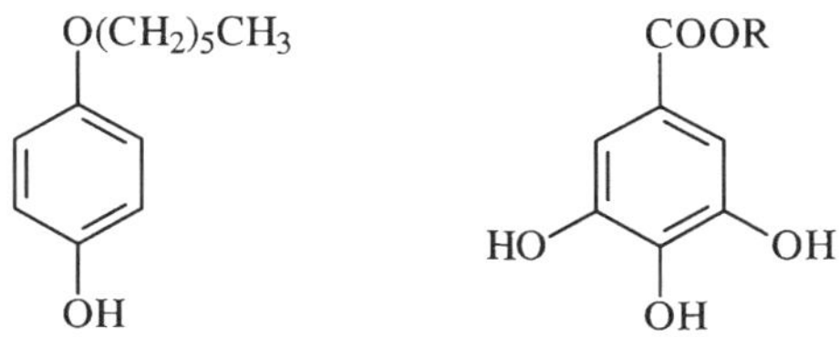

Hexyloxyphenol

Gallates
(R= methyl, propyl, octyl)

More recently, we have begun to examine enzymatic chitosan modification using esters of gallic acid. Gallic acid can be readily obtained from renewable resources (e.g. plant tannins), while gallate esters can be synthesized with a range of properties (depending on the substituent). Further, gallate esters are expected to be reasonably safe since some are used as food additives (e.g. propylgallate). The disadvantage of gallates are that reactions with tyrosinase can sometimes be problematic. We observed that gallate esters can undergo reaction with tyrosinase in ethanol:water mixtures (30:70). Specifically, we monitored the loss of the oxygen co-substrate when gallate esters were incubated with tyrosinase. Figure 2 shows that based on dissolved oxygen consumption, tyrosinase can react with various gallate esters and especially with octyl gallate.

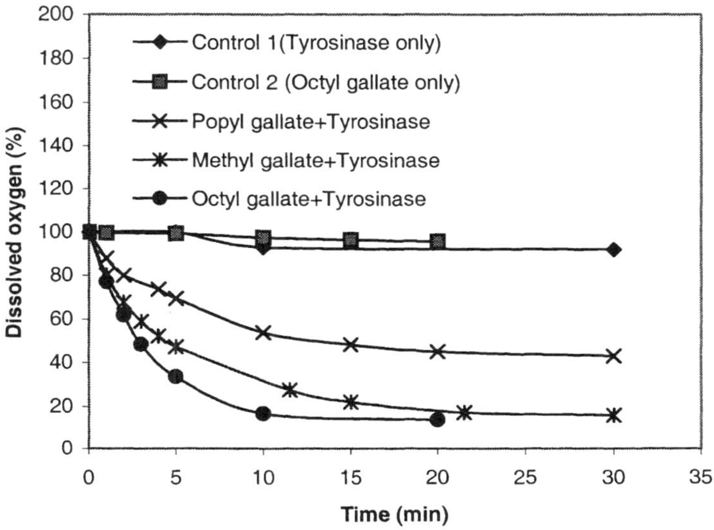

Figure 2. Evidence for tyrosinase-catalyzed oxidation of gallate esters. Reactions were conducted in 30 v/v % ethanol solutions with 6mM ester.

To examine the modification of chitosan with gallate esters, we performed heterogeneous reactions such that the phenolic substrate and tyrosinase were in solution while the quinone reacted at the surface of a chitosan film. After reaction, the films were washed extensively and dried, and the surface hydrophobicity of the film was measured by contact angle. Specifically, a drop of water was placed on the dried films and the contact angle was measured as the drop spread with time (see reference *31* for experimental details). As shown in Figure 3, chitosan films modified by methyl, propyl, or octyl gallates were observed to have larger contact angles (i.e. more hydrophobic surfaces) compared to unmodified chitosan. Also, the film modified with the most hydrophobic ester, octyl gallate, was observed to have the most hydrophobic surface.

Current Research

At this point we believe our research, and that of Muzzarelli (*32*), demonstrate that tyrosinase-generated quinones react with chitosan. Further, these enzymatic modification reactions dramatically alter chitosan's functional properties, and the property changes depend on the phenolic substrate used. Ultimately, the goal of our research is to be able to perform modifications that "tailor" chitosan's properties for specific applications. However, a truly rational

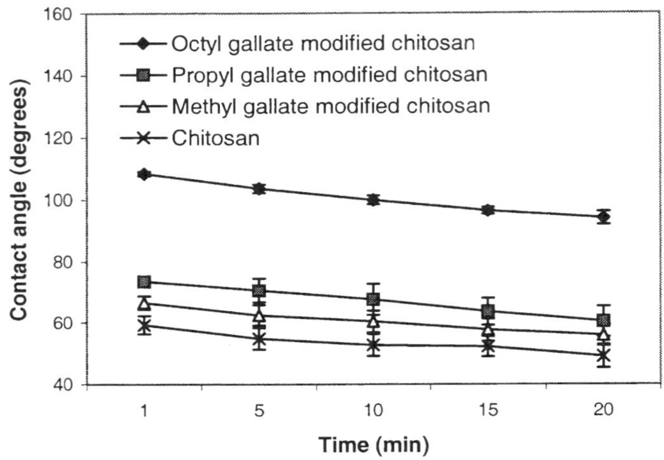

Figure 3. Contact angle measurements for chitosan films, and chitosan films that were modified with different gallate esters. Experimental details of contact angle measurement are reported in reference 31.

approach to "tailor" chitosan's properties will require both a better characterization of quinone chemistry, and a greater understanding of the structure-function relationships for modified chitosans. Both requirements present substantial challenges. To circumvent these requirements we are taking a "discovery-based" approach and are exploring the use of combinatorial screens to efficiently identify phenols and reaction conditions that yield modified chitosans with useful properties (*33*).

Acknowledgement

Financial support for this research was provided by Venture Innovations as a subcontract to NSF grant DMI-9901868.

References

1. *Enzymes in Polymer Synthesis*; Gross, R. A.; Kaplan, D. L.; Swift, G., Eds.; American Chemical Society; Washington, DC, 1998.
2. Argyropoulos, D. S.; Menachem, S. B. Lignin. In *Biopolymers from Renewable Resources*; Kaplan, D.L., Ed. Springer-Verlag, 1998; p. 292-322.

3. *Advances in Lignoncellulosics Characterization*; Argyropoulos, D.S., Ed.; Tappi Press, Atlanta Georgia, 1999.
4. Dordick, J. S.; Marletta, M. A.; Klibanov, A. M. Polymerization of phenols catalyzed by peroxidase in nonaqueous media. *Biotechnol. Bioeng.* **1987,** *30,* 31-36.
5. Akkara, J. A.; Senecal, K. J.; Kaplan, D. L. Synthesis and characterization of polymers produced by horseradish peroxidase in dioxane. *J. Polym. Sci. A: Polym Chem.* **1991,** *29,* 1561-1574.
6. Rao, A. M.; John, V. T.; Gonzalez, R. D.; Akkara, J. A.; Kaplan, D.L. Catalytic and interfacial aspects of enzymatic polymer synthesis in reversed micellar systems. *Biotechnol. Bioeng.* **1993,** *41,* 531-540.
7. Ito, S.; Kato, T.; Shinpo, K.; Fujita, K. Oxidation of tyrosine residues in proteins by tyrosinase. *Biochem. J.* **1984,** *222,* 407-411.
8. Marumo, K.; Waite, J. H. Optimization of hydroxylation of tyrosine and tyrosine-containing peptides by mushroom tyrosinase. *Biochim. Biophys. Acta* **1986,** *872,* 98-103.
9. Shao, L.; Kumar, G.; Lenhart, J. L.; Smith, P.J.; Payne, G. F. Enzymatic modification of the synthetic polymer polyhydroxystyrene. *Enzyme Microbial Technol.* **1999,** *25,* 660-668.
10. Sugumaran, M. Molecular mechanisms for cuticular sclerotization. *Adv. Insect. Physiol.* **1988,** *21,*179-231.
11. Peter, M. G. Chemcial modifications of biopolymers by quinones and quinone methides. *Angew. Chem. Int. Ed. Engl.* **1989,** *28,* 555-570.
12. Andersen, S. O.; Peter, M. G.; Roepstorff, P. Cuticular sclerotization in insects. *Comp. Biochem. Physiol.* **1996,** *113B,* 689-705.
13. Waite, J. H.; Tanzer, M. L. Polyphenolic substance of *Mytilus edulis*: Novel adhesive contiaing L-DOPA and hydroxyproline. *Sci.* **1981**, *212,* 1038-1040.
14. Waite, J. H.; Housley, T. J.; Tanzer, M. L. Peptide repeats in a mussel glue protein: Theme and variation. *Biochem.* **1985**, *24,* 5010-5014.
15. Waite, J. H. Marine adhesive proteins: Natural composite thermosets. *Int. J. Biol. Macromolec.* **1990**, *12,* 139-144.
16. *Biobased Industrial Products: Priorities for Research and Commercialization.* National Research Council, Washington, DC. 1999.
17. Lynd, L. R.; Wyman, C. E.; Gerngross, T. U. Biocommodity engineering. *Biotechnol. Progr.* **1999,** *15,* 777-793.
18. Embree, H. D.; Chen, T.; Payne, G. F. Oxygenated aromatic compounds from renewable resources: Motivation, opportunities, and adsorptive separations. *Chem. Eng. J.* **2001**, *84,* 133-147.
19. Parr, A. J.; Bolwell, G. P. Phenols in the plant and in man: the potential for possible nutritional enhancement of the diet by modifying the phenols content or profile. *J. Sci. Food Agricul.* **2000,** *80,* 985-1012.

20. Fargues, C.; Mathias, A; Rodriques, A. Kinetics of vanillin production from Kraft lignin oxidation. *Ind. Eng. Chem. Res.* **1996,** *35*, 28-36.
21. Bjorsvik, H.-R.; Minisci, F. Fine chemicals from lignosulfonates. 1. Synthesis of vanillin by oxidation of lignosulfonates. *Org. Proc. Res. Dev.* **1999,** *3,* 330-340.
22. Payne, G. F.; Chaubal, M. V.; Barbari, T. A. Enzyme-catalysed polymer modification: Reaction of phenolic compounds with chitosan films. *Polym.* **1996**, *37*, 4643-4648.
23. Lenhart, J. L.; Chaubal, M. V.; Payne, G. F.; Barbari, T. A. Enzymatic modification of chitosan by tyrosinase. In *Enzymes in Polymer Synthesis*; Gross, R. A.; Kaplan, D. L.; Swift, G., Eds.; American Chemical Society; Washington, DC, 1998, p. 188-198.
24. Clifford, M. N. Chlorogenic acids and other cinnamates – nature, occurrence, dietary burden, absorption and metabolism. *J. Sci. Food Agric.* **2000,** *80*, 1033-1043.
25. Kumar, G.; Smith, P. J.; Payne, G. F. Enzymatic grafting of a natural product onto chitosan to confer water solubility under basic conditions. *Biotechnol. Bioeng.* **1999,** *63*, 154-163.
26. Kudsy, M.; Kumazawa, H.; Sada, E. Pyrolysis of Kraft lignin in molten $ZnCl_2$-KCl media with tetralin vapor addition. *Can J. Chem. Eng.* **1995**, *73*, 411-415.
27. Kumar, G.; Bristow, J.F.; Smith, P.J.; Payne, G.F. Enzymatic gelation of the natural polymer chitosan. *Polym.* **2000**, *41*, 2157-2168.
28. Yu, M.; Hwang, J.; Deming, T.J. Role of L-3,4-dihydroxyphenylalanine in mussel adhesive proteins. *J. Am. Chem. Soc.* **1999**, *121*, 5825-5826.
29. Hansen, D.C.; Corcoran, S.G.; Waite, J.H. Enzymatic tempering of a mussel adhesive protein film. *Lang.* **1998**, *14*, 1139-1147.
30. Yamada, K.; Chen, T.; Kumar, G.; Vesnovsky, O.; Topoleski, L. D. T.; Payne, G.F. Chitosan based water-resistant adhesive. Analogy to mussel glue. *Biomacromolec.*, **2000**, *1*, 252-258.
31. Chen, T.; Kumar, G.; Harris, M. T.; Smith, P. J.; Payne, G. F. Enzymatic grafting of a hexyloxyphenol onto chitosan to alter surface and rheological properties. *Biotechnol. Bioeng.* **2000,** *70*, 564-573.
32. Muzzarelli, R. A. A.; Ilari, P.; Xia, W.; Pinotti, M.; Tomasetti, M. Tyrosinase-mediated quinone tanning of chitinous materials. *Carbohydr. Polym.* **1994,** *24*, 295-300.
33. Chen, T.; Vazquez-Duhalt, R.; Wu, C.-F.; Bentley, W.E.; Payne, G.F. Combinatorial screening for enzyme-mediated coupling. Tyrosinase-catalyzed coupling to create protein-chitosan conjugates. *Biomacromolec.*, **2001**, *2*, 456-462.

Chapter 20

Lipase-Catalyzed Grafting Reactions on Polysaccharides

Qu-Ming Gu

Hercules Incorporated Research Center, 500 Hercules Road, Wilmington, DE 19808–1599

Hydrophobic modification of polysaccharides is usually done by chemical means. In this work we have used enzymes, especially lipases, to carry out this hydrophobic modification. One example is the modification of hydroxyethylcellulose (HEC) using lipase-catalyzed transesterification reaction with vinyl stearate. The reaction conditions employed are much milder than those needed for the chemical synthesis. Similarly, this reaction has been applied to cationic guar (galactomannan substituted with 2-hydroxypropyl-3-trimethylammonium chloride), which is being used as an industrial thickening agent. Lipase-catalyzed reaction with vinyl stearate leads to stearoyl-cationic guar, which exhibits a moderate increase in solution viscosity when compared to the unmodified cationic guar. A similar reaction has been used to graft the acrylic functionality on HEC, using vinyl acrylate and lipase. The formation of ester bonds on HEC and on cationic guar has been confirmed by IR and solid state ^{13}C-NMR analysis.

Polysaccharides are sometimes modified with hydrophobic groups to provide surface active and interfacial properties[1-2]. These modifications are usually achieved via chemical reactions[3-5]. Thus, hydrophobic alkyl ether derivatives, which are widely used for industrial applications, are often synthesized under rather harsh caustic conditions[3]. Hydrophobic alkyl ester derivatives can be made by the use of alkyl carboxylic anhydrides[6] and alkyl acyl chloride[7]. In contrast to the traditional chemical approaches, several papers have recently reported the use of enzymes (lipase or protease) to graft alkyl esters on mono- or disaccharides[8-15]. Such reactions on polysaccharides have been much less investigated. Gross and Dordick, *et al*[16] reported the successful grafting of the caproyl group onto amylose using vinyl caprate and subtilisin Carlsberg. In this work, we have extended this reaction and shown that we can graft alkyl esters and acrylic groups onto cationic guar and hydroxyethyl cellulose (HEC).

Guar gum is a carbohydrate polymer containing galactose and mannose as the structural building blocks. It is useful as a thickening agent for water in various industrial applications[17]. Etherification of guar with cationic reagent 2-hydroxypropyl-3-trimethylammonium chloride gives a cationic guar, which has extended the usefulness of the guar gum[18]. Hydrophobic modification of the cationic guar may enhance the rheology of the cationic guar aqueous solution and also provide interfacial properties[2]. As part of our investigation of enzyme-catalyzed modifications of polysaccharides, we attempted the hydrophobic esterification of cationic guar with vinyl alkylate using a lipases or a protease as the catalyst. The resulting material was stearoyl cationic guar that showed both cationic and hydrophobic properties and exhibited a moderate increase in solution viscosity comparing to the unmodified cationic guar. Similarly, HEC was modified hydrophobically through the same enzymatic reaction using vinyl stearate. The reaction conditions employed were much milder than those used for the chemical synthesis. Commercial hydrophobically modified HEC is often used as a thickening agent in paint, construction, and personal care[17].

Experimental

Materials

Vinyl stearate, N,N-dimethylacetamide (DMAc), t-butyl methyl ether, and methanesulfonic acid were obtained from Aldrich. Lipases from *Pseudomonas fluorescens* (lipase AK) and from *Pseudomonas cepacia* (Lipase PS) were obtained from Amano Enzyme USA. Lipase B from *Candida antarctica* (Novozym 435) came from Novozymes. The cationic guar and hydroxyethyl cellulose (HEC) used for the study came from Hercules Incorporated.

Hydrophobic Modification of Cationic Guar in DMAc

A cationic guar (N-Hance 3000, Hercules, 5g) having a degree of substitution of 0.3 was suspended in 100 ml of DMAc. 0.8 g of vinyl stearate was added, followed by 0.4 g of lipase AK. After thorough mixing, a few drop of methanesulfonic acid was added. The resulting solution became very viscous. The mixture was then incubated at 50°C for 24 hours and then treated with isopropyl alcohol (IPA). The resulting precipitate was washed extensively with IPA and dried in air. The yield was 5.1 g. ^{13}C-NMR (solid state, 75.5Hz): δ14.0 (-CH_3), 23 (-CH_2-), 30 (-$OOCCH_2$-); IR: 1740 cm-1 (ester) and 1670 cm-1 (water or acid).

Hydrophobic Modification of Cationic Guar in t-Butyl Methyl Ether

A cationic guar sample (N-Hance 3000, Hercules, 10 g) is suspended in 100 ml of t-butyl methyl ether. 2 g of vinyl stearate was added, followed by 0.5 g of lipase PS. The mixture was stirred at 50°C for 72 hours. The product was recovered by filtration and washed with IPA and hexane. After air-drying, the yield was 10.5 g. The product was a water-soluble material. ^{13}C-NMR (solid state, 75.5Hz): δ14.0 (-CH_3), 24 (-CH_2-); IR: 1735 cm-1 (ester).

Hydrophobic Modification of Hydroxyethylcellulose (HEC) in DMAc

A hydroxyethylcellulose (Natrosol, Pharm-250MR, Hercules, 4 g) was suspended in 20 ml of DMAc, followed by the addition of 1 g of vinyl stearate and 0.5 g of lipase PS. After thorough mixing in 5-10 minutes, the resulting mixture turned into slurry that was subsequently incubated at 50°C for 48 hours. The yellowish slurry was treated with acetone/IPA (1:1) to give precipitates. After washing with IPA and air-drying, 3.8 g of the hydrophobically modified HEC was obtained as a white solid. IR: 1750 cm-1.

Brookfield Viscosity Measurement

As an example, viscosity of the aqueous solutions of the stearoyl-cationic guar was measured at 1% concentration at pH 6.5 and room temperature. A LV type Brookfield viscometer was used for the measurement with the spindle speed set at 30 rpm. The viscosity of the hydrophobically modified cationic guar varies with the degree of esterification (Table I).

Table I. Brookfield Viscosity of Stearoyl-Cationic Guar

Cationic Guar	*Solubility in water*	*pH of the solution*	*Viscosity (cps at 1%)*	*DE**
Unmodified	Yes	6.0	2,000	0
Stearate ester	Yes	6.0	5,600	0.05-0.2
Stearate ester	No	5.0, 6.0	0	>0.5
Stearate ester	Yes	6.0	4,200	0.05-0.1

*Degree of esterification, estimated by IR spectral analysis by comparing the carbonyl signals intensity of a known mixture of methyl palmitate/cationic guar with that of the enzymatically modified products.

Results and Discussion

The reaction of cationic guar with vinyl stearate in the presence of a lipase gave a hydrophobically modified polysaccharide. As shown in Figure 1, the product contains an ester functionality that absorbs at 1740 cm^{-1} in the IR spectrum. The starting material vinyl stearate has a distinct IR absorption at 1775 cm^{-1} that was not observed in the spectrum of the product, indicating all the unreacted vinyl stearate had been washed away during the work-up. Introduction of the ester bonds onto the guar gum has also been confirmed by solid state ^{13}C-NMR analysis, which shows a clear signal at 14.0 ppm, corresponding to $-CH_3$, and signals at 23 and 30 ppm corresponding to the CH_2 of the long chain fatty ester. A control experiment was performed without the use of the enzyme. The product did not show any detectable IR absorption in the region of 1735-1750 cm-1 or higher, suggesting ester formation was low or negligible in the absence of the enzyme.

The degree of substitution (DS) mainly depends on the amount of vinyl stearate used in the starting reaction mixture and reaction time. The more vinyl stearate used, the higher the degree of substitution of the product was. The enzyme use level also has an effect on the DS of the final product. When too much hydrophobe (>0.5 DS) is grafted onto the cationic guar, the product usually becomes water insoluble.

It was found that the enzymatic reaction performed very well when the starting material was dissolved or swollen in DMAc. When using an organic solvent such as t-butyl methyl ether in which cationic guar was merely suspended, the reaction took a longer time and the degree of esterification was also relatively low. However, as the lipases are relatively stable in a non-polar organic solvent environment, a slow but steady increase in the DS of the product could be achieved after long time incubation. Interestingly, when the same enzymatic reaction was applied to guar gum instead of its derivative cationic guar, the ester formation was very low. One interpretation of this result is that transesterification of the cationic guar with vinyl stearate occurs mostly at the 2-hydroxy functionality of the cationic moiety on the polysaccharide (Figure 2). Further study will be carried out to investigate the regioselectivity of this lipase-catalyzed modification of cationic guar.

We screened several commercially available lipases and proteases. The results indicated that the lipases from *Pseudomonus* sp., such as lipase AK and lipase PS from Amano Co., were active in DMAc and t-butyl methyl ether in catalyzing the transesterification of cationic guar and HEC with vinyl stearate. The lipase B from *Candida antarctica* (Novozym 435, Novozymes) was less active. Other lipases and proteases including immobilized Alcalase from Novozymes and subtilisin from Sigma were inactive under the reaction conditions. Suitable suspending media or solvents were hydrocarbons and polar aprotic solvents. The solvent mostly used in this work was DMAc since it solubilized cationic guar in the presence of a small amount of organic acid. It also dissolved HEC easily. The temperature of the reaction was optimally between 40-55°C.

The hydrophobically modified stearoyl-polysaccharides usually have a higher viscosity than the starting material if the products are water-soluble. As shown in Table 1, the viscosity of the cationic guar increased 2-3 times after stearoyl modification. This viscosity increase also confirms covalent attachment of the C18 hydrophobe onto cationic guar.

In order to study the generality of this method, the same lipase-catalyzed reaction has been applied to other polysaccharides. As depicted in Figure 3, cellulose ether hydroxylethylcellulose (HEC) that contains primary alcohol farther away from the polysaccharide backbone, has been modified covalently through transesterification using vinyl stearate as an acyl donor. Both Lipase PS and Novozym 435 have showed activity when DMAc was used as a solvent. In all other solvents such as t-butyl methyl ether, t-butanol and chloroform the formation of the ester material was also observed but with much lower yield according to IR analysis. At a low substrate concentration (<5%) in DMAc, Novozym 435 tended to be more active than lipase PS and lipase AK in catalyzing the grafting reaction.

The enzymatic reaction worked equally well for other polysaccharides. Other vinyl alkylates have also been studied for the transesterification reaction. Dordick, *et al*[11] has shown earlier that vinyl acrylate can be grafted onto sucrose using an enzyme catalyst. We have also successfully grafted the acrylic group onto HEC through enzyme-catalyzed transesetrification. Both the lipase from

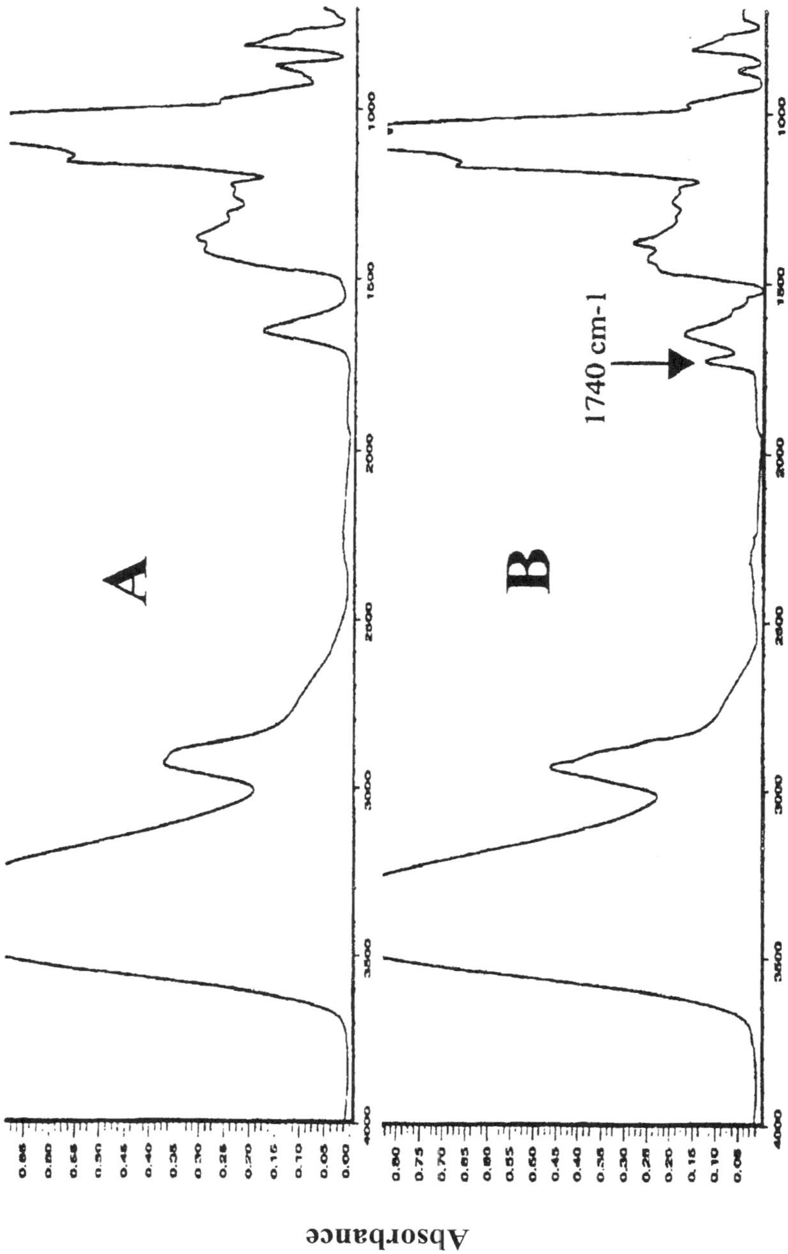
A
B
1740 cm-1
Absorbance
1000
1500
2000
2500
3000
3500
4000

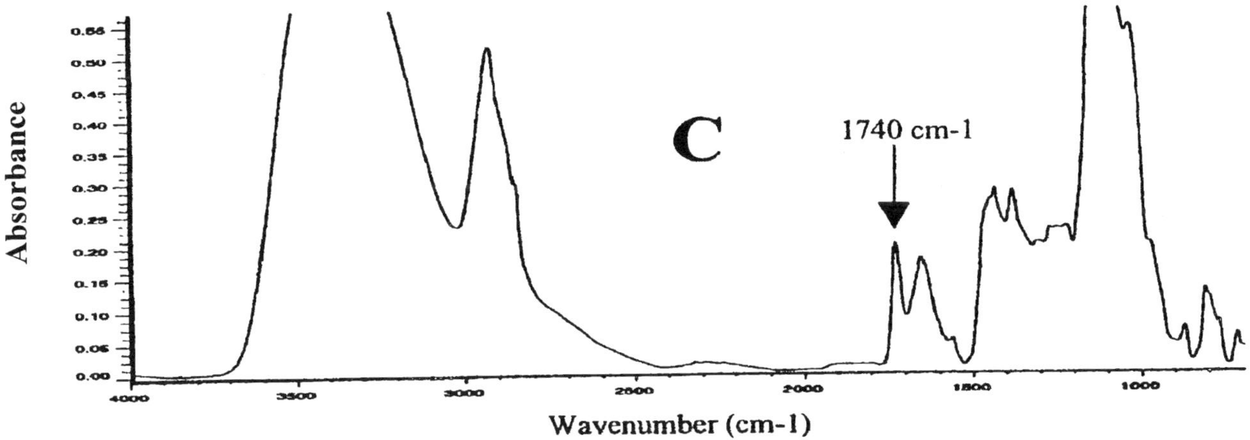

Figure 1. IR spectra of (A) unmodified cationic guar, (B) stearoyl-cationic guar obtained from the reaction with a starting material ratio of vinyl stearate-cationic guar at 1/5 (w/w), and (C) stearoyl-cationic guar obtained from the reaction with a starting material ratio of vinyl stearate-cationic guar at 2/5 (w/w); this material is not water-soluble.

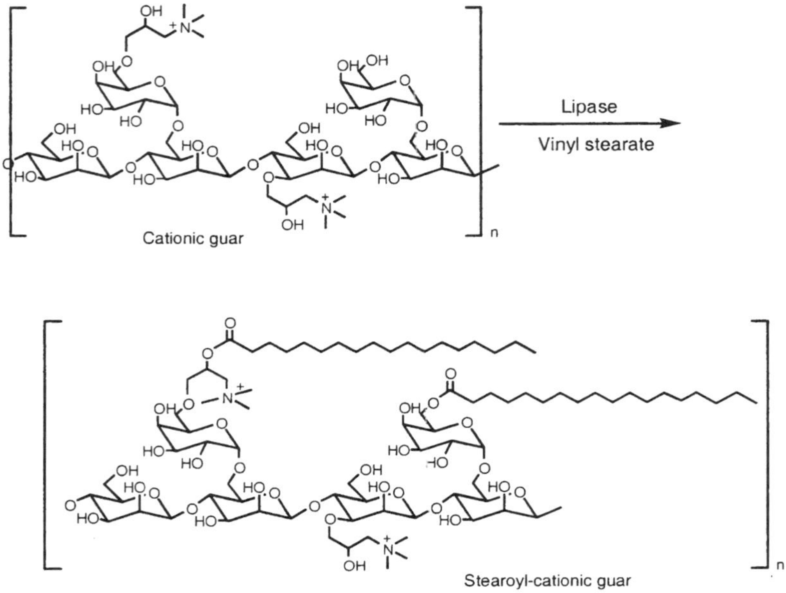

Figure 2. Hydrophobic modification of cationic guar

Pseudomonus cepacia (Lipase PS) and Celite-immobilized bacterial alkaline protease (Alcalase, Novozymes) can be used for this reaction. The structure of acrylic-HEC ester was confirmed by IR and ^{13}C-NMR spectral analysis. Interestingly, the grafted HEC material formed a strong crosslinking gel in DMAc at 5% upon heating to 80°C, indicating covalent bound of acrylate to HEC molecules.

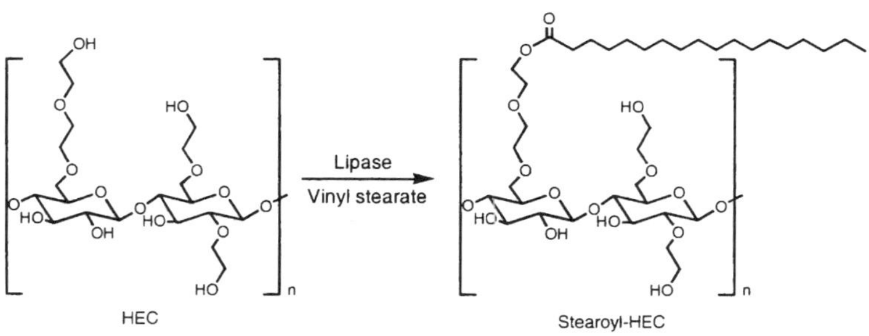

Figure 3. Hydrophobic modification of HEC

Conclusions

This work indicates that it is possible to graft a hydrophobic functionality onto cationic guar and hydroxyethylcellulose using enzymes as catalysts. The hydrophobic polysaccharides exhibit higher solution viscosities than the starting polysaccharides, as expected. The reaction conditions used are mild. These reactions may serve as tools whereby polysaccharid s can be modified in different ways. Furthermore, new structures derived from this method may be conceptualized. For example, the enzymatic reaction of vinyl stearate on any polymers containing 2-hydroxypropyl-3-trimethylammonium moiety may conceivably be used for the preparation of other polymers.

Acknowledgements. The author wishes to thank Dr. H. N. Cheng for his advice and valuable discussion during the work. Thanks are also due to Dr. Robert G. Nickol for his useful suggestions and Mrs. Sadhana Mital for her experimental assistance.

References

1. Denkinger, P.; Kunz, M.; Burchard, W. *Colloid Polymer. Science.* **1990**, 268, 513.
2. Zhang, T.; Marchant, R. E. J. *Coll. Interface Science.* **1996**, 177, 419.
3. Landoll, L. M. *J. Polymer. Science.* **1982**, 20, 443.
4. Zhang, T.; Marchant, R. E. *Micromolecules.* **1994**, 27, 7302.
5. Mintz, R. C., Cai, W.; Marchant, R. E. *Trans. World Biomat. Cong.* **1996**, 5, 828.
6. Malta, C. J. *Analytical. Chemistry.***1953**, 25, 245; Malta, C. J. *Industrial and Engineering Chemistry.* **1957**, 49, 84.
7. Kawaguchi, K.; Matsukawa, K.; Gama, Y. *Carbohydrate Polymers* **1992**, 18, 139.
8. Shibatani, S.; Kitagawa, M.; Tokiwa, Y. *Biotechnology Lett.* **1997**, 19, 511; Hiratake, J.; Yamamoto, K.; Yamamoto, Y.; Oda, J. *Tetrahedron Lett.* **1989**, 30, 1555; Faber, K.; Riva, S. *Synthesis*, **1992**, 895.
9. Uemura, A.; Nozaki, K..; Yamashita, J.-I.; Yasumoto, M.; *Tetrahedron Lett.* **1989**, 30, 248.
10. Yahya, A. R. M.; Anderson, W. A.; Moo-Young, M. *Enzyme and Microbial Technology* **1998**, 23, 438.
11. Chen, X.; Martin, B. D.; Neubauer, T. K.; Linhardt, R. J.; Dordisk, J. S.; Rethwisch, D. G. *Carbohydrate Polymers* **1995**, 28, 15.
12. Cordova, A.; Hult, K.; Iversen, T. *Biotechnology Lett.* **1997**, 19, 15
13. Kitagawa, M.; Tokiwa, Y. *Biotechnology Lett.* **1998**, 20, 627.
14. Degueil-Castaing, M.; De Jeso, B.; Drouillard, S.; Maillard, B. *Tetrahedron Lett.* **1987**, 28, 953.

15. Redmann, I.; Pina, M.; Guyot, B.; Blaise, P.; Farines, M.; Graille, J. *Carbohydrate Research* **1997**, 300, 103.
16. Bruno, F. F.; Akkara, J. A.; Ayyagari, M.; Kaplan, D. L.; Gross, R.; Swift, G.; Dordick, J. S. *Macromolecules* **1995**, 28, 8881.
17. Davidson, R., L. *Handbook of Water-Soluble Gums and Resins*; McGraw-Hill, New York, **1980**; Stephens, A. M. ed., *Food Polysaccharides and their Applications*, Marcel Dekker, New York, **1995**.
18. R. L. Whistler, J. N. BeMiller, eds., *Industrial Gums,* 3rd Ed., Academic Press, San Diego, **1993**; Phillips, G. O.; Wedlock, D. J.; Williams, P. A.. eds., *Gums and Stabilizers for the Food Industry*, Elsevier, London, Vol. 6, **1992**.

Chapter 21

Lipase-Mediated Selective TEMPO Oxidation of Hydroxyethylcellulose

Shanghui Hu[1], Wei Gao[1], Rajesh Kumar[1], Richard A. Gross[1,*], Qu-Ming Gu[2], and H. N. Cheng[2,*]

[1]**NSF Center for Biocatalysis and Bioprocessing of Macromolecules, Department of Chemistry, Polytechnic University, 6 Metrotech Center, Brooklyn, NY 11201**
[2]**Hercules Incorporated Research Center, 500 Hercules Road, Wilmington, DE 19808–1599**

Herein, the lipase-mediated selective TEMPO oxidation of hydroxyethylcellulose was investigated. A system for *in-situ* generation of hypobromite catalyzed by lipase was designed, and was successfully coupled with TEMPO oxidation. Comparison was made with chemically generated hypobromite/TEMPO oxidation and with hypochlorite-mediated TEMPO oxidation. With poly(ethylene glycol) and hydroxyethylcellulose as model polymers, the lipase approach was found to have the least amount of molecular weight degradation and ether cleavage.

Oxidation of polysaccharides leads to enhanced gelatinization strength, increased interaction with other carbohydrates and peptides, increased swelling properties and higher solubilities of the polymer *(1)*. The functionalized polymers can be used in the paper, food, textile, detergent and pharmaceutical industries. The selective oxidation of primary hydroxyl groups in polysaccharides is an important modification reaction, and the polymers produced may be precursors for further derivatization, e.g., cross-linking, esterification, succinate coupling, and reductive amination. Because of the similarity of primary and secondary alcohol groups, this type of oxidation cannot be accomplished with most of the oxidants in common use. However, the nitroxyl radical TEMPO (2,2,6,6-tetramethyl-1-piperidinyloxy) is a highly efficient oxidative catalyst for *regio*selective oxidation of primary alcohol groups in preference over secondary alcohols *(2)*. Recently, De Nooy *et al.* introduced this method to polysaccharide chemistry for the selective oxidation of primary alcohols of aqueous soluble polysaccharides *(3a-c)*.

Hydroxyethylcellulose (HEC) derivatives are widely used as rheology modifiers, thickening agents, protective colloids, and a variety of other applications. HEC is a statistical polymer, where the hydroxyethyl groups are statistically distributed along the polymer backbone and also chained out as substituted poly(ethylene glycol) units. Very recently, enzymatic modifications of HEC have been investigated using β-galactosidases and lipase via transgalactosylation and lipase-catalyzed ring-opening polymerization *(4)*. There have been a few reports on selective oxidation of HEC (Figure 1) by chemical or biological approaches. One method known for the oxidation of primary alcohols of cellulose and its derivatives is the oxidation with nitrogen oxides or halogen oxides *(5)*. Chang and Robyt *(3d)* used TEMPO to oxidize a number of polysaccharides including cellulose, but HEC was not included. However, a patent *(3e)* reported the use of TEMPO to oxidize cellulose to form carboxylic acids and aldehydes, and another patent *(3f)* reported the combined use of TEMPO and an oxidizing enzyme to oxidize HEC.

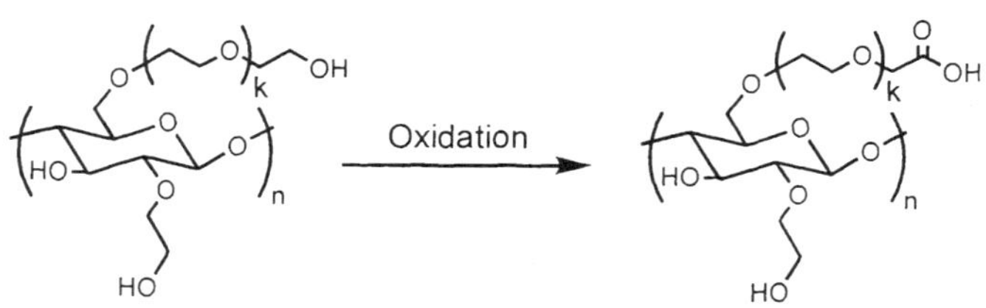

Figure 1. Oxidation of hydroxyethylcellulose (HEC)

Lipases have been extensively used for *regio-* and *stereo*selective synthesis in organic chemistry and for polymerization in polymer chemistry *(6)*. The immobilized lipases can also be applied to generate peroxycarboxylic acids in a suitable organic solvent directly from the parent carboxylic acids and hydrogen peroxide, in which the latter is a nucleophile in the catalytic formation of peroxycarboxylic acids *(7)*.

In this work, we have investigated lipase-mediated selective TEMPO oxidation of HEC. We have designed a novel system for *in-situ* generation of hypobromite catalyzed by lipase and have successfully coupled this oxidative system with TEMPO oxidation of polysaccharides. We have used poly(ethylene glycol) and HEC as model polymers to demonstrate that this oxidative system has a number of advantages over hypochlorite-mediated TEMPO selective oxidation of primary alcohols for polysaccharides and their derivatives.

Results and Discussion

In the lipase approach as shown in Figure 2, lauric acid was used as the starting material. The peroxylauric acid was produced in a yield of 50% from hydrogen peroxide (50%) and lauric acid in a two-phase system of toluene and water, through the catalysis of immobilized *Candida antarctica* lipase. In comparison with traditional acid-catalyzed oxidation of carboxylic acid to peroxycarboxylic acids, the enzyme-catalyzed procedure was a mild and safe alternative. The peroxy acid formed was consumed for *in-situ* oxidation of potassium bromide to bromine, catalyzed by tetrabutylammonium bromide (TBAB) at cold temperature. In an alkaline solution (pH >10), the bromine thus obtained was readily converted to hypobromite at 0°C. The hypobromite was then slowly added to a solution of polysaccharide and a catalytic amount of TEMPO in a two-phase system of toluene and water (1:1) at 0°C during 1 hour. The solution pH was kept at 10.5 by adding NaOH.

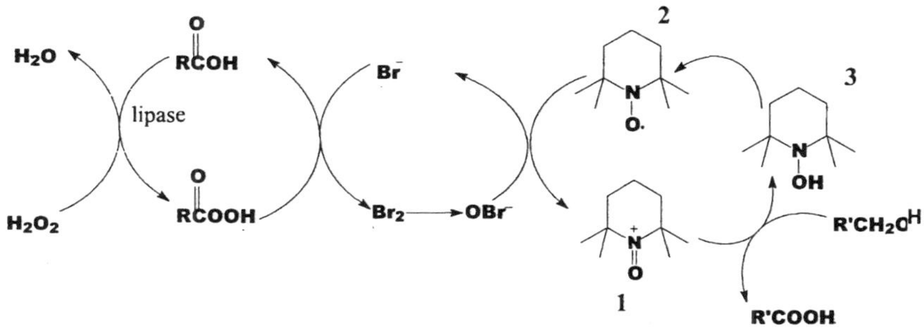

Figure 2.Lipase approach to oxidize HEC

In this oxidative system, TEMPO (**1**) is soluble in the toluene phase, and oxidation occurs at the interface of two phases. The actual oxidant is the nitronium ion **2**, which is formed by one-electron oxidation of TEMPO. During the oxidation of the primary alcohol group of the polysaccharide, **2** is reduced to the hydroxylamine **3** which is reoxidized by HOBr, from which TEMPO is again obtained *(2,3)*. In the course of the reaction, the TEMPO resides primarily in the toluene phase and can be recycled several times. For the alkaline hypohalite oxidation of starch, the use of hypobromite in place of hypochlorite largely reduces polymer degradation *(8)*. Furthermore, we speculate that the long chain lauric acid might play a positive role in limiting the degradation of the hydroxyethyl side chain and the main chain of HEC.

High-resolution ^{13}C NMR spectroscopy is a well-known tool for determining the microstructure of HEC as well as the average degree of substitution of the substituents on the anhydroglucose ring *(9)*. Both ^{13}C NMR and IR spectroscopy have been employed to study the oxidation reaction. In the ^{13}C NMR spectrum (Fig. 3, upper trace) of the oxidized HEC (from the lipase and hypobromite approach), the signals at 179 ppm clearly showed the formation of carbonyl groups in the products. In the IR spectrum (Fig. 3, lower trace), the band absorption at 1739 cm^{-1} further indicated that the oxidized products were mainly carboxylic derivatives of HEC. Furthermore, based on ^{13}C NMR spectral analysis, we confirmed that the oxidation was highly selective on primary alcohol groups (instead of secondary alcohols), and this observation was compatible with the results reported for the TEMPO oxidation of other polysaccharides.

In order to demonstrate the merit of the lipase approach, we also used a modified method from De Nooy's approach *(3)*, in which hypochlorite was employed as an initial oxidant and bromide ion employed as a catalyst in a two-phase system of toluene and water. Alternatively, we used another chemical oxidative system, in which *meta*-chloroperoxybenzoic acid (MCPBA) was the initial oxidant for *in-situ* generation of bromine, which was then coupled to TEMPO oxidation (just as the lipase approach).

In the above two approaches a variety of parameters, such as pH, oxidant concentration (NaOCl, MCPBA and TEMPO) and solvent systems, were optimized for the selective oxidation of HEC (data not shown). We found that *at pH 10.5* the relative molecular weight of the oxidized HEC was higher than at other pH values investigated for both chemical methods. The molecular weight was higher when using two equivalents (eq) of NaOCl and 2.5 eq of MCPBA. These results were similar to that of the hypochlorite-mediated TEMPO oxidation of starch and water-soluble glucans *(3)*. In the case of MCPBA-mediated TEMPO oxidation of HEC, a binary phase system of toluene/H_2O significantly reduced the degradation of the polymer, compared to the single aqueous phase reaction. In this study, the molecular weight of the oxidized polymers was determined by size exclusion chromatography (SEC) as well as by multi-angle laser light scattering combined with SEC (SEC-MALLS). The molecular weights were measured relative to poly(ethylene oxide) to evaluate the efficiencies of the different oxidative systems. The absolute molecular weight and rms radius of gyration were determined by SEC-MALLS.

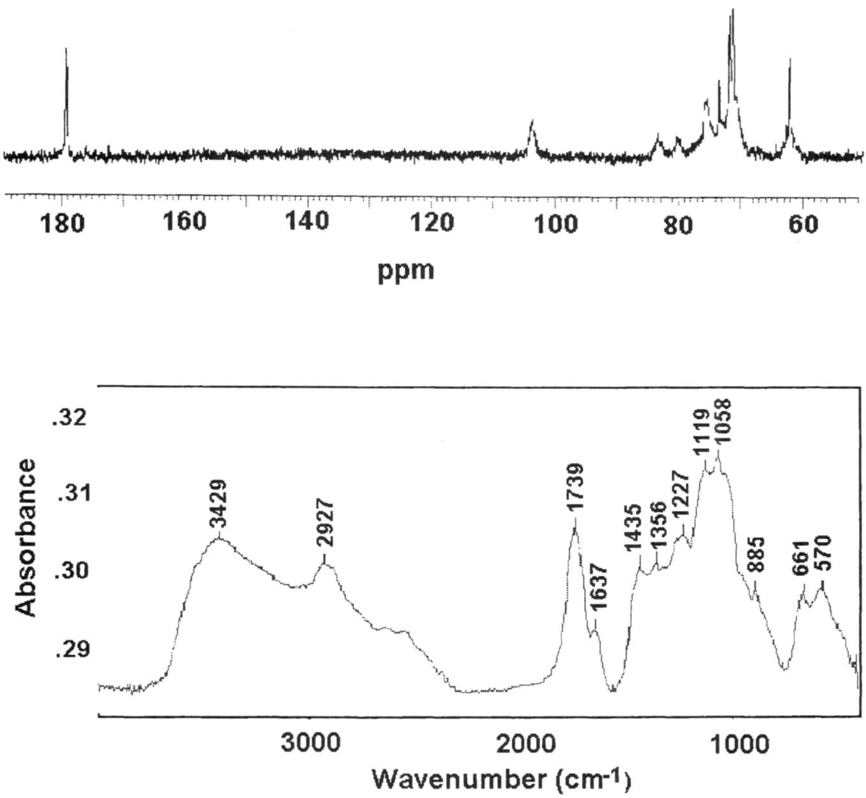

Figure 3. IR and ^{13}C NMR spectra of oxidized HEC from lipase catalysis

We applied the optimized experimental conditions (pH 10.5, 10% TEMPO, and aqueous/toluene two-phase system) to lipase-mediated TEMPO oxidation of HEC (Method **A**), MCPBA (Method **B**) and NaOCl (Method **C**). Under the standard conditions (see the Experimental Section), the *lipase approach produced a polymer with higher molecular weight* when compared to the hypochlorite method (M_w 250,000 versus 105,000, Table 1). In addition, *the lipase approach showed little or no degradation of hydroxyethyl side chains of HEC* when compared with the hypochlorite approach (DS 1.50 versus 1.23, Table 2). Here, the degree substitution (DS) of the hydroxyethyl group on HEC was determined by quantitative ^{13}C NMR analysis on a 600 MHz spectrometer.

Table 1. Absolute molecular weights and rms radii of HEC and its oxidized products

	Molar mass ($x10^{-5}$g/mol)			*RMS radius (nm)*		
Sample	*Mn*	*Mw*	*Mz*	*Rn*	*Rw*	*Rz*
HEC	3.0±0.1	4.8±0.1	6.6±0.5	40±2	45±2	50±2
NaOCl-HEC[1]	0.51±0.03	1.05±0.02	2.5±0.2	18±10	19±5	25±2
Lipase-HEC[2]	1.19±0.05	2.50±0.05	4.8±0.2	27±5	32±2	38±1

[1]NaOCl-mediated oxidation product (NaOCl 2.0 eq, HEC 500 mg in toluene/H_2O, TEMPO 50 mg, pH 10.5).

[2]Lipase-mediated oxidation product (Lipase 2.5g (7000 units/g), HEC 500 mg in toluene/H_2O, TEMPO 50 mg, pH 10.5)

Table 2. Degree of substitution of HEC and its oxidized products based on quantitative ^{13}C NMR analysis

Method[1]	*-COONa*	*-CH_2OH*	*Total DS*
control	0.00	~1.49	~1.49
A (lipase)	0.82	0.68	1.50
B (MCPBA)	0.44	1.06	1.49
C (NaOCl)	0.88	0.35	1.23

[1]Method **A**: Lipase 2.5g (7000 units/g), HEC 500 mg in toluene/H_2O, TEMPO 50 mg, pH 10.5. Method **B**: MCPBA 2.5 eq, HEC 500 mg in toluene/H_2O, TEMPO 50mg, pH 10.5. Method **C**: NaOCl 2.0 eq, HEC 500 mg in toluene/H_2O, TEMPO 50 mg, pH 10.5.

The degradation of the hydroxyethyl side chain was rather surprising. In order to check this reaction, we carried out a study of the effect of different oxidation reactions on the molecular weight of a sample of poly(ethylene glycol) (Figure 4). The results are given in Table 3. It is interesting that TEMPO or NaOBr alone gave no molecular weight degradation. The treatment of PEG with a combination of NaOCl and TEMPO (Method C) gave substantial degradation (reducing the molecular weight by one-third). In contrast, the treatment of PEG with in-situ generated NaOBr (Method B) decreased molecular weight only slightly.

Lipase-mediated TEMPO oxidation of the primary alcohol groups appears to be a general and highly selective reaction. This approach exhibits a number of advantages over the hypochlorite-mediated TEMPO oxidation of polysaccharides. First, the use of lipase as a biocatalyst entails a mild and safe oxidative procedure, and the immobilized lipase can be reused for several cycles. Secondly, owing to the exclusion of hypochlorite, the degradation of the main chain of polysaccharides and the polyether side chain is much reduced. The direct addition of a low concentration of hypobromite also excluded the possible rate limitation in the course of hypochlorite mediated oxidation of KBr, and further enhanced the rate of oxidation at higher alkaline solution (pH >10.5). In an aqueous/toluene two-phase oxidative system, the catalyst TEMPO was soluble in the toluene phase and could be easily recovered and reused for the next reaction.

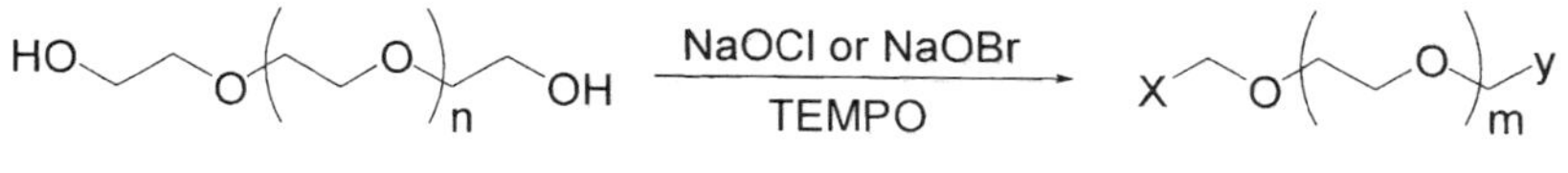

n > m;   x, y = CH_2OH or COOH

Figure 4. Oxidation of poly(ethylene glycol) (PEG)

Table 3. Effect of different oxidation methods on the molecular weights of poly(ethylene glycol) (PEG)

Method	*Treatment*	M_w	M_w/M_n
control	none	31,100	1.22
control	TEMPO only	30,400	1.29
control	NaOBr only	30,200	1.27
B	NaOBr + TEMPO	28,100	1.47
C	NaOCl + TEMPO	13,100	1.99

This synthetic method is highly suited for the selective oxidation of cellulosic derivatives and starch derivatives. In these cases, only modest degradation of the polysaccharides occurs, which makes this type of reaction, in principle, particularly useful for the production of carboxylated polysaccharides. Furthermore, lipase-mediated TEMPO oxidation of primary alcohols might be applied for selective oxidation of other polymers containing primary alcohols.

Experimental Section

Materials

Hydroxyethylcellulose was obtained from Hercules Incorporated (Natrosol®). TEMPO (2,2,6,6-tetramethyl-1-piperidinyloxy), hydrogen peroxide (50% aqueous solution), ~11% hypochlorite aqueous solution, and *meta*-chloroperoxybenzoic acid (MCPBA) were purchased from Aldrich Chemical Company. Poly(ethylene glycol) (PEG 35,000) was from Fluka. The immobilized *C. antarctica* lipase (7000 unit/g) was a gift from Novo Nordisk, which was dried by lyophilization before use. Dialysis membranes with molecular weight cutoff 3,500 and 12,000 were purchased from VWR.

Instrumentation

The FT-IR spectra of polysaccharides were recorded on KBr plates using a Perkin-Elmer 1600 spectrometer at ambient temperature. ^{1}H and ^{13}C NMR spectra were obtained on a Varian 600 MHz NMR spectrometer in D_2O at 70°C. The quantitative ^{13}C NMR spectra were determined in a 10-*mm* tube at 10% w/v in D_2O. Chemical shifts were reported relative to TMS as an internal standard. The relative intensity data were collected with the gated decoupling (suppressed NOE) experiment. The parameters selected to acquire quantitative ^{13}C NMR spectra were: AT = 3.0 sec, □= 90°, and 10-*sec* pulse delay.

The SEC system consisted of a Waters 510 pump, a U6K model injector, and a 410 differential refractometer coupled to a Wyatt DAWN DSP multi-angle laser light scattering photometer. A three-column set consisting of 300x7.5 mm Shodex KB-80M, KB806M and KB802.5 was used to chromatograph the samples. The specific refractive-index increment (dn/dc) of NaOCl-mediated oxidized product was determined by RI detector at 35°C; 0.130 was used for the samples. The solvent used for absolute molecular weight determination was 0.25M NaOAc buffer (pH7.0).

Method A: Lipase-Mediated TEMPO Oxidation of Polysaccharides

Lipase-Catalyzed Production of Peracid

To lauric acid (0.9 g, 4.5 mmol) in toluene (7 ml) was added the immobilized *C. antarctica* lipase (0.5g, Novo Nordisk, 7000 unit/g). Hydrogen peroxide (50%, 6.0 mmol, 390 □l) was added dropwise into the reaction mixture during 1.5 hr. The mixture was then stirred at room temperature (rt) for 3 hrs. After filtration of the lipase the organic layer was washed by distilled water and afforded the peracid solution. The concentration of peracid was measured by HPLC analysis.

In-situ Generation of NaOBr by Peracid

To the peracid solution of toluene/CH_3CN (5ml/3ml) was added NaBr (3.0 mmol, 350 mg) and catalytic amount of *tetra*-butylammonium bromide (TBAB). The mixture was stirred for 30 min at 0°C, and then the pH value of the solution was adjusted up to pH 10.5.

Oxidation of Polysaccharides

To HEC (100 mg) and a catalytic amount of TEMPO (10 mg) in toluene/CH_3CN (3 ml/3ml) was added slowly the above mixture at 0°C- -3 °C for 45 min. The pH was controlled by a pH meter and kept at 10.5 by the addition of 1.0 M NaOH. When the reaction was done, ethanol was added to quench the reaction. The aqueous phase was washed by toluene and precipitated in cold ethanol. The concentrated aqueous solution was dialyzed against the aqueous solution (pH 8.5) and the distilled water (pH 6.5) for 96 hr. After lyophilization the oxidized products were recovered in 80~85% yields, and the degree of oxidation was in the range of 80-90% based on ^{13}C NMR spectral analysis.

Method B: MCPBA-Mediated TEMPO Oxidation of Polysaccharides

In-situ Generation of NaOBr by MCPBA

To NaBr (3.0 mmol, 350 mg) in a mixture of H_2O/toluene/CH_3CN (3ml/3ml/3ml), a catalytic amount of TBAB was added to MCPBA (1.5 mmol,

275 mg, 77%) at 0°C. The mixture was stirred at 0°C for 45 min and the pH of the solution was kept at 10.5 by adding 1M NaOH solution.

Oxidation of Polysaccharides

To HEC (100 mg) and a catalytic amount of TEMPO (10 mg) in toluene/CH3CN (3 ml/3ml) was added slowly the above mixture at 0°C~ -3 °C for 30 min. The pH was controlled by a pH meter and kept at 10.5 by the addition of 1.0 M NaOH. As before, the addition of ethanol quenched the reaction. The aqueous phase was washed by toluene and precipitated in cold ethanol. The concentrated aqueous solution was dialyzed against aqueous solution (pH 8.5) and the distilled water (pH 6.5) for 96 hr. After lyophilization the oxidized product was afforded in yields of 85~90%.

Method C: NaOCl-Mediated TEMPO Oxidation of Polysaccharides

To HEC (100 mg) and a catalytic amount of TEMPO (10 mg) in toluene/CH3CN (3 ml/3ml) was added slowly NaOCl (1.5 mmol, 2.5 eqv) at 0°C- -3 °C for 30 min . The pH was controlled by a pH meter and kept at 10.5 by the addition of 1.0 M NaOH. After 45 min, ethanol was added to quench the reaction. The aqueous phase was washed with toluene and precipitated in cold ethanol. The concentrated aqueous solution was dialyzed against the distilled water for 24 hr. After lyophilization the oxidized product was afforded in yield of 80~85%.

Oxidation of Poly(ethylene glycol)

Method B.

NaBr (6.0 mmol, 700 mg) dissolved in 6 ml water was mixed with 6 ml toluene and 6 ml acetonitrile. TBAB (20 mg) and MCPBA (3.0 mmol, 550 mg) were added at 0°C. The color of the reaction solution turned to yellow. The mixture was stirred at 0°C for 45 minutes and the pH was kept at 10.5 by adding 1.0 M NaOH. To the resulting solution, PEG (35,000 MW, 200 mg) and TEMPO (20 mg) in toluene/acetonitrile (6 ml/6 ml) were added slowly at 0-5°C. The pH was maintained at 10.5 by 1.0 M NaOH. The reaction mixture was stirred at 0-5°C for 30 minutes, and then at room temperature for 30 minutes. The pH was adjusted to 7.0 using 1.0 M HCl. The material was lyophilized and washed with ethyl acetate to give 140 mg of product.

Method C

NaOCl (11.5% available chlorine, 1 ml, 1.54 mmol) was added slowly to PEG (35,000 MW, 200 mg) and TEMPO (20 mg) in toluene/acetonitrile (6 ml/6 ml) at 0-5°C. The pH was maintained at 10.5 by adding 1.0 M NaOH. The reaction mixture was stirred at 0-5°C for 30 minutes, and then at room temperature for 30 minutes. The pH was adjusted to 7.0 using 1.0 M HCl. The material was lyophilized and washed with ethyl acetate to give 150 mg of product.

References

1. (a) Edwards, W. P.; Jackson, E. B. (Ed) *Sugar Confectionary Manufacture*, Chapman & Hall, 1995, p38. (b) Nussinovitch, A. *Hydrocolloid Applications-Gum Technology in the Food and Other Industries*, Chapman & Hall, 1997. (c) Dalgleish, D. G.; Hollocou, A. L. In Dickinson, E.; Bergenstahl, B. (Eds) *Food Colloids, Proteins, Lipids and Polysaccharides*. Royal Society of Chemistry (UK), 1997, p 236.
2. (a) Golubev, V. A.; Rozantsev, E. G.; Neiman, M. B.; Akad, Izv. Nauk SSSR, Ser. Khim., **1965**, 1927. (b) Cella, J. A.; Kelly, J. A.; Keneham, E. F. *J. Org. Chem.* **1975** 40, 1860. (c) Semmelhack, M. F.; Schmid, C. R.; Cortes, D. A. *J. Am. Chem. Soc.* **1984**, 106, 3374-. (d) Anelli, P. L.; Biffi, C.; Montanari, F.; Quici, S. *J. Org. Chem.* **1987**, 52, 2559. (e) Davis, N. J.; Flitsch, S. L. *Tetrahedron Lett.* **1993**, 34, 1181.
3. (a) De Nooy, A. E. J.; Besemer, A. G.; van Bekkum, H. *Recl. Trav. Chim. Pays-Bas* **1994**, 113, 165-166. (b) De Nooy, A. E. J.; Besemer, A. G.; van Bekkum, H. *Carbohydr. Res.* **1995**, 269, 89-98. (c) De Nooy, A. E. J.; Besemer, A. G.; van Bekkum, H.; van Dijk, J. A. P. P.; Smit, J. A. M. *Macromolecules* **1996**, 29, 6541-6547. (d) Chang, P.S.; Robyt, J. F. *J. Carbohydrate Chem.* **1996**, 15, 819-830. (e) Jaschinski, T.; Gunnars, S.; Besemer, A. C.; Bragd, P.; Jetton, J.; Van den Dool, R.; Van Hartingsveldt, W. "Oxidized Cellulose-Containing Fibrous Materials and Products made Therefrom." WO 00/50462, 31 August 2000. (f) Jetton, J.; Van den Dool, R.; Van Hartingsveldt, W.; Van Wandelen, M. T. "Porcess for Selective Oxidation of Primary Alcohols and Novel Carbohydrate Aldehydes." WO 00/50621, 31 August 2000.

4. p(a) Li, J.; Cheng, H. N.; Nickol, R. G.; Wang, P. G. *Carbohydr. Res.* **1999**, 316, 133-137. (b) Li, J.; Xie, W.; Cheng, H. N.; Nickol, R. G.; Wang, P. G. *Macromolecules* **1999**, 32, 2789-2792.
5. (a) Yackel, E. C.; Kenyon, W. O. *J. Am. Chem. Soc.*, **1942**, 64, 121-127. (b) Painter, T. J. *Carbohydr. Res.*, **1977**, 55, 95-103. (c) Painter, T. J.; Cesaro, A.; Delben, F.; Paoletti, S. *Carbodydr. Res.*, **1985**, 140, 61-68. (d) de Nooy, A. E. J.; Pagliaro, M.; van Bekkum, H.; Besemer, A. C. *Carbodydr. Res.*, **1997**, 304, 117-123. (e) Pagliaro, M. *Carbodydr. Res.*, **1998**, 308, 311-317.
6. Gross, R.; Kumar, A.; Kalra, B. *Chemical Rev.* 2001, in press.
7. (a) Bjorkling, F.; Godtfredsen, S. E.; Kirk, O. *J. Chem. Soc., Chem. Commun.* 1990, 1301-1303. (b) Bjorkling, F.; Frykman, H.; Godtfredsen, S. E.; Kirk, O. *Tetrahedron* **1992**, 4587-4592.
8. Besemer, A. C.; van Bekkum, H. *Starch* **1994**, 46, 95-101.
9. DeMember, J. R.; Taylor, L. D.; Trummer, S.; Rubin, L. E.; Chiklis, C. K. *J. Appl. Polym. Sci.*, **1977**, 21, 621-627.

Chapter 22

Reaction-Diffusion of Enzyme Molecules in Biopolymer Matrices

Yu Cheng and Robert K. Prud'homme

Department of Chemical Engineering, Princeton University, Princeton, NJ 08544

In this study, the reaction-diffusion of enzyme molecules in guar galactomannan polymer solutions was investigated. In very dilute polymer solutions, reaction rate increases as first-order kinetics with substrate concentration. In the intermediate concentration regime, the enzyme/polymer binding saturates, and the degradation kinetics becomes zero-order, in agreement with a Michaelis-Menton kinetics model. However, as the solution increases in concentration, the reaction rate decreases and the enzyme diffusion through the concentrated polymer gel becomes the rate–limiting step. A reaction-diffusion model based on polymer scaling theory is presented to express the competition between enzyme reaction and diffusion. The effect of molecular weight and derivatization of the substrate polymer on the degradation kinetics was also explored. The degradation rate was shown to be greatly affected by the type of substituent groups as well as the degree of substitution. A triggered degradation mechanism was found for the enzymatic hydrolysis of cationically derivatized guar solutions.

Guar galactomannan is a naturally occurring polysaccharide which consists of a linear backbone of β-1,4-linked mannose units with α-1,6-lined galactose units randomly attached as side chains (Figure 1). The ratio of mannose to galactose units is approximately 1.8:1 *(1)*. Due to its excellent thickening ability, guar has been widely used as a rheology modifier in food, paper, textile and oil industries *(2)*. Additionally, the architecture of guar can be selectively modified or degraded to tailor properties of guar formulations and open up new opportunities for its applications. For example, the galactose content of guar can be modified so that guar can form synergistic gels with other biopolymers such as xanthan and carrageenan. In oil industry, a guar-based gel needs to be degraded to create the outflow of oil. In both these applications, enzymes offer an efficient and environmentally benign method for modifying the guar structure. Three types of bonds in guar are susceptible to enzymatic hydrolysis: the endo- and exo- β-1,4 linkages on the D-mannose backbone and the α-1,6 linkage between the mannose unit and the galactose side-chain. The enzymes that cleave these bonds are respectively endo- and exo- β-mannanase and α-galactosidase. In order to improve the efficiency of the enzymatic reaction and design enzymes with better properties, it is important to understand the enzymatic degradation kinetics in guar solutions as well as the correlation between the kinetics and the rheological properties of guar solutions during the degradation.

During the past decades, considerable interests have developed in the enzymatic degradation and modification of polysaccharides, due to their great industrial applications ranging from the production of non-fossil energy resources, pulp and paper processing, to drug delivery systems and oil/gas exploration *(3)*. However, most of these studies concentrated on the degradation of cellulose, starch and dextran polymers *(4-6)*. β-D-Mannanases, which are produced by plants, bacteria and fungi, catalyze the hydrolysis of 1,4-β-D-mannosidic linkages in galactomannan polymers. Their ability to degrade these substrates depends on both the degree of polymerization and galactose substitution levels *(7)*. Tayal et al. studied the enzymatic degradation kinetics of high molecular weight native guar *(8)*. Guar solution viscosity was found to be very sensitive to the enzymatic hydrolysis and decreased by several orders of magnitudes during the course of degradation. An empirical zero-order kinetics is recently reported by using Gammanase, a commercial mixed enzymatic system to degrade guar *(9)*. However, they only studied the degradation over a certain range of polymer concentrations. There is no report in the literature on the degradation kinetics at very low and high polymer concentration ranges. Furthermore, even though a variety of substituted guars are used widely in industrial applications, there has been no published data on the influence of substitution on enzyme reactivity *(10, 11)*.

In this study, the enzymatic degradation process was followed over a wide range of substrate concentrations using gel permeation chromatography (GPC) and steady shear viscometry. A Michaelis-Menton model is used to explain the results based on previous studies of Suga et al. *(5)*. We also proposed a reaction-diffusion model to further discuss the competition between enzyme reaction and diffusion in all concentration ranges. In addition, guar can be modified by grafting side-groups (-R) onto the hydroxyl sites. Three commercially available derivatized guars were used in this study: hydroxypropyl guar (HPG), carboxymethyl guar (CMG) and hydroxypropyl trimethylammonium guar (HPTMAG) (Figure 2). HPG is a neutral polymer, while both CMG and HPTMAG are ionized in solution. The β-mannanase enzyme we used in the experiment has an isoelectric point (pI) of 3.5 *(12)*, so in a polymer solution at pH 7 the enzyme is negative charged. Therefore, by adding either neutral or charged groups on guar, we studied both the steric and columbic interactions between the enzyme and the substrate for three derivatized guars, and explored the effect of side groups on the degradation kinetics.

Experimental Section

Materials

Polymers used in this study, guar, HPG (MS=0.18), HPG (MS=1.53), CMG (DS=0.1), CMG (DS=0.6) and HPTMAG (DS=0.14), are provided by Rhone-Poulenc, Inc. (Cranbury, NJ) as a gift. For HPG, the addition of hydroxypropyl groups to guar is described by the molar substitution (MS) which is defined as the average number of moles of hydroxypropyl groups substituted per mole of sugar units. It is a measure of the total number of moles of propylene oxide which have been added to the guar polymer chain. For CMG and HPTMAG, the addition of substituting groups to the guar polymer chain is described by the term degree of substitution (DS) which is the average number of carboxymethyl groups substituted per sugar unit *(13)*.

Polymer solutions were prepared by adding appropriate amount of polymer powder into deionized water under agitation. Endo-β-mannanase extracted from *Aspergillus niger* was obtained from Megazyme Inc. (Ireland). The enzyme was supplied as an ammonim sulphate suspension in 0.02% sodium azide. 1.35 μl of this suspension was diluted in 2 ml 0.1M sodium acetate (EM Science) / acetic acid (Glacial, FisherChemical) buffer solution with pH adjusted to 6.

Enzymatic Degradation

The enzymatic degradation reaction was run at room temperature and under pH 7. Into 200 ml guar solution, 0.2 ml enzyme buffer solution was injected

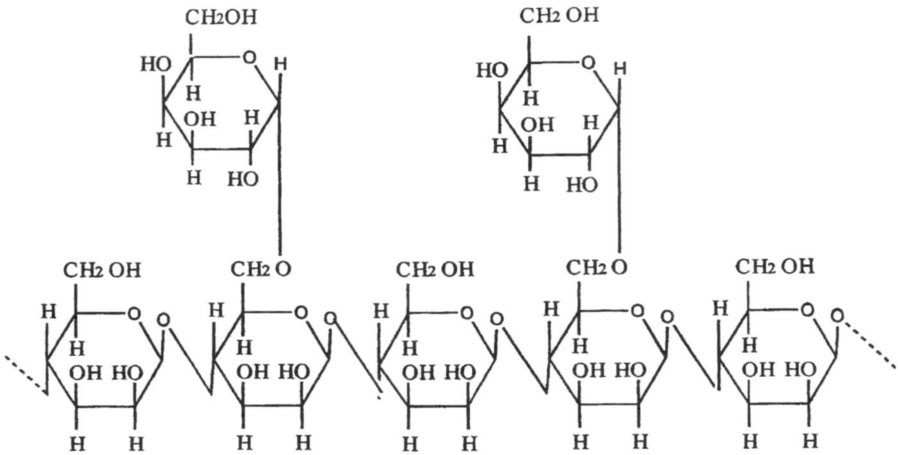

Figure 1: The structure of guar.

Guar Derivative	R =
Hydroxypropyl Guar (HPG)	$-CH_2-CH(OH)-CH_3$
Carboxymethyl Guar (CMG)	$-CH_2-COOH$
Hydroxypropy Trimethylammonium Guar (HPTMAG)	$-CH_2-CH(OH)-CH_2-N^{+}(CH_3)_3$

Figure 2: Side groups of derivatized guars. The hydrogen atom of hydroxyl group of guar can be substituted by a side group (-R) to form different derivatized guars: hydroxypropyl guar (HPG), carboxymethyl guar (CMG) and hydroxypropyl trimethylammonium guar (HPTMAG)

using a microsyringe. The mixture was magnetically stirred during the reaction. After the reaction began, aliquots of the guar and enzyme mixture were taken out at various times. Each aliquot was immediately heated to 100 °C for 20 minutes to denature the enzyme and stop the reaction. Biochemical analyses reveal that the half-life of endo β-mannanase is 6 min at 85 °C *(9)*. The viscosity of the solution does not change after the enzyme is denatured. Experiments were also done to show that this denaturing protocol (i.e. heat treatment) did not produce a viscosity reduction for polymer solutions in the absence of enzymes (data not shown). In order to compare results from different experiments, equal enzyme concentration was used as a basis.

At high guar concentrations (>1.5 wt%), the high viscosity of the polymer solution makes (rapid) homogenization of the guar and enzyme difficult. As a result, high substrate concentration samples were made by concentrating the 1 wt% guar solution and enzyme mixture by dialysis at 0 °C where enzyme activity is low. A control experiment was run to show there is a negligible molecular weight reduction at this low temperature (data not shown). For example, to study the degradation experiment at 3 wt% substrate concentration, 5 μl enzyme buffer solution was added to 15 g 1 wt % guar solution and mixed well. The mixture was immediately transferred into a Spectra/Por dialysis bag with a molecular weight cut-off of 3,500. Then the dialysis bag was concentrated by placing it in a bed of absorbent silica gel (Fisher Scientific). After around two hours, the guar solution was concentrated to 3 wt% with a total mass of 5g. Finally, the substrate-enzyme mixture was heated so that the temperature quickly increased from 0 to 25 °C and the reaction began.

Steady Shear Viscometry

Steady shear rheological tests were applied to determine the viscosity of guar samples using a strain-controlled rheometer (RFS-II, Rheometrics, Piscataway, NJ). A Couette geometry, with inner bob and outer cup radii of 16mm and 16.925mm respectively and a bob length of 33.3mm, was chosen. Samples without enzyme were also tested as controls to assess the initial rheology of the solutions. All viscosity measurements were made at 25 °C.

GPC Measurement

The molecular weight of degraded guar samples were analyzed using GPC. The HPLC system consisted of two columns in series (TSK G3000PW_{XL} and TSK G6000PW_{XL}), a pump (Waters M510), a differential refractometer (Waters 410) and an injector (Valco SSA C12P). Both columns were thermostated at

40°C to decrease solvent viscosity and peak broadening. The mobile phase was 55mM Na_2SO_4 and 0.02% NaN_3 aqueous solution. A flow rate of 0.6 ml/min was used. Degraded guar samples were diluted to 0.05 wt% and filtered through 0.45 μm filter (Whatman Autovial) before injection. Both pullulan and guar standards were used to calibrate the columns. Nine fractions of pullulan standards with average molecular weight (MW) ranging from 5,900 to 1.6 million were used (Shodex Corp., Japan). Two fractionated guar standards, Meyprogat 7 (Mp=58,000) and CSAA 200 (Mp=2,000,000), were used as secondary standards *(10)*. The Mark-Houwink-Sakurada relationship $[\eta]=K(M_w)^a$ of guar and pullulan are: $K=3.8\times10^{-4}$ dL/g, a=0.723 (guar); $K=1.9\times10^{-4}$ dL/g, a=0.67 (pullulan) *(14, 15)*. A universal calibration curve was plotted for guar and pullulan and all points fall on a single straight line *(16)*. The molecular weight (MW) and molecular weight distribution (MWD) of degraded guar samples were determined based on the universal calibration curve.

Results and Discussion

Enzymatic Degradation Kinetics

Figure 3 shows viscosity versus shear rate plots of six guar solutions upon exposure to β-mannanase enzyme at room temperature and a pH of 7. For each sample we observed a Newtonian region at low shear rates and a shear thinning region at higher shear rates, as has been observed in previous studies *(9)*.The β-mannanase, which cleaves the mannose backbone, is very effective in reducing the viscosity of guar solutions. Under enzymatic hydrolysis, the solution viscosity decreased by over two orders of magnitude after twenty hours.

The reaction between enzyme and substrate is asssumed to be of the common Michaelis-Menton type *(17)*:

$$S + E \underset{k_2}{\overset{k_1}{\rightleftharpoons}} ES \xrightarrow{k_{cat}} P + E$$

The enzyme (E) first attaches to the polymer substrate (S) to form an enzyme-substrate complex (ES). Then the enzyme cleaves the substrate, followed by a release of the product (P) and the enzyme itself. If we consider the substrate as every breakable bond in the system and the scission of these bonds on a polymer chain to be random, the degradation kinetics are *(5)*

$$\frac{dL}{dt} = -\frac{k_{cat} C_e L}{K_m + L} \quad (1),$$

where L represents the molar concentration of every cleavable bonds in the system, C_e is the concentration of the enzyme, k_{cat} and K_m are the rate constant and the Michaelis-Menton constant respectively.

When $L << K_m$, Eq. (1) reduces to a first-order reaction

$$\frac{dL}{dt} = -\frac{k_{cat} C_e L}{K_m} \quad (2);$$

when $L >> K_m$, Eq. (1) reduces to a zero-order reaction

$$\frac{dL}{dt} = -k_{cat} C_e \quad (3).$$

In order to determine the reaction rate experimentally, we can let

$$k = \frac{k_{cat} C_e}{K_m + L} \quad (4).$$

Therefore Eq (1) can be written in a first order form as

$$\frac{dL}{dt} = -kL \quad (5),$$

where k stands for the apparent rate constant.

If the number average molecular weight of the polymer, M_n, is much greater than the molecular weight of a monomeric unit, m (for guar, m ≈ 266), during the degradation, the following relation between molecular weight and degradation time can be derived *(18)*:

$$\frac{1}{M_n(t)} = \frac{1}{M_{no}} + \frac{kt}{m} \quad (6),$$

in which M_{no} is the initial M_n of the polymer. Therefore, we followed the change of M_n with reaction time by GPC and plotted $1/M_n$ as a function of degradation time at different guar concentrations (Figure 4). The inverse M_n was shown to change linearly with time, consistent with Eq. (6). The apparent rate constant k was obtained from the slope of the straight line, and it decreases with guar concentration. According to Eq. (5) and the obtained rate constant, the

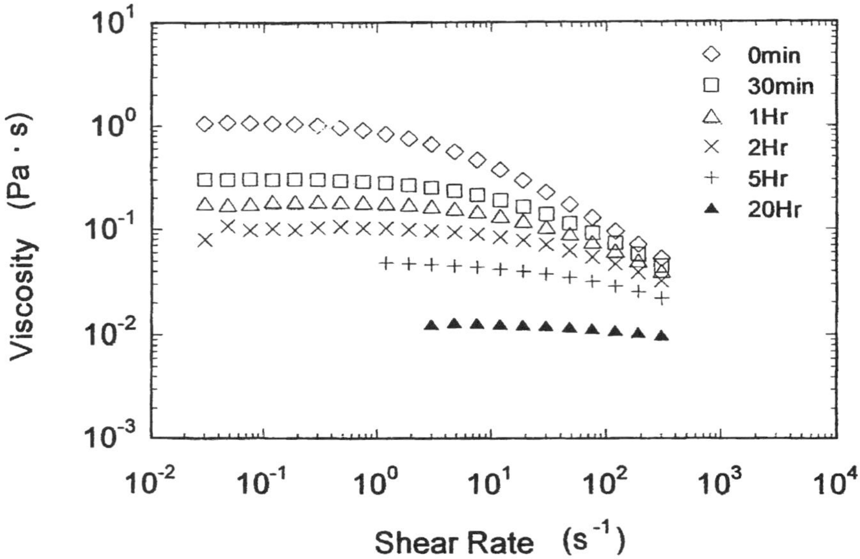

Figure 3: Viscosity of 0.5wt% guar solution as a function of shear rate plotted at different periods during enzymatic degradation. The reaction is run at ambient temperature and pH of 7. The concentration of β-mannanase is 0.0002 units / ml polymer solution.

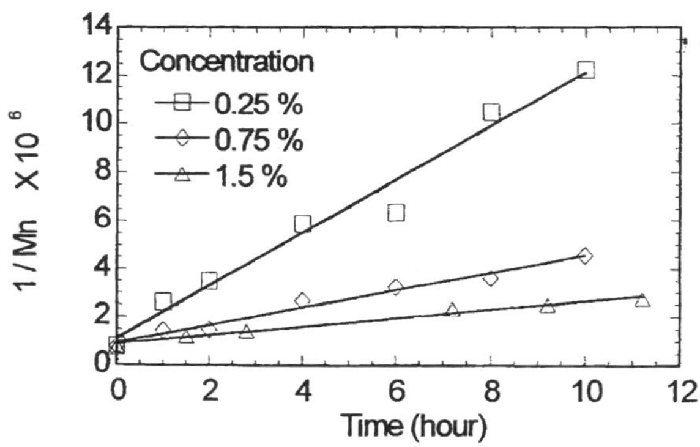

Figure 4: Inverse of number average molecular weight, M_n, plotted vs. enzymatic degradation time at different guar concentrations: 0.25, 0.75, 1.5 wt%.

enzymatic degradation rate can be calculated and plotted versus polymer concentration (Figure 5). At very low guar concentration (<0.03wt%), the reaction rate increases with substrate concentration. This is the first-order regime predicted by the Michaelis-Menton kinetics. In this regime, polymer concentration is much lower than the overlap concentration of native guar (C* ~ 0.1wt%) *(19)*. Guar appears as random coils in solution *(15)*, and its molar concentration is extremely low. At intermediate concentrations (0.05 ~ 3 wt%), the degradation rate becomes independent of substrate concentration: this is the predicted zero-order regime. This arises because at high polymer concentrations, there are sufficient substrate sites for the enzyme to bind. Therefore, the increasing of substrate concentration does not increase the number of enzyme-substrate complexes. The reaction rate is limited by the enzyme cleavage kinetics and not by the enzyme diffusion to the substrate: the reaction becomes zero-order in polymer substrate. The experimental results in the first two regimes were fitted according to the Michaelis-Menton model to yield K_m and V_{max} *(20)*. In this study, V_{max} is measured in terms of the molar concentration of glycosidic bonds broken per unit time. K_m and V_{max} were determined to be 0.6 mM and 7.8×10^{-10} mol/ml-s. At very high polymer concentrations (> ~ 5 wt%), the degradation rate begins to decrease with substrate concentration, corresponding to a diffusional resistance to the enzyme mobility in the concentrated polymer mesh. To our knowledge, this study is the first to show the three transitions from substrate limited, to enzyme limited, to diffusion controlled.

In figure 4, the inverse MW changes linearly with time during the first ten to twelve hours of the reaction, through which the apparent rate constant can be determined. However, after twelve hours, the apparent reaction rate appears to decrease with reaction time. This phenomena could be attributed to several reasons. First, it has been found that the distribution of galactose along the mannose backbone is not homogeneous *(1)*. The enzymatic reaction rate is faster in regions unsubstituted, or lightly substituted by galactose (higher reactivity) than in the galactose-rich regions of guar chain. Therefore, enzyme molecules would prefer to attack the unsubstituted regions at the beginning of the reaction. During the first ten hours, there are plenty of these highly reactive sites in the system, and the reaction rate stays constant. At longer times, the number of high-reactivity sites decreases, and the apparent reaction rate slows down accordingly. Furthermore, the reduction of reaction rate could also be due to the fouling of enzymes and possible feed back inhibition by the degradation product.

Reaction-Diffusion Model

The enzymatic degradation of polymers can be divided into a two-state process, a mobile state and a bound state. The mobile state is the free diffusion of

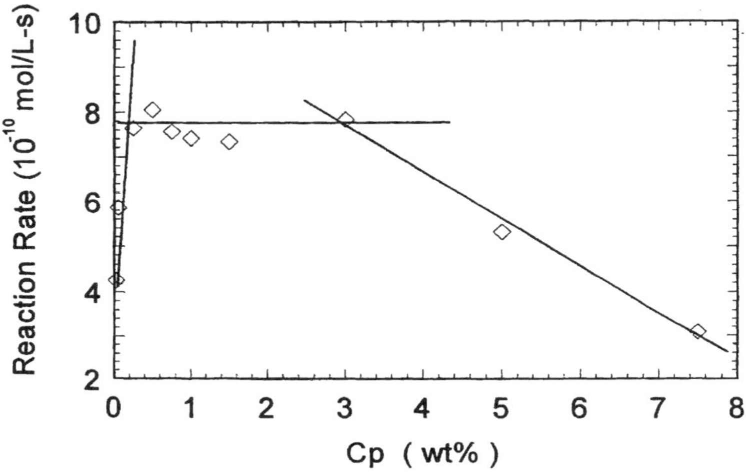

Figure 5: The enzymatic degradation rate at pH 7 is plotted versus substrate concentration. At very low concentration regime, reaction rate increases first-orderly with polymer concentration. In the intermediate regime, the reaction rate is independent of substrate concentration and the kinetics is zero-order. At high concentration regime, the reaction becomes diffusion limited. The concentration of β-mannanase is 0.0002 units / ml polymer solution.

enzyme molecules through a polymer solution; the bound state consists of the binding of enzyme to polysaccharide chain, the cleavage of glycosidic bonds and the release of the enzyme from the binding site. The time of the mobile state and the bound state are denoted as t_d (diffusion) and t_r (reaction) respectively. The total time per enzyme molecule per cut of bond is $(t_d + t_r)$. Since the polymer is uncharged, we assume the interactions between enzyme and the polymer substrate are short-ranged. As a result, the diffusion of an enzyme molecule can be treated as a random diffusive process, where the time to diffuse a distance between polymer strands is:

$$t_\xi \sim (\xi^2/D) \qquad (7),$$

where ξ is the correlation length between polymer chains *(21)* and D is the self-diffusion coefficient of the enzyme molecule. However, the diffusion time t_d between binding / reaction steps is greater than t_ξ, the time to diffuse the distance between polymer strands. Most collisions between the enzyme and polymer chains do not result in binding because the molecular recognition for binding requires that the collision event occurs at the enzyme active site, and has the proper spatial configuration. Therefore, the time for free diffusion, t_d, is longer than t_ξ by a "collision efficiency factor", α, such that *(22)*

$$t_d = \alpha\,(\,\xi^2\,/\,D\,) \qquad (8).$$

The polymer mesh size ξ defines the network through which the enzyme diffuses. In the entanglement regime and good solvents the mesh size ξ decreases with polymer concentration C_p by *(21)*,

$$\xi = R_g\left(\frac{c}{c^*}\right)^{-\frac{3}{4}} \qquad (c > c^*)\ (9),$$

where R_g and c^* are the radius of gyration and the overlap concentration of guar respectively.

Enzymes are compact globular macromolecules. The diffusion of solid spherical particles in polymer solutions has been widely studied *(23-26)*. In many systems, the diffusion coefficient decreases with concentration as a stretched exponential: $D/D_o=\exp(-\alpha c^\gamma)$, in which D_o is the diffusion coefficient of the probe in pure solvent, and γ is a constant which varies from 0.5 to 1 *(25)* (Figure 6a). De Gennes also argued that, when the hydrodynamic diameter of the enzyme, d, is much smaller than the mesh size ξ, we expect the enzymes to move rather easily, with a friction coefficient related to the viscosity of the pure solvent; when

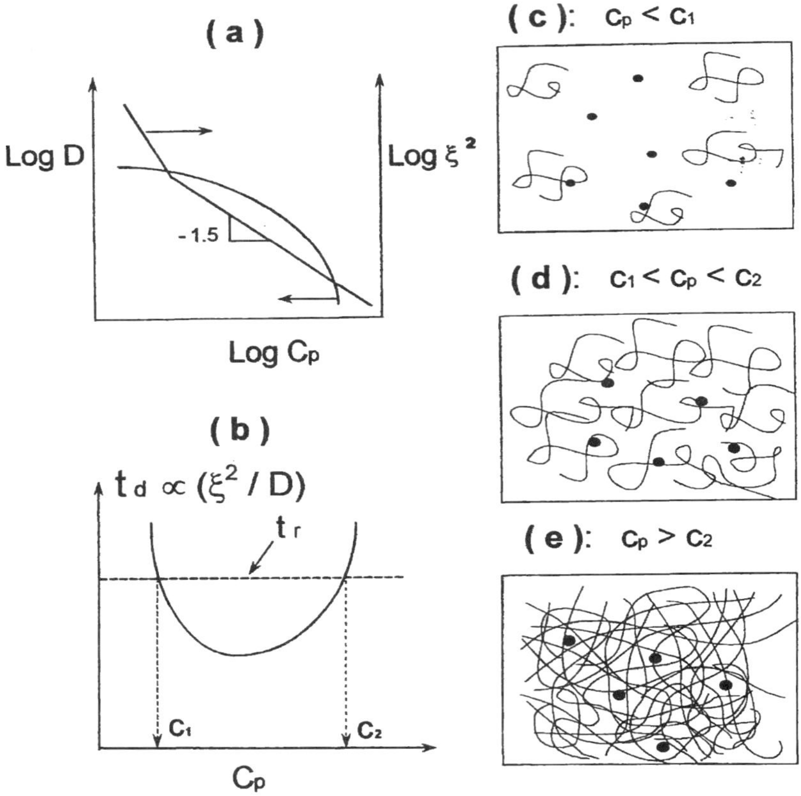

Figure 6: (a) Changes of the enzyme diffusion coefficient (D) and polymer mesh size (ξ) with the matrix polymer concentration (c_p). (b) Changes of the enzyme diffusion time with polymer concentration. The reaction time has a constant value at given conditions, and is shown as the dashed line. (c) Dilute regime ($c<c_1$): diffusion time $t_d >> $ reaction time t_r. The black dots denote enzyme molecules. (d) Intermediate regime ($c_1<c<c_2$): $t_r >> t_d$. (e) Concentrated regime ($c>c_2$): $t_d >> t_r$.

d is greater than ξ, the enzymes experience high friction from the polymer and the friction is related to the viscosity of the entangled solution *(27)*.

Figure 6a and Figure 6b qualitatively explain the competition between diffusion and reaction. The diffusion coefficient decreases with polymer concentration c_p as a stretched exponential, while mesh size ξ changes with c_p as a power law through Eq. 9. In the dilute range (Figure 6c), the average distance between polymer chains is so large that the diffusion time t_d is long for the enzyme to diffuse from one chain to another. At low enough concentration, t_d is bigger than t_r (an intrinsic property of the enzyme) and the reaction is enzyme diffusion-controlled. As the concentration increases, the average distance between chains decreases as a power law, as described by Eq. 9. However, the diffusion coefficient of the enzyme, D, does not decrease significantly because the polymer concentration is low and the enzyme molecule diffuses through essentially solvents. As a result, t_d decreases (Eq. 8), and reaction rate increases with concentration. With increasing concentration t_d decreases until $t_d \sim t_r$ which we denote as concentration c_1. For concentrations greater than c_1, t_d becomes smaller than t_r, and the reaction is reaction-controlled. Therefore in the intermediate range (Figure 6d), the reaction rate is determined by the enzyme cleavage kinetics and is independent of polymer concentration (zero order). As the polymer concentration continues to increase, the stretched exponential function drops faster than a power law (Figure 6a). So D decreases faster than ξ^2, and t_d begins to increase again. Eventually, at some high concentration c_2, t_d will reach the value of t_r. When $c>c_2$, the reaction rate will decrease, and the reaction becomes diffusion-limited again (Figure 6e).

According to Figure 5, the degradation rate begins to decrease at a guar concentration of 5 wt %. This is approximately the concentration c_2 where $t_d \sim t_r$. The reaction time t_r can be evaluated from the catalytic constant of the enzyme, k_{cat}, which is defined as,

$$k_{cat} = V_{max} / c_e \tag{10}$$

This quantity is also known as the turnover number of an enzyme because it is the number of reaction processes (turnovers) that each active site catalyzes per unit time. The reaction time of the enzyme molecule, $t_r \sim 1/k_{cat}$. The enzyme concentration c_e used in the reaction is approximately 10^{-10} M. Based on Eq. 10, t_r is calculated to be ~ 0.1 second. At 5 wt%, the mesh size $\xi \sim 10$ nm, and the diffusion coefficient of the enzyme molecule is on the order of 10^{-8} cm^2/sec, according to diffusion experiments *(26)*. The diffusion time for one collision between the enzyme molecule and polymer chain is approximately $\xi^2/D = 10^{-4}$ second. Therefore, we can estimate the collision efficiency, α, to be on the order of 1000. This suggests that there are thousands of collisions between the enzyme and the polymer chain before an effective collision (reaction) occurs.

Effect of substrate molecular weight

Effect of substrate molecular weight on the rate of degradation has also been studied. Guar samples with various MWs were prepared by acid hydrolysis. All enzymatic reactions were run in the intermediate concentration range (0.5 wt %), where the enzymes were saturated. Figure 7 shows the degradation rates of guar substrates with different initial MWs. In the range that we are interested (the range where most of the viscosity reduction of guar solution occurs is M_n from $10^6 \sim 5\times10^4$), the reaction rate is insensitive to the molecular weight of guar. In this range, the MW of guar is still very high, and the polymer solution is in the semi-dilute regime. There are plenty of breakable bonds in the system which the enzyme can cleave, while the enzyme still does not see the change in polymer MW. Furthermore, it also suggests that feedback inhibition by the degradation products is probably not important in this study.

Effect of Substrate Structure

During the enzymatic hydrolysis process, the binding event involves molecular recognition between the enzyme and the substrate. A change of structure of the substrate may change the ability of enzyme to recognize and attach to the substrate, which can cause an alteration in the degradation rate. As previously discussed, HPG is a neutral polymer in solution. At pH 7, HPTMAG is positively charged, and CMG is negatively charged. The β-mannanase enzyme we used in the experiment has a pI of 3.5. At neutral pH, the enzymes are negatively charged.

Since MW determination of polyelectrolyte polymers is complicated, steady shear rheometry was used to follow the degradation. Morris et al. *(2)* have shown that random coil polymers obey a universal log (η_{sp}) ~ log (c[η]) master curve, where η_{sp} is the specific viscosity, c is the solution concentration and [η] denotes the intrinsic viscosity. Due to the "hyperentanglements" between polymer chains, they showed that galactomannan polymers fall on a master curve with a lower overlap concentration and somewhat greater concentration dependence thereafter. Therefore, based on the master curve by Morris et al and the Mark-Houwink-Sakurada equation given by Robinson et al. *(15)*, we can calculate the molecular weight change with the reaction from the zero shear viscosity of the solution. Since the initial molecular weights of different polymers are not the same, normalized molecular weight, which is defined as the ratio of the molecular weight of the polymer solution incubated with enzyme to that of the initial, non-degraded polymer, was chosen as the basis of comparison. Figure 8 shows the normalized molecular weight changes with time during the degradation reaction. After incubation with enzyme for the same time period, the average molecular

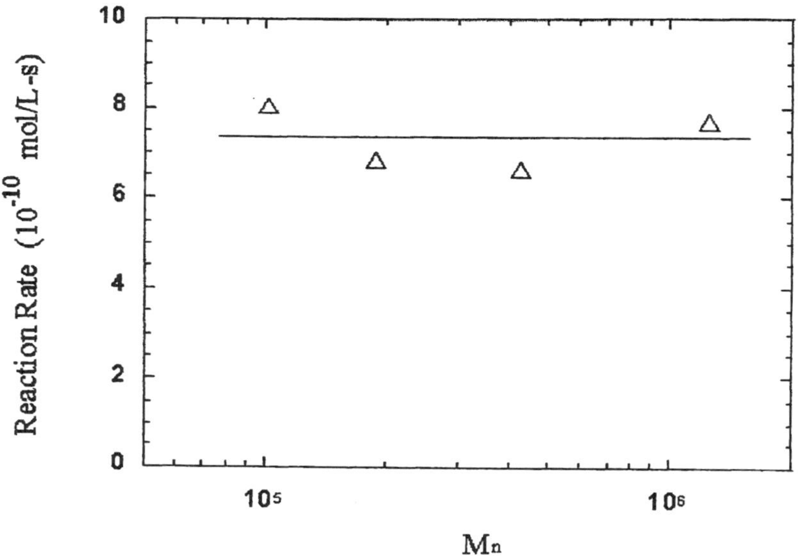

Figure 7: The enzymatic degradation rate at intermediate substrate concentrations (the plateau region in Figure 5) is plotted as a function of substrate molecular weight. The reaction rate is independent of substrate molecular weight in the range of this study. The concentration of β-mannanase is 0.0002 units / ml polymer solution.

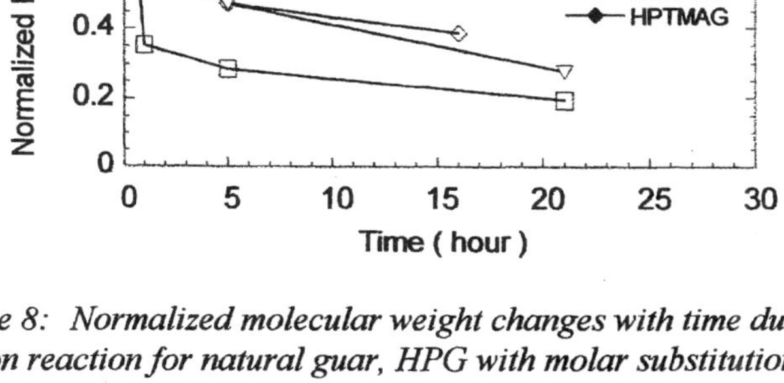

Figure 8: Normalized molecular weight changes with time during the degradation reaction for natural guar, HPG with molar substitution of 0.18 and 1.53, CMG with degree of substitution of 0.1 and 0.6, and HPTMAG. The concentration of β-mannanase is 0.0002 units / ml polymer solution (pH=7).

weights of HPG and CMG are both higher than that of guar. The addition of hydroxypropyl groups on the polymer chain sterically hinder the attachment of enzymes to the polymer molecules, which decreases the total enzymatic degradation rate. For CMG at pH 7, the electrostatic repulsion between the enzyme and the carboxyl groups on the polymer backbone also slows down the degradation.

Triggered enzymatic degradation

For cationic guar, the molecular weight did not change with time: there was no enzymatic degradation. It has been known that proteins can interact strongly with both synthetic and natural polyelectrolytes to form protein-polyelectrolyte complexes (PPCs) *(28, 29)*. Since the cationic polymer and the enzyme have opposite charges, the ionic attractions immobilize the enzyme molecules on the polymer chain to form an enzyme-polymer complex. This interaction is the basis for ion-exchange separations of proteins *(30)*. Because enzymes have both carboxyl and amine groups, their net charges can be altered by changing the solution pH. In Figure 9, as the pH is lowered to the isoelectric point of β-mannanase (pH=3.5), the enzyme is released from the complex, and degradation begins immediately. The viscosity of cationic guar solution decreased dramatically by almost two orders of magnitude over twenty hours. In many applications, it is desirable to initiate the polymer degradation in a controlled manner. For example, making a homogeneous enzyme / substrate mixture is difficult in very viscous polymer solutions. Inactivating the enzyme during dispersion, and then triggering degradation would lead to a more uniformly modified product. This experiment demonstrates a novel triggering mechanism by which we can turn on and off the degradation reaction by shifting the solution pH. Since the acetal linkages between the mannose can hydrolyze at low pH, a control curve was run to see the effect of acid-catalyzed cleavage of the polymer backbone at pH 3.5 without added enzyme. The control solution viscosity dropped from 0.75 Pa·s to 0.45 Pa·s due to the acid catalyzed cleavage of polymer molecules over a forty-hour period.

Another experiment was run to further explore the charge complex triggering. Figure 10 shows the effectiveness of the β-mannanase in degrading cationic guar solutions at different ionic strengths but a constant pH of 7. At low ionic strength, the enzymatic degradation was stopped. When the ionic strength was increased to 0.1 M, the solution viscosity decreased substantially. As the solution ionic strength increases, the electrostatic attraction between the enzyme molecule and the polymer chain is screened, and the complex is disrupted. A control experiment was run on native guar to show that salt concentration has little effect on enzyme activity. Therefore, this offers us another trigger to control

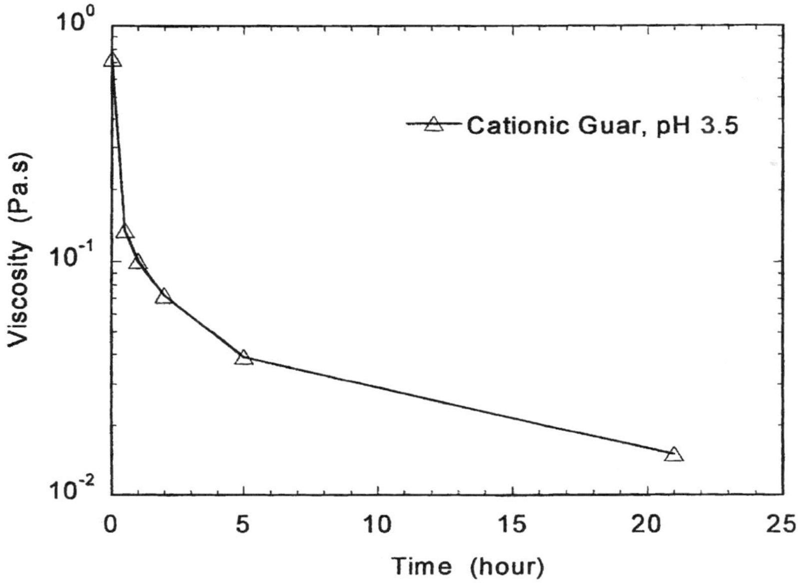

Figure 9: Viscosity of 0.5 wt % cationic guar solution is plotted as a function of enzymatic degradation time at pH 3.5.

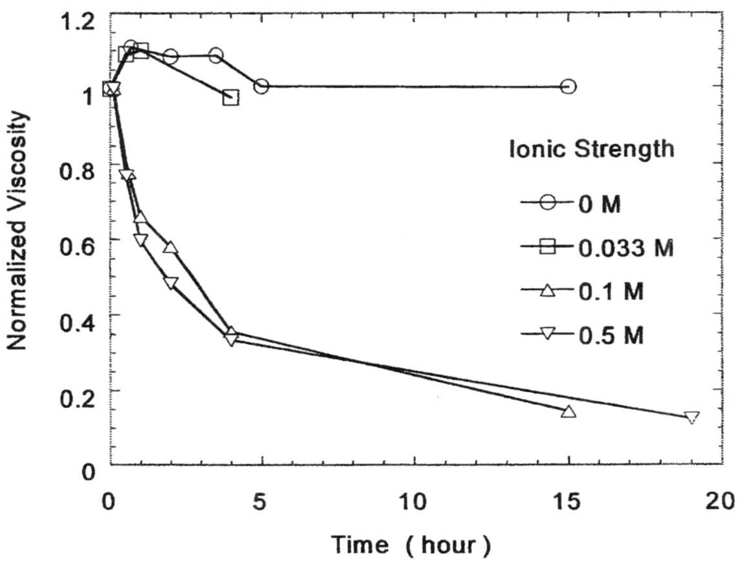

Figure 10: Normalized viscosity of 0.5 wt % cationic guar solution vs. reaction time at different solution ionic strengths. All reactions were run at ambient temperature and pH of 7.

the enzymatic degradation through adjusting the ionic strength of the cationic guar solution.

Conclusion

Understanding the reaction-diffusion of enzyme molecules in polymer matrices is important for the control of enzymatic reactions with biopolymers. In this study, we investigated the reaction kinetics and mechanism of enzymatic degradation of guar and substituted guar over a wide range of polymer (i.e. substrate) concentrations. In dilute and semidilute guar solutions, a Michaelis-Menton kinetics model was used to analyze the results, and for native guar at pH=7, the Michaelis-Menton parameters are: $K_m = 0.6$ mM and $V_{max} = 7.8\times10^{-10}$ mol/ml-s. As predicted by the kinetic model, in the very dilute regime, the reaction rate is first-order with substrate concentration. In the intermediate regime, the enzyme is saturated and the degradation kinetics is zero-order. Our new observation is the onset of diffusion controlled reaction at higher polymer concentrations. The reaction rate decreases by 50% between 5 % and 7.5 % of substrate polymer. A novel reaction-diffusion model is proposed to describe the enzymatic degradation in all concentration ranges. Based on this model, the boundaries of different polymer concentration regimes can be defined.

The effect of substrate structure on the degradation kinetics was also explored. The degradation rate was found to be greatly affected by the type of substituent groups added and the degree of substitution. Increasing either carboxymethyl or hydroxypropyl substitution causes a decrease in enzymatic degradation rate due to steric and electrostatic blocking of enzyme binding onto the polymer chain. The triggering mechanism was found for the enzymatic hydrolysis of cationically derivatized guar solutions. At high pH and low salt concentration, the enzyme is complexed with the guar and inactive; at low pH or high salt concentration, the enzyme is released and activated. This mechanism holds promise for the controlled degradation of electrolyte biopolymers in industry, and provides a model for studying the enzyme-polymer interactions.

Reference

1 B. V. McCleary, A. H. Clark, I. C. M. Dea, and D. A. Rees, *Carbohydrate Research* **139,** 237-260 (1985).

2 E. R. Morris, A. N. Cutler, S. B. Ross-Murphy, and D. A. Rees, *Carbohydrate Polymers* **1,** 5-21 (1981).

3 J. Rollings, *Carbohydrate Polymers* **5,** 37-82 (1985).

4 A. Marc, J. M. Engasser, M. Moll, and R. Flayeus, *Biotechnology and Bioengineering* **25,** 481-496 (1983).
5 K. Suga, G. van Dedem, and M. Moo-Young, *Biotechnology and Bioengineering* **17,** 185-201 (1975a).
6 S. Wald, C. R. Wilke, and H. W. Blanch, *Biotechnology and Bioengineering* **26,** 221-230 (1984).
7 B. V. McCleary and N. K. Matheson, *Carbohydrate Research* **119,** 191-219 (1983).
8 A. Tayal and S. A. Khan, in *Abstracts of Papers of the American Chemical Society*, Vol. 209 (AMER CHEMICAL SOC, WASHINGTON, 1995), pp. 28-CELL.
9 A. Tayal, R. M. Kelly, and S. A. Khan, *Macromolecules* **32,** 294-300 (1999).
10 Rhodia Inc., Personal Communication.
11 V. Constein, Dowell-Schlumberger Inc., Personal Communication.
12 Megazyme International Ireland Ltd.
13 Y. Cheng, Ph.D. Thesis, Princeton University (2001).
14 T. Kato, T. Okamoto, T. Tokuya, and A. Takahashi, *Biopolymers* **21,** 1623-1633 (1982).
15 G. Robinson, S. B. Ross-Murphy, and E. R. Morris, *Carbohydrate Research* **107,** 17-32 (1982).
16 Y. Cheng and R. K. Prud'homme, *ACS polymer preprints* **41(2),** 1711 (2000).
17 L. Michaelis and M. L. Menten, *Biochemische Zeitschrift* **49,** 533 (1913).
18 H. H. G. Jellinek, *Degradation of vinyl polymers* (Academic Press Inc., New York, 1955).
19 F. M. Goycoolea, E. R. Morris, and M. J. Gidley, *Carbohydrate Polymers* **27,** 69-71 (1995).
20 A. Cornish-Bowden, *Fundamentals of enzyme kinetics* (Portland Press, London, 1995).
21 P. G. de Gennes, *Scaling concepts in Polymer Physics* (Cornell University Press, 1979).
22 C. R. Cantor and P. R. Schimmel, *Biophysical Chemistry* (W. H. Freeman and Company, San Francisco, 1980).
23 A. G. Ogston, B. N. Preston, and J. D. Wells, *PROCEEDINGS OF THE ROYAL SOCIETY OF LONDON SERIES A-MATHEMATICAL PHYSICAL AND ENGINEERING SCIENCES* **333,** 297-316 (1973).
24 A. R. Altenberger and M. Tirrell, *Journal of Chemical Physics* **80,** 2208-2213 (1984).
25 G. D. J. Phillies, *Journal of Physical Chemistry* **93,** 5029-5039 (1989).
26 Y. Cheng, J. L. Thomas, and R. K. Prud'homme, Submitted to *Macromolecules* (2001).

27 D. Langevin and F. Rondelez, *Polymer* **19,** 875 (1978).

28 K. W. Mattison, Y. F. Wang, K. Grymonpre, and P. L. Dubin, *Macromolecular Symposia* **140,** 53-76 (1999).

29 S. Dumitriu and E. Chornet, *Advanced Drug Delivery Reviews* **31,** 223-246 (1998).

30 A. A. Garcia, M. R. Bonen, J. Ramirez-Vick, M. Sadaka, and A. Vuppu, *Bioseparation Process Science* (Blackwell Science, Inc., 1999).

Other Examples of Polymer Biocatalysis

Chapter 23

Enzymatic Degradation of Polyester Urethanes Studied by Multi-Angle Laser Light Scattering

D. Himel[1], D. P. Norwood[1,*], and W. F. Reed[2]

[1]Department of Chemistry and Physics, Southeastern Louisiana University, Hammond, LA 70402–0878
[2]Department of Physics, Tulane University, New Orleans, LA 70118

We report an investigation of the degradation of polyester urethanes using enzymes purified from members of the genus Pseudomonas. Using multi-angle laser light scattering [MALLS], we observe the decrease in substrate weight-average molecular weight as enzymes purified from Comomonas acidovirans attack and degrade polyurethanes suspended in unsalted water. We obtain an enzymatic degradation rate of $(8.50\pm1.56)\cdot10^{-4}$ min^{-1}. Using a typical model of enzyme catalysis, we conclude that $K_M \ll [E]_0 = 6 \cdot 10^{-7}$ M.

Introduction

Despite its xenobiotic origins, polyurethane has long been known to be susceptible to attack by various bacteria and fungi (*1-3*). Recently, enzymes have been purified from several Pseudomonads that show enzymatic activity against polyesterurethanes (*4*). Static laser light scattering has been shown to be a valuable technique in the characterization of polymers (*5-9*), providing the molecular weight, radius of gyration, and second virial coefficient of polymers in dilute solution. In particular, light scattering is most effective for large molecular weight polymers where most other methods become ineffective, and thus complements those other techniques. Furthermore, by monitoring, as a function of time, the molecular weight and other parameters during degradation of polymers, one can deduce other properties of the polymer system such as strandedness (*10*) and branching (*11*), and properties of the degrading agent, such as absolute depolymerization rates (*12-14*) and method of attack (*11, 15, 16*). Finally, it is known that laser light scattering, when coupled with high performance liquid chromatography (HPLC) (*17*) and on-line viscometry (*18*), is a powerful characterization technique that emphasizes the advantages of each individual technique (*19*). The fractionation produced by the HPLC system overcomes the ambiguity polydispersity causes in the interpretation of light scattering, and the absolute molecular weight determinations provided by light scattering overcome uncertainties present in HPLC used alone. The information provided by these techniques proves invaluable in understanding the kinetics and mechanism of the enzymes' attack on the polyurethane substrate. Here, we describe time-resolved static light scattering measurements of the polyurethane substrate while it is under attack by the pseudomonas enzymes. These data are reduced to provide directly the enzymatic degradation rate. A series of such measurements would provide the Henri-Michaelis-Menten parameters in the standard way.

Experimental Protocol

The light scattering measurements unambiguously provide the molecular weight, radius of gyration (or size) and second virial coefficient of the dissolved polymer, as pointed out by Zimm (*9*). Experiments of this type are carried out by measuring the Rayleigh scattering of incident laser light as a function of the scattering angle. The Rayleigh ratio is the fraction of light scattered from the incident beam, per unit scattering volume. The inverse of the Rayleigh ratio is a linear function of the polymer concentration and the square of the scattering wavevector (*9*). That is,

$$\frac{Kc}{R_\Theta} = \frac{1}{M_W}\left(1 + \frac{R_g^2}{3}q^2\right) + 2A_2c \qquad (1)$$

where the independent quantities are c, the polymer mass concentration in solution, and q, the scattering wavevector (=$(2\pi/\lambda)\sin(\theta/2)$ with λ the laser vacuum wavelength and θ the scattering angle). The constant K is an instrument constant given by $K = 4\pi^2 n^2 (dn/dc)^2/N_A\lambda^4$ (where n is the refractive index of the solvent at the laser wavelength, dn/dc is the change in refractive index per unit solute concentration and N_A is Avogadro's number). By plotting the inverse of the Rayleigh ratio as a function of the wavevector squared and the polymer concentration (the so-called "Zimm plot"), one obtains the weight average molecular weight, M_W, the radius of gyration, Rg, and the second virial coefficient, A_2. Figures 1 and 2 show such Zimm plots, which will be discussed in the results section.

One may make such measurements in real time as a polymer substrate is being digested by an enzyme, and by monitoring the consequent decrease in molecular weight as a function of time, one obtains the enzymatic degradation rate constant. The molecular weight as a function of time should satisfy (*16*):

$$\frac{1}{M(t)} = \frac{1}{M_n}(1 + rt) \qquad (2)$$

and so the ratio of the slope to the intercept of a linear fit to the reciprocal of the molecular weight as a function of time gives the enzymatic degradation rate, r. Such results are shown as Figure 3, and discussed in the results section.

Materials

Polyurethanase enzymes used in the study, which were purified from Comomonas acidovirans, were a gift from Dr. Gary Howard (Biological Sciences, Southeastern Louisiana University), as was the polyurethane substrate. All solutions were aqueous solutions, prepared using purified water from a Modulab Analytical Research Grade UF/UV water-polishing system (Continental Water Systems Corporation), terminated by a 0.2 μm capsule filter. Ionic strength was not adjusted. The polyurethane substrate was received as a thick paste, approximately 40% polyurethane - 60% water by weight. Dilutions of the substrate were made to concentrations appropriate for light scattering measurements; precise final concentration varied, but was typically about $5 \cdot 10^{-6}$ g/mL which corresponds to a density of $\sim 10^{-10}$ M. After gentle stirring overnight, samples were filtered using either 0.22 micron (Millex-GV) or 0.1 micron (Millex-VV) sterile filters obtained from Millipore Corp. Polyurethanase enzyme was received as an aqueous solution of concentration $0.83 \cdot 10^{-3}$ g/mL. Precise experimental details varied, but typically, ~0.2 mL of this enzyme solution was added to ~3.8 mL of substrate solution, resulting in an experimental enzyme concentration of $\sim 4 \cdot 10^{-5}$ g/mL, corresponding to a density of about $6 \cdot 10^{-7}$ M. Note that the enzyme density is typically about 6000 times that of the substrate density. That is, we should be safely in a regime for which the enzyme is unsaturated by substrate (i.e., a pseudo-first order reaction in substrate concentration).

Light scattering measurements were made using a DAWN EOS multi-angle laser light scattering [MALLS] detector (Wyatt Technology Corporation) operating in batch mode. Measurements of the polyurethane substrate and

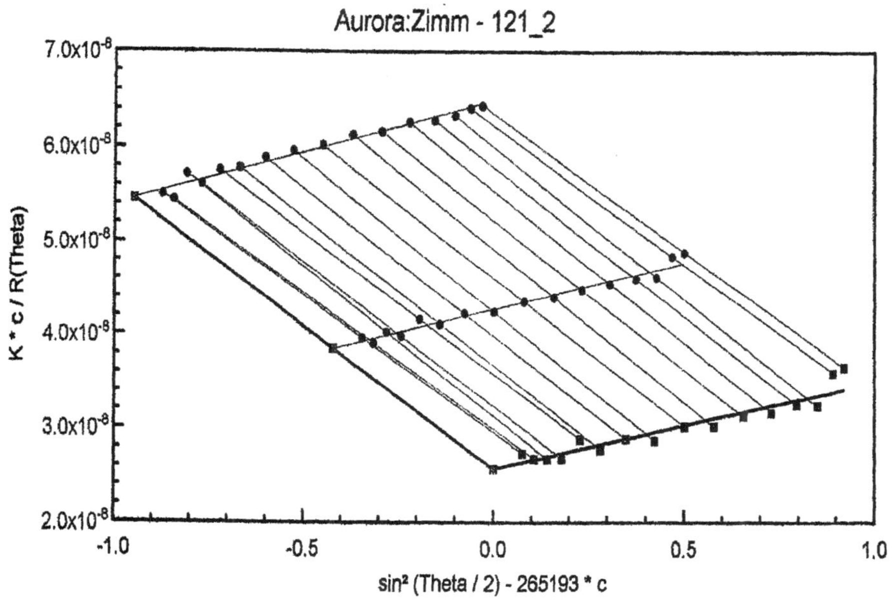


```
RMS Radius         : 42.9 +/- 1.9 nm
Molecular Weight   : (3.89 +/- 0.09)e7 g/mol
2nd Virial Coef.   : (4.03 +/- 0.09)e-3 mol ml/g**2
```

Figure 1 - Zimm plot of light scattering data for polyurethane substrate. Fit to equation (1) gives molecular weight of $39 \cdot 10^{6}$ g/mol, radius of gyration of 43 nm, and a virial coefficient of $4 \cdot 10^{-3}$ mol-mL/g^2.

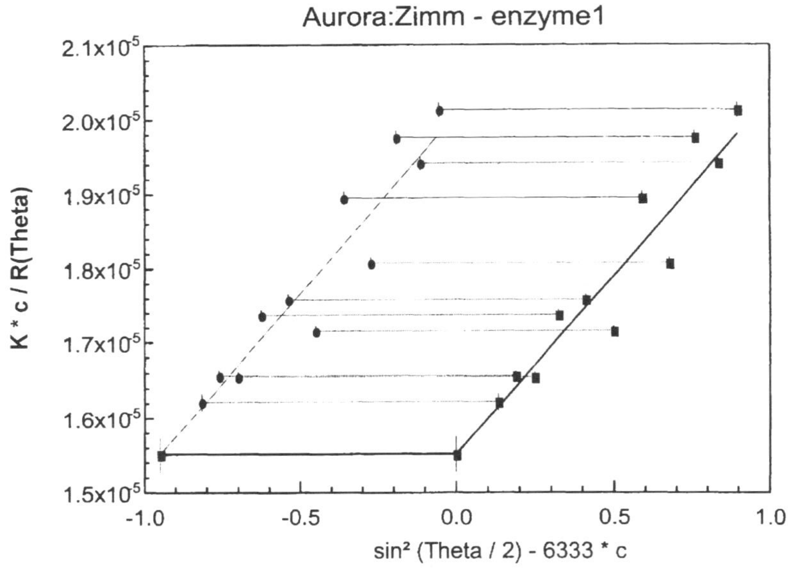


```
RMS Radius         : 39.7 +/- 1.9 nm
Molecular Weight : (6.45 +/- 0.1)e4 g/mol
2nd Virial Coef.  : 0.00e+00 mol ml/g**2
```

Figure 2 - Zimm plot results for the purified enzyme. Molecular weight results are in excellent agreement with SDS-PAGE measurements. Radius of gyration suggests the molecule is expanded in its structure.

enzyme separately (shown in Figs. 1 and 2) were acquired and analyzed using the DAWN for Windows software, ver. 3.31 (Wyatt Technology Corporation). Time resolved data for the enzymatic degradation experiments was acquired and analyzed using the ASTRA for Windows software, ver. 4.73 (Wyatt Technology Corp.).

Results and Discussion

Figure 1 shows a Zimm plot of light scattering data taken for the polyurethane substrate. Parameters from a fit to equation (1) are: molecular weight of $39 \cdot 10^6$ g/mol, radius of gyration of 43 nm, and a virial coefficient of $4 \cdot 10^{-3}$ mol-mL/g^2. Using this molecular weight and radius of gyration, we can investigate the geometry of the polyurethane substrate molecule. We estimate the molecular weight and length of the repeat unit as m=150 g/mole and l=1.5 nm (precise details of this proprietary material are not known). From this, we can estimate the persistence length, L_T, of the polyurethane substrate, assuming a linear, random coil. Using $L_T = 3\ Rg^2/L$, with the measured radius of gyration and the contour length estimated as $L = (M_W / m)\ l = 390{,}000$ nm, we estimate $L_T = 0.01$ nm. This clearly unlikely result suggests strongly that the polyurethane substrate is NOT linear, but is in fact a very compact molecule; either highly branched or perhaps in the form of extremely small, heavily cross-linked pellets. This is not surprising, as isolated polyurethane chains would not be expected to be water soluble, as this proprietary material is. We can also compute a straightforward density (mass/volume = $m/v = (M_W/N_A)/(4/3\ \pi\ Rg^3) \cong 0.3$ g/mL) for the substrate and use it to predict the intrinsic viscosity, whose relationship to the density is as predicted by Einstein's expression for spherical particles: $[\eta] = 2.5\ v_H/m = 2.5\ N_A\ (4/3\ \pi\ R^3 / M_W) = 7.7$ mL/g. This is in good agreement with our measured intrinsic viscosity for this material of $[\eta] = 3.8$ mL/g. Differences here are probably due to our assumption that the radius of gyration is the precisely the same as the hydrodynamic radius - these data suggest they differ by about 25%. The close agreement reinforces the idea that the substrate material consists of tightly packed molecules.

The enzyme light scattering data, shown in Figure 2, gives a molecular weight of $64.5 \cdot 10^{+3}$ g/mole and a radius of gyration of 40 nm. This agrees well with other results, obtained by other methods (primarily SDS-PAGE) which give $65 \cdot 10^{+3}$ g/mole (these other methods give no information on size of the molecule). A calculation similar to that above gives a persistence length of L_T = 10 nm. Although the enzyme in question is almost certainly NOT a random coil, this calculation still suggests that the molecule is quite expanded (as opposed to compact) in its structure. Finally, the relative dimensions of the enzyme and the substrate suggest that degradation occurs when one enzyme attaches to one substrate molecule. That is, since the radii of gyration of the two molecule are comparable, it is unlikely that more than one enzyme can attach to a given substrate molecule.

Finally, Figure 3 shows the inverse of the molecular weight as measured by MALLS plotted as a function of time. The slope of this graph gives the enzymatic degradation rate (enzyme cuts per unit time) divided by the initial molecular weight, and the intercept gives the inverse of the initial molecular weight. Thus, the degradation rate is obtained directly from the ratio of the slope to the intercept. Using the parameters obtained from the fitted straight line, we calculate a degradation rate of $(8.503\pm1.558)\cdot10^{-4}$ min^{-1}. The very small value for the enzymatic degradation rate constant is not unexpected considering the very small concentration of both substrate and enzyme used in this particular experiment.

We now consider the effect of diffusion on the rate we extract. The substrate molecules are approximately 10^{-4} cm apart (obtained by setting the reciprocal of the number density of the substrate molecules equal to $4/3\pi R^3$). Using a diffusion coefficient typical of macromolecules ($D\sim10^{-7}$ cm^2/s), enzyme molecules will diffuse near a substrate molecule on a time scale of $\tau = L^2/D \sim$ 1 sec. That the observed time for enzyme action is so much longer ($\sim10^{+5}$ sec) suggests that either the bond between enzyme and substrate is extremely weak (requiring many encounters to form the complex) or that the enzyme scission itself is extremely slow.

With only one degradation experiment, we cannot determine Michaelis-Menten parameters. However, since we are in a regime for which [E] >> [S], we can make some statements.

We assume a reaction of the typical form:

$$E + S \underset{k_{-1}}{\overset{k_1}{\rightleftharpoons}} ES \underset{k_{-2}}{\overset{k_2}{\rightleftharpoons}} E + P$$

We assume, given the condition $[E]_0 >> [S]_0$, that entropy will drive the first reaction to the right. That is, we assume that $[E]_0 = [E] + [ES] \cong [E]$ and that $[S]_0 = [S] + [ES] \cong [ES]$. Also, the large disparity in enzyme versus substrate molecules suggests that $(1/[E])\ d[E]/dt = d \ln [E]/dt \cong 0$. Then $(k_{-1}+k_2)[ES] = [E]\ (k_{-2}[P] + k_1 [S])$, and making the standard assumption that $k_{-2}[P]$ is small, we have the typical result that $[ES] = k_1 [E][S] / (k_{-1} + k_2) = [E][S] / K_M$. With the above assumptions, this can be put in the form $[S]/[S]_0 = K_M/[E]_0$. Now, assuming that the majority of substrate is bound to enzyme (i.e., that $[S]/[S]_0 << 1$), we conclude that $K_M << [E]_0 = 6 \cdot 10^{-7}$ M.

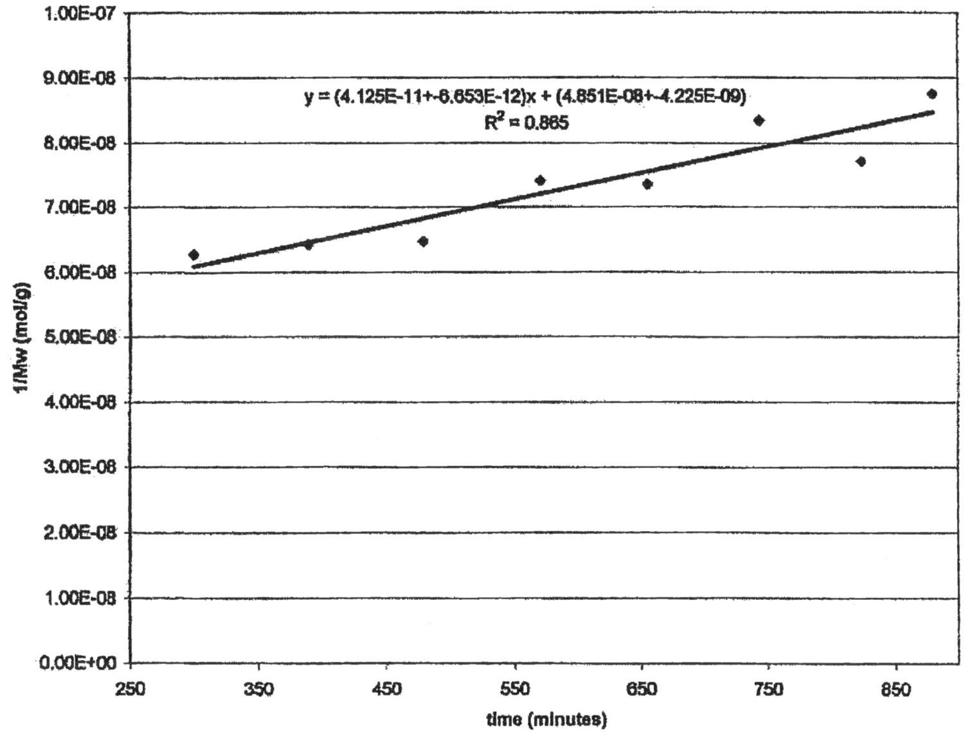

Figure 3 - Light scattering measurements of the molecular weight of the polyurethane substrate while under attack by polyurethanase enzyme. The ratio of the slope to the initial molecular weight gives the degradation rate of the enzyme (enzyme cuts per unit time).

Conclusions

We have used multi-angle laser light scattering [MALLS] to measure the degradation of polyesterurethanes by an enzyme purified from Comomonas acidovirans, and to characterize the substrate and the enzyme. Light scattering results for the substrate and enzyme separately show that, while the enzyme and substrate differ in molecular weight by a factor of 1000, they are comparable in spatial extent. This suggests that enzyme action occurs when a single protein attaches to a single substrate molecule. The time dependence of the measured molecular weight allows us to calculate an enzyme degradation rate of $(8.503\pm1.558)\cdot10^{-4}$ min^{-1}. This rate is much slower than is suggested by a simple diffusive encounter model, implying that the affinity of enzyme for the substrate is very small or that the scission process itself is very slow.

Acknowledgements

The researchers gratefully acknowledge the support of the SLU College of Arts and Sciences OSCAR Program, the Louisiana Board of Regents grant # LEQSF (1999-02) - RD - A - 33, and the SLU Department of Chemistry and Physics.

References

[1] Kaplan, Arthur M., Richard T. Darby, Marvin Greenberger, and Morris R. Rogers, *Microbial Deterioration of Polyurethane Systems, Developments in Industrial Microbiology*, Vol 9 (Society for Industrial Microbiology, 1968)

[2] Phua, B. K., E. Castillo, J. M. Anderson and A Hiltner, *Journal of Biomedical Materials Research*, **21**, 231-246 (1987)

[3] Santerre, J. P., R. S. Labow, D. G. Duguay, D. Erfle, and G. A. Adams, *Journal of Biomedical Materials Research*, **28**, 1187-1199 (1994)

[4] Ruiz, C., T. Main, N. Hilliard, and G.T. Howard. *International Biodeterioration and Biodegradation*. **43,** 43-47 (1999)

[5] Arakawa, T., *Cell Technology,* **15** (5) 679-688 (1996).

[6] Benmouna, M., and Reed, W.F.; "Theoretical developments in static light scattering from polymers," in *Light Scattering, Principles and Development Ch. 1*, (Clarendon Press, Oxford, England 1996).

[7] Kamide, K., M. Saito, and Y. Miyazaki, "Molecular weight determination", in *Polymer Characterization Ch. 5*, (Blackie Academic and Professional Press, London 1993).

[8] Norwood, D. P.; Benmouna, M.; and Reed, W. F.; "Static Light Scattering from Mixtures of Polyelectrolytes in Low Ionic Strength Solution", *Macromolecules*, **29**, 4293-4304 (1996)

[9] Zimm, Bruno H., "The Scattering of Light and the Radial Distribution Function of High Polymer Solutions", *J. Chem. Phys.*, **16**, 1093-1116 (1948)

[10] Norwood, D. P.; and Reed, W. F.; "Determination of Polymer Solution Structure by Time Resolved Static Light Scattering", *Polymer Preprints*, **37**(1), 752 (1996)

[11] Ghosh, S., and W. F. Reed, "New characteristic signatures from time-dependent static light scattering during polymer depolymerization, with application to proteoglycan subunit degradation," *Biopolymers*, **35**, 435-450 (1995).

[12] Reed, W. F.; Reed, C. E.; and Byers, L. D.; "Random Coil Scission Rates Determined by Time-Dependent Total Intensity Light Scattering: Hyaluronate Depolymerization by Hyaluronidase", *Biopolymers*, **30**, 1073-1082 (1990)

[13] Ghosh, S., I. Kobal, D. Zanette, and W. F. Reed, "Conformational contraction and hydrolysis of hyaluronate in sodium hydroxide solutions," *Macromolecules*, **26**, 4685-93 (1993).

[14] Singh, S.K., S.P. Jacobssen, "Kinetics of acid hydrolysis of k-carrageenan as determined by molecular weight (SEC-MALLS-RI), gel breaking strength, and viscosity measurements," *Carbohydrate Polymers,* **23** 89-103 (1994).

[15] Catalani L.H.; Rabello, A.M.; Florenzano, F.H.; Politi,M.J.; and Reed,W.F.; "Real-time Determination of Ultraviolet Degradation Kinetics of Polymers in Solution", *Int'l. J. of Polymer Characterization and Analysis*, **3** (2), 231-247 (1997)

[16] Reed,W.F.; "Time dependent light scattering from single and multiply stranded polymers undergoing random and endwise scission", *J. Chem. Phys.*, **103**, 7576-7584 (1995)

[17] Jackson, C., L.M. Nilsson and P.J. Wyatt, "Characterization of biopolymers using a multi-angle light scattering detector with size exclusion chromatography," *J. Appl. Polym. Sci.* **43** 99 (1989).

[18] Jackson, C.; and Barth, Howard G.; "Size Exclusion Chromatography with Multiple Detectors: Inherent Problems in Data Analysis", *Trends in Polymer Science*, **2**, 203-207 (1994)

[19] Reed, W.F., "Data evaluation for unified multi-detector size exclusion chromatography-molar mass, viscosity and radius of gyration distributions", *Macromol. Chem. Phys.*, **196**, 1539-1575 (1995).

Chapter 24

Peroxidase-Mediated Free Radical Polymerization of Vinyl Monomers

B. Kalra and R. A. Gross*

NSF-I/UCRC for Biocatalysis and Bioprocessing of Macromolecules, Polytechnic University, 6 Metrotech Center, Brooklyn, NY 11201

Peroxidase-mediated polymerizations of methyl methacrylate, acrylamide, and sodium acrylate were studied. Horseradish peroxidase (HRP), hydrogen peroxide, and 2,4-pentanedione functioned as the oxido-reductase, oxidant, and reducing substrate, respectively, for the free radical polymerizations. The polymerizations were conducted in solution (aqueous and water/co-solvent mixtures) and emulsions (oil/water with and without surfactant, concentrated). The poly(acrylamide)s and poly(acrylate)s formed were atactic. However, poly(methylmethacrylate) with high syndiotactic was prepared as a result of the ability of enzymes to function effectively at ambient temperature and below.

Introduction

The ability of enzymes to provide an advantage over traditional catalysts for monomer and polymer synthesis is gaining in recognition (*1*). Although enzyme catalysis has been known for well over a century (*2*), there use has been largely restricted to aqueous systems. However, over the past 15 years, non-aqueous enzymology has continued to grow in importance (*3-5*) In some cases, non-aqueous enzyme-catalysis has provided synthetic routes to chemical transformations that are difficult, or even impossible, with conventional chemical catalysts. Important examples of reactions catalyzed by enzymes in organic media include: *i*) the lipase-catalyzed synthesis of optically active polyesters (*6*), *ii*) lipase-catalyzed interesterification of triglycerides and fatty acids (*7*), *iii*) the regioselective oxidation of phenols by phenol oxidase (*8*), *iv*) peroxidase-mediated reactions that yield polymers with useful electro-optical properties such as polyphenols and polyaromatic amines (*9,10*), and polysaccharide modification to regulate their properties (*11*).

Horseradish peroxidase (HRP) is an oxido-reductase that acts on hydrogen peroxide and/or alkyl peroxide as an oxidant (*12*) and on several reducing substrates such as phenol, hydroquinone, pyrogallol, catechol, aniline and p-aminobenzoate (*13*). The oxidative coupling of a variety of substrates such as phenols and aromatic amines catalyzed by HRP in the presence of hydrogen peroxide have been reported in aqueous (*14*), nonaqueous (*3-5, 8-10, 15-19*) and interfacial systems (*20*).

Recently, HRP has been used to catalyze free-radical polymerizations of commodity vinyl monomers. The potential of such reactions with vinyl monomers including methyl methacrylate, acrylamide, 2-hydroxyethyl methacrylate, and acrylic acid was first recognized and reported by Derango *et al. (21)*. In a related study, Kobayashi and coworkers (*22*) reported the HRP-catalyzed polymerization of phenylethyl methacrylate. Later, HRP-mediated free-radical polymerizaton of acrylamide was shown to take place in the presence of β-ketones as initiators (*23,24*). Our group investigated HRP-mediated methyl methacrylalte polymerization in a mixture of water and water-miscible solvents at ambient temperature (*25, 26*).

This chapter provides an overview of recent contributions by our laboratory towards enzyme-mediated vinyl monomer polymerizations. These studies addressed how various reaction parameters can be adjusted to regulate: *i*) the HRP-mediated polymerization of water-soluble acrylamide and sodium acrylate in aqueous medium (*27*), *ii*) polymerizations of water-insoluble methyl methacrylate, and *iii*) polymerizations of acrylamide in concentrated emulsions (*27*). In addition, we have carried out experiments to probe whether the mechanism of the polymerization involved a direct electron transfer from the oxoiron(IV) π-radical cation to monomer, or whether chain initiation takes place through the radical species generated by 2,4-pentanedione.

Experimental

Materials.

Horseradish peroxidase (Type II, activity 235 purpullogallin units/mg), Horserdish peroxidase (Type I, activity 100 purpullogallin units/mg), Soybean peroxidase (90 purpullogallin units/mg), *Arythomyces ramosus* peroxidase [(75 purpullogallin units/mg), Acrylamide, sodium acrylate, dioctyl sulfosuccinate sodium salt, cetyl trimethyl ammonium bromide, sorbitane monooleate, hydrogen peroxide (30% (wt/vol)] were all obtained from the Sigma Chemical Company. Methanol, dioxane, acetone, tetrahydrofuran and dimethyl formamide were all of analytical grade and were used as received. Methyl methacrylate (MMA), obtained from the Aldrich Chemical Co., was fractionally distilled over calcium hydride under reduced pressure with a nitrogen atmosphere. 2,4-Pentanedione from Aldrich was distilled prior to use.

Instrumentation

The NMR data were recorded on a Bruker DPX300 and Bruker AMX500. The chemical shifts in parts-per-million (ppm) for proton (^{1}H) NMR spectra were referenced relative to tetramethylsilane (TMS, 0.00ppm) as the internal reference. The stereochemistry of the polymer backbone was calculated by observing the NMR signals (*28*) due to the backbone methyl groups: syndiotactic triad, 0.81 ppm; atactic triad, 0.97 ppm; isotactic triad, 1.14 ppm. The standard deviation for the syn dyad fractions was calculated by taking the mean for three replicates of the NMR integration values. The distribution of repeat unit sequences that differ in stereochemistry was analyzed for poly(acrylamide) by observing the NMR signals due to the methine carbon region in the ^{13}C-NMR.

The number average molecular weights (M_n) of the polymer samples (PMMA) were determined by gel permeation chromatography (GPC) using a Waters HPLC system equipped with model 510 pump, Waters model 717 autosampler, model 410 refractive index detector, and model T-50/T-60 detector from Viscotek Corporation with 500, 10^3, 10^4 and 10^5 Å ultrastyragel columns in series. Trisec GPC software Version 3 was used for calculations. Chloroform was used as the eluent at a flow rate of 1.0 mL/min. Sample concentrations of 0.2 percent wt/vol and injection volumes of 100 µL were used. Molecular weights were determined based on conventional calibration curve generated by polystyrene standards of low dispersity (Aldrich Chemical Company). Since the

PMMAs formed in this work were not completely soluble in chloroform, the molecular weight values reported are relevant to only part of the product.

The weight average molecular weights (M_w) of the poly(acrylamide) and poly(sodium acrylate) were determined by gel permeation chromatography (GPC). Studies by GPC were carried out using a Waters, Inc. Model 510 pump, two Shodex KB 80m and one Shodex KB 802.5 columns, and a waters 410 differential refractometer. The software used for molecular weight calculations was millennium chromatography manager version 2.15. Sodium dihydrogen phosphate, 20 mM, pH 7.0, was used as the eluent. Analyses were carried out at 35 °C, flow rate 1 mL/min and with injection volumes of 10 μL. Polyethylene glycol standards with narrow polydispersity were used to generate a calibration curve.

Differential Scanning Calorimetry (DSC) was performed on TA instrumental model 2920 DSC. The temperature program used was to increase the sample temperature from room temperature to 150 °C, cool to 30 °C and then to reheat to 150 °C. The heating rate was 20 °C min^{-1} and the purge gas was helium. The cooling was performed using TA instruments refrigerated cooling systems using an equilibration step that gives rapid cooling. The glass transition temperature (T_g) was determined by the midpoint at half-height. The T_g results were taken from the second heating scan.

Enzyme-catalyzed polymerization of MMA in binary solvents

Methyl methacrylate (MMA) (5.6 mmol) was added to a solution of distilled water (0.7 mL) and organic solvent (0.3 mL) in a dual inlet ampule under nitrogen atmosphere. An example of a typical reaction is the successive addition under a nitrogen atmosphere of 0.2 mL of HRP (80 mg/mL, 16 mg of enzyme), hydrogen peroxide (0.092 mmol) and 2,4-pentanedione (0.136 mmol). The reaction mixture was maintained under nitrogen with stirring at room temperature for a predetermined time period. Then, the reaction mixture was poured into a large excess of methanol. The precipitate obtained was separated by filtration, washed with methanol and dried (*in vaccuo*, 50 °C, 30 mm Hg, 24 hours).

Polymerization of acrylamide in aqueous medium

Acrylamide (2.92 mmol in 4 mL water) in a dual inlet ampule was purged with nitrogen for 10 min. Into the above solution, HRP (8 mg in 0.2 mL water), hydrogen peroxide (0.046 mmol) and 2,4-pentanedione (0.068 mmol) were successively injected while stirring. The reactions were carried out at room temperature for a predetermined time period while maintaining both stirring and a nitrogen atmosphere. After the predetermined time, the reaction mixture was poured into an excess of methanol. The resulting precipitate was filtered off, washed with methanol and dried *in vacuo* (50 °C, 30mm Hg, 24 h). The enzyme was soluble in methanol and thus removed from the polymer.

Polymerization of acrylamide in concentrated emulsion

Toluene (42 μL) and sorbitane monooleate (17.2 μL) were degassed in a dual inlet ampule for 10 min. Acrylamide (2.92 mmol) dissolved in water (0.45 mL) was added to the surfactant solution with stirring. To the above was added HRP (8 mg in 0.3 mL water), hydrogen peroxide (0.046 mmol) and 2,4-pentanedione (0.068 mmol) while vigorously stirring. The polymerization was carried out under a nitrogen stream for different time periods. The polyacrylamide was isolated by precipitation in methanol and then dried in a vacuum oven (50 °C, 30mm Hg, 24 h).

Polymerization of sodium acrylate in aqueous medium

Sodium acrylate (4 mmol) was added to a solution of distilled water (1.5 mL) in a dual inlet ampule under nitrogen atmosphere. An example of a typical reaction is the successive addition under a nitrogen atmosphere of HRP (11.4 mg in 0.5 mL water), hydrogen peroxide (0.064 mmol) and 2,4-pentanedione (0.097 mmol). The reaction mixture was maintained under nitrogen with stirring at room temperature for a predetermined time period. Then, the reaction mixture was poured into a large excess of methanol. The precipitate obtained was separated by filtration, washed with methanol and dried (*in vacuo*, 50 °C, 30 mm Hg, 24 h).

Results and Discussion

HRP-mediated polymerizations were conducted with 2,4-pentanedione as the reducing substrate and hydrogen peroxide as the oxidant (stoichiometric ratio 1.5:1). The acrylate to 2,4-pentanedione ratio was maintained at about 42-44:1. MMA polymerization was conducted in a mixture of water and a water-miscible co-solvent so that MMA is solubilized. Co-solvents (dioxane, acetone, THF and DMF) were all useful for the HRP-mediated polymerization of MMA.

Effect of co-solvents

The room-temperature polymerization of MMA was catalyzed by HRP in the presence of hydrogen peroxide and 2,4-pentanedione in binary solvent mixtures consisting of water and different co-solvents. The polymer formed, based on analyses by high field proton (^1H) NMR, was predominantly syndiotactic. When the binary mixture was water:DMF (3:1), the polymerization was very slow (<5% monomer conversion in 24 h). The low yields in water/DMF are due to unfavorable interactions between HRP and DMF that cause enzyme deactivation. The yields of PMMA increased when less polar solvents were used. Specifically, for 24h polymerizations in water/acetone, water/dioxane and water/THF (all 3:1 v/v), the PMMA yields were 8, 45 and 90%, respectively.

Table I. Effect of the Water-Miscible Co-Solvent on MMA Polymerizations Carried out for 24 h at 25 °C

Co-solvent	*%-yield*	*M_n (10^4) (g/mol)*	*M_w/M_n*	*T_g (°C)*	*Syn-dyad fraction*
25% DMF	<2.0	N.D.	N.D.	N.D.	N.D.
25% Acetone	7	1.8	4.7	128	0.85
25% Dioxane	45	6.3	3.0	131	0.86
50% Dioxane	38	1.9	3.8	126	0.86
75% Dioxane	7	2.8	4.3	N.D.	0.85
25%THF	85	7.2	3.1	128	0.84
50%THF	58	6.4	2.6	N.D	0.84

Study of the repeat unit sequence distribution for MMA polymerization using HRP as the catalyst showed that the syndiotacticity of the product was highest when conducting the polymerization in water:dioxane (3:1 v/v). Increasing the ratio of dioxane to water resulted in decreased polymer yield and lower syndiotacticity (Table I). This is attributed to that HRP normally looses its activity at high non-aqueous solvent concentration. Analysis of the syndio-PMMA products by GPC established that they were of moderate molecular weight (M_n between 9,000 and 75,000 g/mole). Measurements by DSC confirmed that PMMA was indeed highly syndiotactic (glass transition temperatures of about 131°C).

Effect of reaction temperature and concentration of HRP, hydrogen peroxide, and 2,4-pentanedione

HRP-mediated free radical polymerization of MMA was performed in water/dioxane (3:1 v/v) for 24 h at 5, 15, 20, 25, 34 and 40°C. The polymer yield at 25°C was 45%. Increase of the MMA polymerization temperature to 34°C resulted in a decrease in the polymer yield as well as the degree of syndiotacticity. Further increase in temperature to 40°C gave PMMA in 4% yield. Decrease of the MMA polymerization temperature to 5°C resulted in both higher yields and polymer syndioregularity (Figure 1).

A study was performed to determine how variations in the concentrations of HRP, hydrogen peroxide, and 2,4-pentanedione effect the outcome of MMA (5.6 mmoles) polymerizations. This work showed that higher chain syndioregularity resulted by using 16 mg of HRP, 0.092 mmol of hydrogen peroxide, and 0.136 mmol of 2,4-pentanedione (Figures 2a and 2b).

Acrylamide polymerizations mediated by HRP in the presence of hydrogen peroxide and 2,4-pentanedione were conducted in aqueous media and concentrated emulsions. In aqueous media, yields of poly(acrylamide) were up to 98% within 3 h. The rate of the polymerization was further enhanced by adding anionic or cationic surfactants. By using surfactants, an isolated polymer yield of 94% was obtained by 75 min. When a concentrated emulsion formed with sodium monooleate was used in place of a homogeneous aqueous solution, the polymerization increased in rate giving nearly quantitative yields of polymer in 1.25 h (Figure 3).

Further comparison of product molecular weights showed that, by carrying out the polymerization in aqueous solution, M_n was higher and polydispersity was lower. The weight average molecular weight (M_w) of the poly(acrylamide) products ranged between approximately 128 K and 210 K in the aqueous medium and 106 K and 154 K in concentrated emulsions. For both reaction conditions the polymer formed has repeat unit sequence distributions that are best described as random. HRP-mediated polymerization of sodium acrylate in the presence of hydrogen peroxide and 2,4-pentanedione was also carried out to give poly(sodium acrylate) in 88% yield, in 24 h, with M_w ca. 219 K, M_w/M_n = 5.2. Analysis of this product by NMR showed that it too was atactic.

Mechanism of HRP-Mediated MMA Polymerization

For HRP-mediated polymerizations of MMA in the absence of 2,4-pentanedione, with an MMA to H_2O_2 ratio of 61:1 mol/mol, MMA polymerization did not occur. If the reaction did occur in the absence of 2,4-pentanedione, then the oxoiron (IV) π-radical cation would likely be the

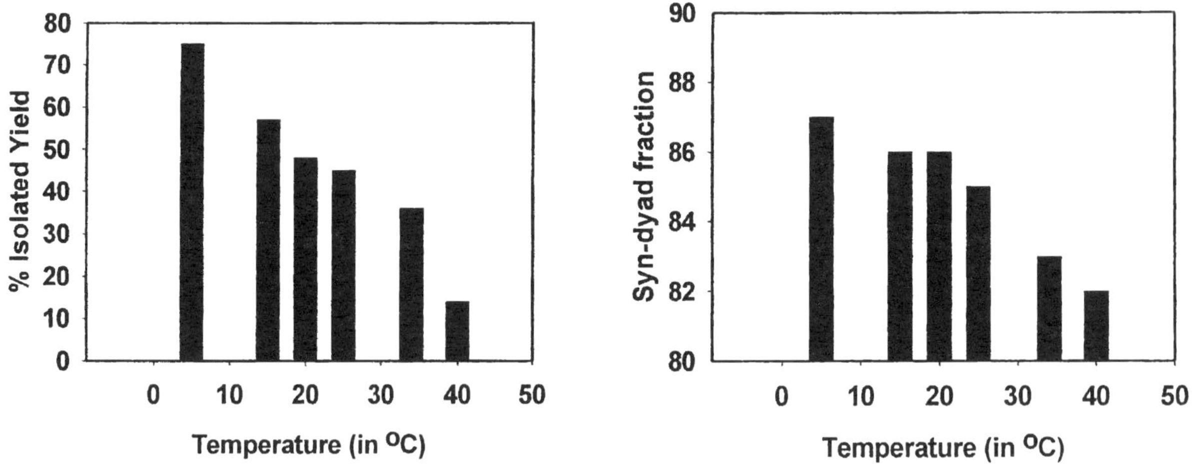

Figure-1. Effect of temperature on % isolated Yield and syndiotacticity of HRP II mediated MMA polymerization.

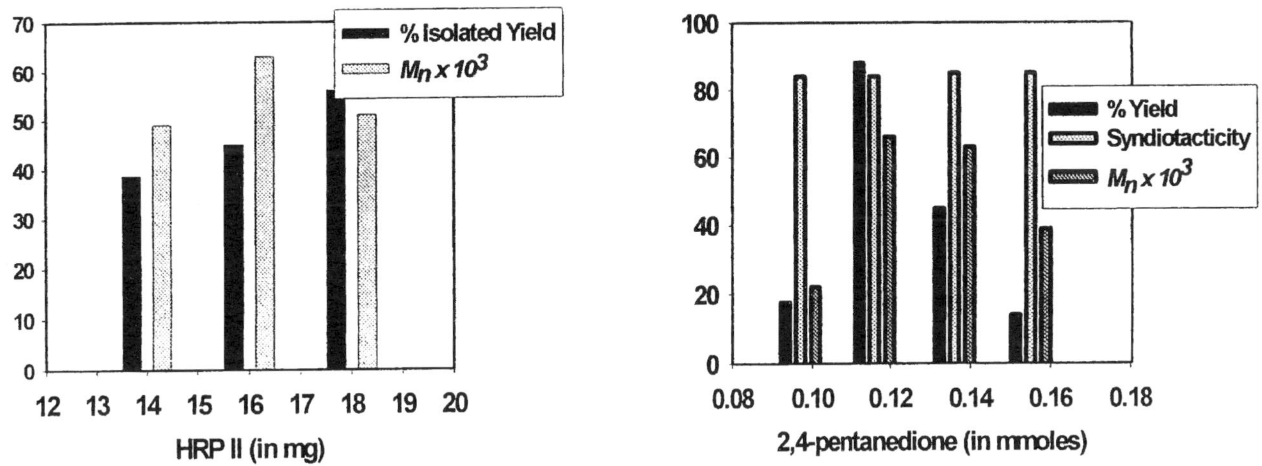

Figure-2. Effect of HRP-II (a) and 2,4-pentanedione (b) concentration on MMA yield, M_n and syndioregularity.

intermediate responsible for initiation of MMA free-radical polymerization. Furthermore, in the absence of enzyme but with 2,4-pentanedione and H_2O_2, no polymer was obtained. This shows that MMA polymerization does not takes place due to the homolytic cleavage of hydrogen peroxide. Therefore, we believe that the formation of primary radicals arises from the second and third steps of the enzymatic cycle. The keto-enoxy radicals are generated as the first radical species. Then, initiation may take place through this radical or by the radical transfer to any other molecule, i.e. enzyme impurity, amino acid residue or hydrogen peroxide itself.

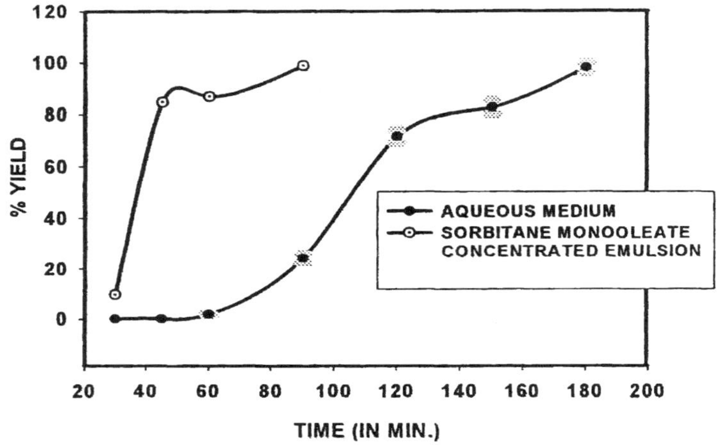

Figure 3. Horseradish peroxidase mediated free radical polymerization of acrylamide in aqueous medium and in sorbitane monooleate concentrated emulsion (The error bars were generated by determining the mean deviation for two replicate reactions)

Conclusions

The HRP-mediated free radical polymerization of MMA was found to be useful for the preparation of highly syndiotactic PMMA. Normally, to attain high degrees of PMMA syndioregularity, it is necessary to use toxic metals or extreme temperatures. For example, ionic and conventional free-radical polymerizations are often conducted at about –80 °C and ≥70 °C, respectively. HRP-mediated MMA polymerizations provide an'environmentally friendly' route to stereoregular PMMA, poly(acrylamide) and poly(sodium acrylate) at ambient temperatures. Comparison of acrylamide polymerizations in aqueous media without surfactant versus in concentrated emulsions showed that the latter gave more rapid polymerizations but products with broader polydispersity. In addition, the above strategy for HRP-mediated free-radical polymerizations was found useful with sodium acrylate. Hence, after a 24 h polymerization in aqueous medium without surfactant, poly(acrylate) was prepared in 88% yield, M_w ca. 191 K, and PD=5.2. The chain stereochemistry of the polymers formed was found to be similar to that for free-radical polymerizations carried out without enzyme. This suggests that the propagating chain and the incoming monomer do not associate in such a way that would regulate chain stereoregularity.

Acknowledgements

The authors of this paper would like to thank Rohm and Haas Company and Novozymes for financial support of this research

References

1. Gross, R. A.; Kaplan, D. L.; Swift, G., Eds. *Enzymes in Polymer Synthesis*. ACS symposium series 684. American Chemical Society, Washington, DC, 1998.
2. Dixon, M.; Webb, E. C. *Enzymes*, 3rd edn., Academic Press 1979.
3. Zaks, A.; Klibanov, A. M. *Proc. Nat. Acad. Sci., U.S.A.* **1985**, *82*, 3192.
4. Klibanov, A. M. *Trends Biotechnol.* **1997**, *15*, 97.
5. Cooney, C. L.; Hueter, J. *Biotechnol Bioeng.* **1974**, *16*, 1045.
6. Wallace, J. S.; Morrow, C. J. *J. Polym. Sci., Part A* **1989**, *27*, 2553.
7. Basheer, S.; Mogi, K.; Nakajima, M. *Biotechnol Bioeng.* **1995**, *45*, 187.
8. Kazandjian, R. Z.; Klibanov, A. M. *J. Am Chem Soc.* **1985**, *107*, 5448.
9. Akkara, J. A.; Senecal, K.; Kaplan, D. L. *J Polym Sci.* **1991**, *29*, 1561.
10. Akkara, J. A.; Salapu, P.; Kaplan, D. L. *Indian J Chem.* **1992**, *31B*, 855.
11. Bruno, F. F.; Akkara, J. A.; Ayyagari, M.; Kaplan, D. L.; Gross, R. A.; Swift, G.; Dordick, J.S. *Macromolecules* **1995**, *28*, 8881.
12. Schonbaum, G. R.; Lo, S. *J Biol. Chem.* **1972**, 10, 3353.
13. Barman, T.E. *Enzyme Handbook*, Springer-Verlag: New York, **1985**, Vol I, p-234.
14. Saunders, B. C.; Holmes-Siedle, A. G.; Stark, H. P. *Peroxidase*. Butterworth publishers 1942.
15. Akkara, J. A. et al. *Polymeric Materials Encyclopedia* (Vol 3) (J C Salamone, ed), CRC press, 1996, 2115.

16. Dordick, J. S.; Marletta, M. A.; Klibanov, A. M. *Biotechnol Bioeng.* **1987**, *30*, 31.
17. Ryu, K.; Stafford, D. R.; Dordick, J. S. *Am Chem Soc. Sýmp. Ser.* **1989**, *389*, 408.
18. Ayyagari, M.; Akkara, J. A.; Kaplan, D. L. *Acta Polym.* **1996**, *47*, 193.
19. Ayyagari, M; Akkara, J. A.; Kaplan, D. L. *Mater. Sci. Eng.* **1996**, *C4*, 169.
20. Bruno, F. F.; Akkara, J. A.; Kaplan, D. L.; Sekher, P.; Marx, K. A.; Tripathy, S. K. *Ind. Eng. Chem. Res.* **1995**, *34*, 4009.
21. Derango, A. R.; Chiang, L-C.; Dowbenko, R.; Lasch, J. G. *Biotechnol. Tech.* **1992**, *6*, 523.
22. Uyama, H.; Lohavissavapanich, C.; Ikeda, R.; Kobayashi, S. *Macromolecules* **1998**, *31*, 554.
23. Emery, O.; Lalot, T.; Brigodiot, M.; Marechal, E. *J Polym. Sci.,Part A, Polym. Chem.* **1997**, *35*, 3331
24. Teixeira, D.; Lalot, T.; Brigodiot, M.; Marechal, E. *Macromolecules* **1999**, *32*, 70.
25. Kalra, B.; Kumar, A.; Gross, R.A. *Polymer Preprint* **2000**, 41(2), 1804.
26. Kalra, B.; Gross, R.A. *Biomacromolecules,* **2000**, 1(3), 501.
27. Kalra, B.; Gross, R. A. *Polymer Preprint* **2000,** 41(2), 1935.
28. a. Bovey, F. A. '*High resolution NMR of Polymers*' Academic Press, New York, 1972, Chapters III and IV.
    b. Schilling, F. C.; Bovey, F. A.; Bruch, M. D.; Kozlowski, S. A. *Macromolecules* **1985**, 18, 1418.

Indexes

Author Index

Al-Azemi, Talal F., 156
Andreana, Peter R., 188
Bernard, Laetitia, 80
Bisht, Kirpal S., 156
Brandstadt, Kurt F., 141
Bulone, Vincent, 65
Bylina, Edward J., 50
Chanzy, Henri, 65
Chen, Tianhong, 231
Cheng, H. N., 1, 203, 253
Cheng, Yu, 265
Coleman, William J., 50
Dordick, Jonathan S., 34
Fan, H., 93, 99
Gao, Wei, 253
Govar, C. Justin, 231
Gross, Richard A., 1, 107, 172, 253, 297
Gu, Qu-Ming, 203, 243, 253
Harris, Michael T., 231
Hasegawa, Yoshie, 80
Him, Joséphine Lai Kee, 65
Himel, D., 286
Hsieh, You-Lo, 217
Hu, Shanghui, 253
Iwaki, Hiroaki, 80
Kalra, Bhanu, 172, 297
Kelley, Aaron S., 124
Kim, Dae-Yun, 34
Kitagawa, M., 93, 99
Kobayashi, Shiro, 128
Kumar, Ajay, 172
Kumar, Rajesh, 107, 253
Lau, Peter C. K., 80
Liu, Nai-Chi, 231
Norwood, D. P., 286
Payne, Gregory F., 231
Pelosi, Ludovic, 65
Prud'homme, Robert K., 265
Putaux, Jean-Luc, 65
Raku, T., 93, 99
Reed, W. F., 286
Saam, John C., 141
Sharma, Ajit, 141
Srienc, Friedrich, 124
Steinbüchel, Alexander, 120
Tanaka, Keiji, 57
Teraoka, Masahiro, 80
Tokiwa, Y., 93, 99
Tokuyama, Tai, 80
Uyama, Hiroshi, 128
Wang, Peng George, 188
Wu, Xiaoqiu, 34
Xie, Jiangbing, 217
Xie, Wenhua, 188
Youvan, Douglas C., 50
Yu, Hyuk, 57

Subject Index

A

Acetobacter xylinium, polysaccharides, 16
Acinetobacter sp. NCIMB 9871
cyclohexanol degradation, 81, 82*f*
See also Cyclohexanol (chn) biodegradation genes
Acrylamide
horseradish peroxidase mediated free radical polymerization, 306*f*
polymerization in aqueous medium, 300–301
polymerization in concentrated emulsion, 301
See also Free radical polymerization, peroxidase-mediated
Acrylation
general procedure, 115
lipases for selective, 110*f*
stereoselective, 109
See also Macromers around sugar core
Activity. *See* Screening enzymes
Acylation
effect of dimethylsulfoxide (DMSO) on enzymatic, 102–103
small molecules, 220
vinyl esters as acylating agents, 219
See also Transesterification
Adenosine
synthesis of vinyl sugar esters, 96*t*
See also Nucleoside branched poly(vinyl alcohol)
Agrobacterium faecalis
model enzyme system, 51
See also Screening enzymes
Alcaligenes eutrophus, polyhydroxyalkanoates, 144
Alginate industry, growth, 189
Allylamine, oligomerization using epoxidase, 211*f*, 214
Amino acid sequence, open reading frames (ORFs) of genes, 85, 87
Amylose, acylation, 220
Aniline, polymerization, 13–14
Antibodies, catalytic, 8
Arabidopsis thaliana
characterization of β-glucan synthase, 73, 76
polysaccharides, 16
See also Plant β-glucans

B

Bacillus licheniformis, acylation of thymidine, 98
Bacillus subtilis
conversion of uridine, 95
synthesis of vinyl sugar esters, 96*t*
Beverages, opportunities for natural polymers, 189
Biocatalysis
applications, 1
polymer science use lagging, 2
problems, 7
Biocatalysis in polymer science
catalytic antibodies, 8
chiral or enantio-enriched monomers, 19
combined condensation and ring opening polymerization, 12
condensation from different monomers, 10

crosslinked enzyme crystals, 8–9
development of novel methodologies, 2–9
directed evolution technique, 2–3
enrichment culture and screening, 5–6
enzymatic condensation polymerization, 9–10
enzymatic peptide synthesis, 23
enzymatic ring opening polymerization, 11–12
enzymatic transesterification, 12
enzyme-catalyzed polymer modifications, 16–17
enzyme-related developments, 2–5
ester interchange reaction, 19
extremophile organisms, 3
genetic engineering, 6
immobilization techniques, 3–4
improved enzymes, 2–3
microbial approaches to polyesters, 13
microzyme, 9
monomers from specific biocatalytic reactions, 20
new bioprocesses, 7–8
new reactions or products, 7
optimization of enzyme reactions and processes, 5
phenolic and vinyl polymers, 13–15
plant cells and cell extracts, 6–7
polycarbonate synthesis, 11–12
polyesters and polycarbonates, 9–13
polymer degradation, 17–18
polymerization of lactones, 11
polymerization of phenol and aniline, 13–14
polymerization of vinyl polymers, 14–15
polynucleotide mimetic, 23
polytransesterification, 10
reactive oligomers, 22–23
self-condensation, 9–10
silicon bioscience, 24
silicone acrylate, 21–22
solubilization techniques, 4–5
synthesis of monomers and reactive oligomers, 18–23
synthesis of oligosaccharides and polysaccharides, 15–16
synzyme, 9
vinyl acrylate, 21
vinyl monomer synthesis, 20–22
whole-cell approaches, 5–7
Biocatalysts, definition of immobilized, 3
Biocatalytic plastics, synthesis, 4
Biodegradable polymers
polyesters, 125
textiles, 189
See also Polyhydroxyalkanoates (PHA)
Biodegradation genes. *See* Cyclohexanol (chn) biodegradation genes
Biopolymer, chitosan for grafting, 234–235
Biopolymer matrices. *See* Reaction-diffusion of enzymes in biopolymer matrices
Bioprocesses, new, 7–8
Biosynthesis. *See* Plant β-glucans
Biotechnology
clean industrial products and processes, 80–81
See also Carbohydrates; Cyclohexanol (chn) biodegradation genes
Block copolymers
reactions between polymer chains, 172
See also Polytransesterification
Butyrate kinase, synthesis of coenzyme A thioesters of hydroxyalkanoic acids, 121
β-Butyrolactone, ring opening polymerization by thermophilic lipases, 193–196

C

Candida antarctica. See Enzymatic polymerizations
Candida rugosa, stereoregular polyesters, 142
δ-Caprolactone
- copolymerization with achiral lactones, 134*f*
- ring structure and optical resolution, 133

ε-Caprolactone (ε-CL)
- polymerization with multifunctional initiator, 110–111
- ring opening polymerization, 146–147
- *See also* Macromers around sugar core; Polytransesterification

Carbohydrates
- CLONEZYME™ library, 190, 191
- degree of substitution (DS) and degree of polymerization (DP) from lipase-catalyzed ε-caprolactone hydroxyethylcellulose (HEC) backbone, 197*t*
- enzymatic assay via peroxidase-chromogen test, 192
- enzymatic synthesis with recombinant thermophilic glycosidase CLONEZYME™ library and conventional mesophilic β-galactosidase, 191*t*
- lipase catalyzing ring opening polymerization (ROP) of β-butyrolactone, 194*t*
- lithium chloride as catalyst for ROP of lactide, 197–200
- macroinitiators in LiCl-catalyzed ROP of lactide, 199–200
- modification, 189–190
- modification of HEC by transgalactosylation with β-galactosidases, 190, 192–193
- polycaprolactone-modified HEC by lipase-catalyzed ROP, 196–197
- polymer degradation by controlled methanolysis in determination of enantiomeric excess, 195
- polymers containing, 108
- ROP by LiCl, 198
- ROP of β-butyrolactone by thermophilic lipases, 193–196
- ROP of lactide by LiCl, 199*t*
- synthesis of vinyl sugar, 99
- synthetic potential of sugar-modifying enzymes, 190
- thermophilic glycosidase library, 190, 191
- thermophilic lipase catalyzing ROP, 194
- transgalactosylation between lactose and HEC, 192
- transgalactosylation of HEC with galactosidases, 193*t*
- *See also* Macromers around sugar core; Vinyl sugar esters

Carbonates. *See* Polycarbonates
Carboxymethylcellulose (CMC)
- addition of cationic charge, 206
- aminolysis of CMC and CMC esters, 207*f*
- enzyme-catalyzed reactions, 212*t*
- hydrolysis by cellulase, 18
- *See also* Polysaccharides

Catalysts
- lithium chloride for ring opening polymerization, 197–200
- *See also* Enzymes

Cell extracts, polymer reactions, 6–7
Cellulose
- enzyme-catalyzed transesterification, 224–225
- FTIR spectra of cellulose, cotton fibers, cotton fibers, and filter paper, 226*f*
- method for enzyme-catalyzed transesterification, 223
- water contact angles, 226*t*

See also Transesterification
Cellulose ethers, growth, 188
Cellulose solids
characterization, 223
enzyme-catalyzed transesterification, 221–222
method for enzyme-catalyzed transesterification, 223
regioselectivity of enzyme-catalyzed transesterification, 223
See also Transesterification
Cellulose synthase
activity, 69–70
See also Plant β-glucans
Charge, addition to polysaccharide, 205–206
Chemical heterogeneity, supramolecular structures, 157
Chitosan
biopolymer for grafting, 234–235
conferring water-resistant adhesive properties, 237
current research, 239–240
grafting hydrophobic phenols onto chitosan, 237–239
modification by conferring base solubility, 235–236
modification to confer base solubility, 235–236
modification using esters of gallic acid, 238–239
solution viscosity for, and hexyloxyphenol-modified, 238*f*
tyrosinase-catalyzed, crosslinking, 236
Cholesterol, monolayers, 58
Chromatium vinosum, polyhydroxyalkanoate synthase, 121
Chromobacterium viscosum, stereoregular polyesters, 142
α-Chymotrypsin, enzyme transfer from aqueous buffers to isooctane, 224*t*
CLONEZYME™, glycosidase library, 190, 191
Cloning. *See* Cyclohexanol (chn) biodegradation genes
Closed reactions, water for polyesterifications, 149
Clostridium acetobutylicum, synthesis of coenzyme A thioesters of hydroxyalkanoic acids, 121
Clostridium propionicum, propionyl-CoA transferase, 121
Cofactors, addressing problem, 8
Collagen, growth, 188
Collision efficiency factor, 275
Colorimetric solid phase assay
enzymatic hydrolysis, 52*f*
See also Screening enzymes
Combinatorial strategy. *See* Polymer libraries
Condensation
combining with ring opening polymerization, 12
enzymatic polymerization, 9–10
self-condensation, 9–10
See also Enzymatic polymerizations
Condensation polymerization, dicarbonates monomers and diols, 158
Contact angles
chitosan films and different gallate esters, 240*f*
water wetting in cellulose derivatives, 225, 226*t*
Cotton
water contact angles, 226*t*
water wetting, 225, 226*t*
See also Cellulose
Crosslinkable polyesters, lipase catalysis, 138, 139*f*
Crosslinked enzyme crystals, improvements, 8–9
Crosslinking, tyrosinase-catalyzed chitosan, 236
Cyclic carbonates
polymerization, 158

See also Polycarbonates
β-Cyclodextrin, acylation, 220
Cyclohexanol (chn) biodegradation genes
amino acid sequence features of enzymes ChnA, ChnC, and ChnD, 85, 87
bacterial strains and plasmids in study, 83*t*
biochemical pathway of chn degradation, 81, 82*f*
biological materials, genetic manipulation, and sequence analysis, 81, 83*t*
chemicals, 81
cloning and sequencing of chnBER flanking regions, 84–85
enzyme assays and protein expression, 84
expression of ChnA, ChnB, ChnC, ChnD, and ChnE in *E. coli* and activity assays, 87, 88*f*
features of chn gene, 90–91
green opportunities and oxidations with chn genes, 89–90
homology of open reading frames (ORFs) with proteins in databases, 86*t*
new insertion sequence of chn gene, 90–91
ORFs up- and downstream of chnBER cluster, 85, 86*t*
potential β-oxidation genes, 90
specific activities of native and clones enzymes, 87*t*
subcloning in *E. coli*, 83–84
Cytidine
synthesis of vinyl sugar esters, 96*t*
See also Nucleoside branched poly(vinyl alcohol)

D

Degradable plastics, growth, 188
Degradation
biochemical pathway of cyclohexanol, 81, 82*f*
enzymatic, kinetics, 270–273
enzymatic, of polysaccharides, 266
enzymatic, reaction method, 267, 269
Michaelis–Menton model, 266–267
polycarbonates, 157–158
polymer, by enzymes, 17–18
relation between molecular weight and, time, 271
1,2,3-trichloropropane, 20
triggered enzymatic, 280, 282
See also Reaction-diffusion of enzymes in biopolymer matrices
Degradation genes. *See* Cyclohexanol (chn) biodegradation genes
Dendrimers, research, 108
Detergent extractions
cryo-transmission electron micrographs of *Arabidopsis thaliana* plasma membranes in presence of, 74*f*
effect on β-glucan synthesis, 72–73
plant glucan synthases, 68
transmission electron micrographs of β-glucans by callose synthase in *Rubus fruticosus*, 75*f*
See also Plant β-glucans
N,N'-Diacetyl chitobiose, oxazoline derivative, chitin formation, 15–16
Diffusion. *See* Reaction-diffusion of enzymes in biopolymer matrices
Diffusion control. *See* Enzyme kinetics on monolayers
Dihydroxyphenylalanine (DOPA), conferring water-resistant adhesive properties, 237
L-α-Dilauroylphosphatidylcholine (DLPC)
enzyme catalysis on DLPC monolayers, 58
See also Enzyme kinetics on monolayers

Dimethyl sulfoxide (DMSO), effect on enzymatic acylation, 102–103
Directed evolution technique, enzyme technology, 2–3
Direct polyesterification, enzyme catalyzed
aqueous media, 142
Candida rugosa and *Chromofacterium viscosum* lipases, 142
chemoenzymatic synthesis of poly(sucrose adipamide), 144, 145*f*
closed reactions, 148
closed reactions and water, 149
conditions and reaction times, 143
diester-diol polyesterification, 146*f*
dispersion, 151
evaluation of molecular structure, 152
evaluation of monomer to enzyme weight ratio, 151, 152*f*
evaluation of solvent, 149
evaluation of use of recycled enzyme, 152
experimental, 147–148
feasibility, 148–149
instrumentation, 148
nonpolar organic solvent, 144, 146*f*
open reactions, 148
optically active poly((meth)acrylic) materials, 144, 145*f*
organic solvents as media, 142–143
physical use of sonication, 151
poly(β-hydroxyalkanoates) (PHA), 143–144
poly(12-hydroxydodecanoic acid) [poly(12-HDA)] gel permeation chromatography, 154*t*
poly(ε-caprolactone) by ring opening polymerization, 146–147
porcine pancreatic lipase, 146
reactivity of primary hydroxy acid equilibrium reaction, 150–151
relationship between thermal history of enzyme and its activity, 151
ring opening polymerization of ε-caprolactone, 146*f*
solvent-free transesterification, 147
stereochemical repeat unit of PHA, 143*f*
temperature, 150–151
time study of condensation of 12-HDA, 152, 153*f*
water, 149–150
water and dependence on pH, 149–150
Dispersion, polyesterification reaction, 151
Divinyl adipate
library from reaction with diols, 37, 38*f*, 39*f*
vinyl sugar ester from sugar and, 100–102
See also Enzymatic polymerizations
Divinyl sebacate
lipase-catalyzed polymerization with sorbitol, 136, 137*f*
lipase-catalyzed polymerization with triols, 134–135
See also Regioselective polymerization

E

Electrochemical regeneration, cofactors, 8
Enantioselective polymerization
δ-caprolactone, ring-structure and optical resolution, 133
catalytic site of lipase, 130
copolymerization of β-butyrolactone with achiral lactones, 131*f*
copolymerization of δ-caprolactone with achiral lactones, 134*f*
lipase catalysts, 129–130
lipase-catalyzed ring opening copolymerization, 132*t*

microstructural analysis of copolymer by ^{13}C NMR, 133–134
monitoring copolymerization by gas chromatography (GC), 131
optically active polyesters using *Candida antarctica* lipase, 130
postulated mechanism of lipase-catalyzed polymerization of lactones, 130–131
relationships between reaction time and conversion, 133*f*
relationships between reaction time and enantiomeric excess, 133*f*
See also Regioselective polymerization
Enrichment culture, techniques, 5–6
Enzymatic acylation, effect of dimethylsulfoxide (DMSO), 102–103
Enzymatic degradation. *See* Degradation; Reaction-diffusion of enzymes in biopolymer matrices
Enzymatic polymerizations
acyl donors and acyl acceptors for polymer libraries, 41*f*
analytical measurements, 36–37
array from diverse monomers showing application range, 42*f*
Candida antarctica lipase, 35
catalysis of *C. antarctica* lipase in polycondensation, 37, 38*f*, 39*f*
catalysis of porcine pancreatic lipase in polycondensation, 37, 38*f*, 39*f*
catalysis of protease Protex-6L in polycondensation, 37, 38*f*, 39*f*
chemoenzymatic methods with, 35
divinyl adipate (DVA) reaction with diols, 37
histograms of metal ions from polyphenol array, 46*f*
hydrolysis and depolymerization phenomena, 43–44
materials and enzymes, 35
maximum emission wavelength (MEX) of polyphenol array and fluorescence intensity change of metal ions, 47*t*
measurement of fluorescence response of polymer array to metal ions, 36–37
molecular weight of polycondensation polymers, 40, 43
molecular weight of polymers from DVA/sorbitol and DVA/octanediol, 45*f*
normalized deviation of fluorescence response, 46
phenolic polymer array for metal-ion sensing, 44, 46–47
polycondensation of activated esters with acyl acceptors, 40–44
polycondensation procedures, 35–36
polycondensation reactions, 37–44
polyphenol array of homo- and copolymers from phenolic monomers, 45*t*
polyphenol reaction procedures, 36
polyphenol reactions, 44
See also Polymer libraries
Enzymatic transesterification, polyesters, 12
Enzyme attack. *See* Polyurethanes
Enzyme engineering
limitation, 51
screening systems, 51, 55
Enzyme kinetics on monolayers
t-butylmethacrylate (tBMA)
composition dependence of diffusion coefficients in L-α-dilauroylphosphatidylcholine (DLPC)/PtBMA monolayers, 60, 61*f*
correlating, with monolayer dynamics, 60
enzyme catalyzed hydrolysis reaction, 62, 64
examining chemical reactions on surfaces, 57–58
experimental, 59

fraction (tBMA) dependence of hydrolysis reaction rate on DLPC/PtBMA mixed monolayers, 63*f*
lateral diffusion, 59–62
lipid monolayers at air/water interface, 58
normalized initial hydrolysis rates, 63*f*
surface pressure-area isotherms, 59
surface pressure dependences of diffusion coefficient at surface mass fractions, 60, 61*f*
umbelliferone stearate on DLPC/cholesterol binary monolayers, 58
Enzyme mimetics, synzyme, 9
Enzyme regeneration, cofactors, 8
Enzymes
advantages for synthesis of polycarbonates, 160
advantages in organic syntheses, 218
crosslinked enzyme crystals, 8–9
degrading cellulosic derivatives, 204–205
developments, 2–5
diffusion, 275, 277
directed evolution technique, 2–3
enhancing activity, 3
ester interchange reaction, 19
immobilization techniques, 3–4
improving, 2–3
industrial applications, 218
optimization of reactions and processes, 5
polymer degradation, 17–18
polymer modifications, 16–17
polynucleotide mimetic, 23
production of drugs and vitamins, 189–190
screening mutants, 3
solubilization, 4–5
thermal history and activity, 150
See also Reaction-diffusion of enzymes in biopolymer matrices; Screening enzymes
Enzyme transfer
aqueous buffers to isooctane, 224*t*
method, 221, 223
Epoxidase, oligomerization of allylamine, 211*f*, 214
Escherichia coli
bacterial strains, 83*t*
enzyme assays and protein expression, 84
recombinant strain, 121–122
subcloning, 83–84
See also Cyclohexanol (chn) biodegradation genes
Ester interchange reaction, enzymes, 19
Extremophile organisms, extreme conditions, 3

F

Feasibility, polytransesterification reaction, 148–149
Fermentation products, growth, 188
Filter paper
water contact angles, 226*t*
See also Cellulose
Fluorescence
intensity, 37
measurement of response of polymer array to metal ions, 36–37
normalized deviation of response, 46
phenolic polymer array for metal-ion sensing, 44, 46–47
polyphenol sensor array, 47*t*
Food, opportunities for natural polymers, 189
Free radical polymerization, peroxidase-mediated
effect of co-solvents, 302
effect of horseradish peroxidase (HRP) and 2,4-pentanedione

concentration on methyl methacrylate (MMA) yield, M_n and syndioregularity, 305*f*
effect of reaction temperature and concentration of HRP, hydrogen peroxide, and 2,4-pentanedione, 303
effect of temperature on yield and syndiotacticity of HRP-mediated MMA polymerization, 304*f*
effect of water-miscible co-solvent on MMA polymerization, 302*t*
enzyme-catalyzed polymerization of MMA in binary solvents, 300
experimental, 299–301
HRP, 298
HRP-mediated, of acrylamide, 306*f*
instrumentation, 299–300
materials, 299
mechanism of HRP-mediated, 303, 306
polymerization of acrylamide in aqueous medium, 300–301
polymerization of acrylamide in concentrated emulsion, 301
polymerization of sodium acrylate in aqueous medium, 301
potential vinyl monomers, 298

G

Galactose side chains, polysaccharide, 205
β-Galactosidases
modification of hydroxycellulose by transgalactosylation with, 190, 192–193
synthesis of glycidyl galactose using, 210*f*
Gallate esters
chitosan modification, 238–239
contact angle measurements for chitosan films, 240*f*
evidence for tyrosinase-catalyzed oxidation of gallate esters, 239*f*
Gene shuffling, libraries, 3
Genetic engineering, techniques, 6
Genetic manipulation. *See* Cyclohexanol (chn) biodegradation genes
Glass transition temperature, polycarbonate copolymer, 168
β-Glucan synthase
activity, 68–69
See also Plant β-glucans
Glycosidases, thermophilic, 190, 191
Grafting
chitosan, biopolymer for, 234–235
hydrophobic phenols onto chitosan, 237–239
renewable phenolic substrates for grafting, 234
See also Chitosan; Polysaccharides
Green technology
industrial products and processes, 80–81
oxidations with cyclohexanol genes, 89–90
See also Cyclohexanol (chn) biodegradation genes
Guanosine
synthesis of vinyl sugar esters, 96*t*
See also Nucleoside branched poly(vinyl alcohol)
Guar
Brookfield viscosity of stearoyl-cationic, 246*t*
degree of substitution (DS), 246
effect of substrate molecular weight, 278
effect of substrate structure, 278, 280
enzymatic degradation kinetics, 270–273
enzyme-catalyzed reactions, 212*t*
hydrophobic modification, 244
hydrophobic modification of cationic, 245, 250*f*

infrared spectra of unmodified cationic and stearoyl-cationic guars, 248*f*, 249*f*
inverse molecular weight vs. enzymatic degradation time, 272*f*, 273
low molecular weight polysaccharides, 204
Michaelis–Menton type reaction, 270
preparation of maleated, 213
reaction between enzyme and substrate, 270
reaction of cationic, with vinyl stearate, 246
side groups of derivatized, 268*f*
solvent effects, 246–247
structure, 268*f*
synthesis of maleated, 208*f*
use, 244
viscosity vs. shear rate during enzymatic degradation, 272*f*
See also Polysaccharides; Reaction-diffusion of enzymes in biopolymer matrices
Guar galactomannan, polysaccharide, 266

H

Horseradish peroxidase (HRP)
description, 298
experimental, 299–301
free radical polymerization of acrylamide, 306*f*
potential vinyl monomers for polymerization, 298
syndiotactic poly(methyl methacrylate) (PMMA), 306–307
See also Free radical polymerization, peroxidase-mediated
Hydrolysis reactions
enzymatic, of indolyl derivative, 52*f*
kinetics of lipase catalyzed, of umbelliferone stearate (UMB-C_{18}) on monolayers, 62, 64
normalized initial hydrolysis rates, 63*f*
polycarbonates, 157–158
UMB-C_{18} on monolayers, 58
Hydrophobic modification, polysaccharides, 206, 209*f*
Hydroxyalkanoic acids, copolyesters, 122
12-Hydroxydodecanoic acid (12-HDA)
gel permeation chromatography, 154*t*
molecular structure, 152
recycled enzyme, 152
time study of condensation of 12-HDA, 152, 153*f*
See also Direct polyesterification, enzyme-catalyzed
Hydroxyethylcellulose (HEC)
absolute molecular weights and rms radii of HEC and oxidized products, 258*t*
acylation, 220
^{13}C NMR spectroscopy, 256
degradation of side chains, 258–259
degree of substitution of HEC and oxidized products, 258*t*
enzymatic assay via peroxidase-chromogen test, 192
enzyme-catalyzed reactions, 212*t*
galactose side chains, 205
hydrophobic modification, 206, 209*f*, 214, 245, 247, 250*f*
instrumentation, 260
IR and ^{13}C NMR spectra of oxidized HEC from lipase catalysis, 257*f*
lipase-mediated TEMPO oxidation method, 261
macroinitiators for polymerization of lactide, 199–200
materials, 260
MCPBA-mediated TEMPO oxidation method, 261–262

modification by transgalactosylation with β-galactosidases, 190, 192–193
NaOCl-mediated TEMPO oxidation method, 262
optimized experimental conditions, 257–258
oxidation, 254*f*
oxidation methods, 261–262
parameter optimization for selective oxidation of HEC, 256–257
polycaprolactone-modified HEC by lipase-catalyzed ring opening polymerization, 196–197
preparation of acetylated, 213–214
preparation of succinated, 213
synthesis of acetylated HEC, 205
synthesis of succinated HEC, 206, 209*f*
transgalactosylation between lactose and, 192
uses, 254
See also Carbohydrates; Polysaccharides

I

Immobilization, biocatalysis techniques, 3–4
Indolyl derivatives, enzymatic hydrolysis, 52*f*

K

Kcat technology
distinguishing small differences, 55
feasibility, 51
graphic user interface, 54–55
See also Screening enzymes
Kinetics
analysis of enzyme model system mutants, 54–55
enzymatic degradation, 270–273
See also Enzyme kinetics on monolayers; Reaction-diffusion of enzymes in biopolymer matrices

L

Laccases
polymerization of phenol and aniline, 13–14
polymerization of vinyl polymers, 14–15
Lactide
copolymerization with trimethylene carbonate (TMC), 159
LiCl as catalyst for ring opening polymerization of, 197–200
macroinitiators for polymerization of, 199–200
Lactones
copolymerization of δ-caprolactone with achiral, 133, 134*f*
polymerization, 11
postulated mechanism for lipase catalysis in, polymerization, 130–131
ring opening polymerization, 22
See also Enantioselective polymerization
Libraries
recursive ensemble mutagenesis (REM), construction, 54–55
thermophilic glycosidase, 190, 191
See also Polymer libraries
Light scattering. *See* Polyurethanes
Lipases
bulk polymerization of trimethylene carbonate (TMC), 158–159
catalytic site, 130
catalyzing pentadecalactone (PDL)/TMC copolymerization, 182
enantioselective copolymerization, 131, 132*t*
enzyme transfer, 224

esterification or transesterification reactions, 218–219
hydrophobic modification of guar and hydroxyethylcellulose (HEC), 247, 250
lactones with enantio-enrichment, 19
lipid/water interface, 58
optically active polyesters, 129–130
optically active polyesters by copolymerization, 130, 131*f*
oxidation of HEC, 255*f*
polycaprolactone-modified HEC by ring opening polymerization, 196–197
postulated mechanism, 130–131
production of peracid, 261
regio- and stereoselective synthesis, 254–255
regioselective polymerization, 134, 136, 138
ring opening polymerization of β-butyrolactone by thermophilic, 193–196
selective acrylation, 110*f*
selective TEMPO oxidation of HEC, 255
synthesis of acetylated HEC, 205
synthesis of polyesters, 129
thermophilic for ring opening polymerization (ROP), 193–196
See also Enantioselective polymerization; Enzymatic polymerizations; Polycarbonates; Polytransesterification; Regioselective polymerization
Lithium chloride, ring opening polymerization of lactide, 197–200
Lyases, degrading polymers, 18

M

Macroinitiators, LiCl-catalyzed polymerization of lactide, 199–200
Macromers, reactive oligomers, 22–23
Macromers around sugar core
attachment of different substituent, 111, 114
enzyme catalysis of monomers and polymers, 108
experimental, 114–116
general procedure for acrylation of triol, 115
^{1}H NMR spectrum of poly(ε-caprolactone) [poly(ε-CL)] substituted acryl sugar, end-capped product, and chemically acetylated product, 112*f*, 113*f*
homopolymerization of product using AIBN, 114
4'-hydroxymethylmethacryl, 4-C-hydroxymethyl-1,2-O-isopropylidene-α-D-pentofuranose (HMG) as multifunctional initiator, 108–109
instrumental methods, 114
ketal deprotection procedure, 116
lipases for selective acrylation, 110*f*
materials, 114
molecular weight measurements, 115
regioselective end capping of HMG oligo(ε-CL), 116
research towards stars and dendrimers, 108
ring opening polymerization procedure, 116
sugar acryl derivative HMG as initiator for ε-CL polymerization, 110–111
Mechanism
horseradish peroxidase mediated methyl methacrylate polymerization, 303, 306
postulated, for lipase catalysis in lactone polymerization, 130–131
ring opening polymerization (ROP) of cyclic carbonates, 160
transacylation reactions for chains of mixed ester-carbonate linkages, 182, 184

Metabolic engineering, molecular biology, 6
Metal ion sensing, phenolic polymer array, 44, 46–47
5-Methyl-5-benzyloxycarbonyl-1,3-dioxan-2-one (MBC)
co-monomer diad fractions, 167*t*
copolymerization with trimethylene carbonate (TMC), 161, 164–165
homopolymerization, 161, 162–163
microstructure analysis of poly[MBC-*co*-TMC], 165–167
molecular weight to monomer conversion relationship, 163–164
number-average molecular weight vs. MBC conversion, 164*f*
percent conversion vs. time, 163*f*
regeneration of pendant carboxylic acid groups, 169
regeneration of pendant carboxylic acid groups in polyMBC and poly[MBC-*co*-TMC], 169
relationship between monomer conversion and reaction time, 163
ring opening polymerization of MBC in bulk at 80°C, 162*t*
thermal analysis of poly[MBC-*co*-TMC], 167–169
thermal properties of poly[MBC-*co*-TMC], 167*t*
Methyl methacrylate (MMA)
effect of co-solvents, 302
effect of horseradish peroxidase (HRP) and 2,4-pentanedione concentration on yield, M_n, and syndioregularity, 305*f*
effect of reaction temperature and concentration of HRP, hydrogen peroxide, and 2,4-pentanedione, 303
effect of temperature on yield and syndiotacticity of HRP-mediated polymerization, 304*f*
enzyme-catalyzed polymerization in binary solvents, 300
mechanism of HRP-mediated polymerization, 303, 306
See also Free radical polymerization, peroxidase-mediated
Michaelis–Menten reaction, polyurethane, 292
Michaelis–Menton model, enzymatic degradation, 266–267
Microbial approach
monomer syntheses, 20
polyesters, 13
Microbiology, enrichment culture techniques, 5–6
Microstructure
analysis of copolymer by ^{13}C NMR, 133–134
analysis of polycarbonate copolymer, 165–167
polyester analysis by nuclear magnetic resonance (NMR) techniques, 136, 137*f*
Microzyme, reducing size, 9
Mimetic
polynucleotide, 23
synzyme, 9
Model
enzyme system from *Agrobacterium faecalis*, 51
Michaelis–Menton, of enzymatic degradation, 266–267
Molecular weight reduction, polysaccharides, 204–205
Monolayers. *See* Enzyme kinetics on monolayers
Monomers
biocatalytic reactions, 20
chiral or enantio-enriched, 19
enzyme catalysis, 108
microbial approaches, 20
silicone acrylate, 21–22
synthesis of vinyl, 20–22
vinyl acrylate, 21

Monomer to enzyme weight ratio, polyesterification, 151, 152*f*
Multi-angle laser light scattering (MALLS)
molecular weight vs. time, 292
See also Polyurethanes
Multifunctional initiators. *See* Macromers around sugar core
Mutation, enrichment culture, 5–6

N

Natural polymers, opportunities, 189
Nonpolar organic solvent, lipase catalyzed transesterification, 144, 146*f*
Nucleobases, synthetic polymers containing, 93
Nucleoside branched poly(vinyl alcohol)
conversion rate of uridine by alkaline protease, 95
enzymatic synthesis of vinyl nucleoside derivatives, 96*f*
enzymatic synthesis of vinyl thymidine and its polymerization, 97*f*
polymerization of vinyl nucleosides, 95
synthesis of vinyl nucleoside esters, 95
synthesis of vinyl sugar esters by proteases from *Streptomyces* sp. or *Bacillus subtilis*, 96*t*
transesterifications, 95
transesterification with divinyl adipate by *Streptomyces* sp. protease, 98

O

Oligomerization, allylamine using epoxidase, 211*f*, 214
Oligosaccharides, synthesis, 15–16
Open reading frames (ORFs)
homology of, with proteins, 86*t*
upstream and downstream of cyclohexanol cluster, 85
See also Cyclohexanol (chn) biodegradation genes
Optically active polymers
applications, 129
chemoenzymatic synthesis of poly((meth)acrylic) materials, 144, 145*f*
enantioselective synthesis, 130
lipase catalysis, 129–130
polyesters by lipase catalyzed copolymerization, 130, 131*f*
See also Enantioselective polymerization; Lipases
Optimization, enzyme reactions and processes, 5
Organic syntheses, advantages of enzyme catalysis, 218
Oxazoline derivative of *N,N'*-diacetyl chitobiose, chitin formation, 15–16
Oxidation
experimental, 260–263
hydroxyethylcellulose (HEC), 254*f*
lipase approach to oxidize HEC, 255–259
poly(ethylene glycol) (PEG), 259*f*, 262–263
polysaccharides, 253–254
primary alcohol groups, 259
See also Hydroxyethylcellulose (HEC)

P

Packaging market, growth, 189
Pentadecalactone (PDL)
copolymerization with ε-caprolactone (CL), 173–176
See also Polytransesterification
Peptide synthesis, enzymatic, 23

Peroxidases
polymerization of phenol and aniline, 13–14
polymerization of vinyl polymers, 14–15
See also Free radical polymerization, peroxidase-mediated; Horseradish peroxidase (HRP)
Phenol
polymerization, 13–14
See also Enzymatic polymerizations
Phenolic substances, renewable, substrates for grafting, 234
Phenolic substrates, renewable, for grafting, 234
Phosphotransbutyrylase, synthesis of coenzyme A thioesters of hydroxyalkanoic acids, 121
Phylogeny, active site, 53
Plant β-glucans
Arabidopsis thaliana and *Rubus fruticosus*, 66
biosynthesis, 66
callose synthase, 65
cellulose synthase activity, 69–70
characterization of β-glucan by callose synthase from *R. fruticosus*, 72*t*
characterization of β-glucans synthesized in vitro, 70, 72
characterization of synthase from *A. thaliana*, 73, 76
cryo-transmission electron microscopy (TEM) in vitreous ice after incubation of *A. thaliana* plasma membranes with detergents, 74*f*
detergent extractions of synthases, 68
determination of conditions for in vitro synthesis, 68–70
effect of detergents on glucan synthesis, 72–73
glucan synthase activity, 68–69
glucan synthases in plants, 65
hypothetical model for biosynthesis, 67*f*
importance of synthases, 66
lack of mechanistic understanding, 66
purification and characterization of glucan synthases, 66, 68
TEM micrographs of cellulose after extraction of *R. fruticosus* cellulose synthase with Brij, 71*f*
TEM micrographs of cellulose from cell walls of *R. fruticosus* cells, 71*f*
TEM micrographs of glucans in vitro by callose synthase from *R. fruticosus*, 75*f*
Plant cells, polymer reactions, 6–7
Polar substituents, addition to polysaccharides, 205
Poly(*N*-acryroyl piperidine)
nucleoside containing, 94*f*
reversibly soluble-insoluble against water, 93
Polycaprolactone-modified hydroxyethylcellulose, lipase-catalyzed, 196–197
Poly(ε-caprolactone) (PCL)
^{1}H NMR spectrum, 112*f*, 113*f*
multifunctional initiator, 110–111
ring opening polymerization procedure, 116
See also Macromer with sugar core
Polycarbonates
advantages of enzymatic synthesis, 160
bisphenol A polycarbonate, 156
^{13}C NMR spectrum of poly[(5-methyl-5-benzyloxycarbonyl-1,3-dioxan-2-one)-*co*-(trimethylene carbonate)] poly[MBC-*co*-TMC], 166*f*
chemical heterogeneity for supramolecular structures, 157
condensation polymerization, 158
copolymerization of MBC and TMC, 161, 164–165

copolymers with esters, 12
cyclic carbonate monomers, 158
degradation, 157–158
enzymatic ring opening polymerization (ROP), 11–12
enzymes for ROP, 158
experimental, 161–162
experimental and calculated co-monomer diad fractions, 167*t*
glass transition temperature, 168
graph of $1/T_g$ vs. weight composition of MBC, 168*f*
homopolymerization of MBC, 161, 162–163
hydrolytic chain cleavage, 157–158
industrial process, 156–157
introduction of functional groups, 157
lipase-catalyzed copolymerization of TMC and lactide, 159
lipases for bulk polymerization of TMC, 158–159
materials, 161
mechanism for lipase-catalyzed ROP of cyclic carbonates, 160
microstructure analysis, 165–167
molecular weight determination, 161
molecular weight to monomer conversion relationship, 163–164
NMR spectroscopy method, 161
number-average molecular weight and weight distribution vs. MBC conversion, 164*f*
percent monomer conversion (MBC) vs. time, 163*f*
polymerization of cyclic dicarbonates by lipase, 158–159, 159
polymerization of dicarbonate monomers and diols by condensation, 158
polymerization procedure, 162
poly(TMC), 157
regeneration of pendant carboxylic acid groups, 169
regio control in lipase-catalyzed polymerization of TMC, 159
relationship between monomer conversion and reaction time, 163
ring opening copolymerization of MBC and TMC by lipase, 165*t*
ROP of MBC in bulk at 80°C, 162*t*
ROP of TMC by lipase, 159
synthesis, 11–12
thermal analysis, 167–169
thermal properties of poly[MBC-*co*-TMC], 167*t*
Polycondensation. *See* Enzymatic polymerizations
Poly(ester-*co*-carbonate), synthesis, 12
Polyesters
combined condensation and ring opening polymerization, 12
condensation from different monomers, 10
enzymatic condensation polymerization, 9–10
enzymatic ring opening polymerization, 11
enzymatic synthesis of crosslinkable, 138, 139*f*
enzymatic transesterification, 12
microbial approaches, 13
polymerization of lactones, 11
polytransesterification, 10
self-condensation, 9–10
stereoregular, with enzyme catalysts, 142
synthesis of coating resins, 142
See also Direct polyesterification, enzyme-catalyzed; Polycarbonates; Polyhydroxyalkanoates (PHA)
Poly(ethylene glycol) (PEG)
effect of different oxidation methods on molecular weights of PEG, 259*t*
lipase-mediated TEMPO oxidation of primary alcohols, 259–260

macroinitiator for LiCl-catalyzed polymerization of lactide, 199–200
oxidation method, 262–263
oxidation of PEG, 259*f*
See also Hydroxyethylcellulose (HEC)
Poly(γ-glutamic acid), microbial approach, 13
Polyhydroxyalkanoates (PHA)
approach to synthesizing controlled PHA microstructures, 125, 126*f*
bacterial fermentation, 143–144
biodegradable polyesters, 125
biodegradable thermoplastic and elastomers, 121
biosynthesis pathway, 121–122
controlling microstructure of polymer, 125
in vitro biosynthesis methods, 121
polyhydroxybutyrate (PHB), 125
polyhydroxyvalerate (PHV), 125
Poly(hydroxyalkanoic acid), microbial approach, 13
Polylactic acid, biodegradable polymer, 189
Polylactide, LiCl as catalyst for ring opening polymerization of lactide, 197–200
Polymer degradation
controlled methanolysis in determination of enantiomeric excess, 195
enzymes, 17–18
Polymer libraries
activated esters reacting with acyl acceptors, 40–44
divinyl adipate reacting with diols, 37
phenolic polymer array, 44, 46–47
See also Enzymatic polymerizations
Polymer mesh size
defining network, 275
See also Reaction-diffusion of enzymes in biopolymer matrixes
Polymer microstructure, polyester analysis by nuclear magnetic resonance (NMR) techniques, 136, 137*f*
Polymer modifications, enzyme catalysis, 16–17
Polymer modulation. *See* Enzyme kinetics on monolayers
Polymer processing, cell-free biosynthesis, 232
Polymers
containing nucleobases, 93, 94*f*
enzymatic degradation, 17–18
enzyme catalysis, 108
Polymer science
new reactions or products, 7
See also Biocatalysis in polymer science
Poly((meth)acrylic) materials, chemoenzymatic synthesis, 144, 145*f*
Polynucleotide mimetic, enzymatic approach, 23
Polyphenol
enzyme-catalyzed reactions, 36, 44
histograms of metal ions from polymer array, 46*f*
maximum emission wavelengths of elements in sensor array, 47*t*
nucleoside branched, 94*f*
polymer array for metal-ion sensing, 44, 46–47
See also Enzymatic polymerizations
Polysaccharides
addition of charge, 205–206
addition of polar substituent, 205
aminolysis of carboxymethylcellulose (CMC) and CMC esters, 207*f*
applications, 203
Brookfield viscosity measurements, 213, 245–246
Brookfield viscosity of stearoyl-cationic guar, 246*t*
degradation, 18

degree of substitution (DS), 246
enzymatic degradation and modification, 266
enzymatic degradation process, 266–267
enzyme-catalyzed reactions, 212*t*
enzymes on degradation of, 7
experimental, 212–214, 261–262
formation of reactive oligomers, 206, 209, 212
generality of lipase-catalyzed method, 247, 250
guar galactomannan, 266, 268*f*
guar gum, 244
hydrolase and oxidoreductase enzymes, 212
hydrophobic modification, 206, 243–244
hydrophobic modification of cationic guar, 250*f*
hydrophobic modification of cationic guar in DMAc, 245
hydrophobic modification of cationic guar in *t*-butyl methyl ether, 245
hydrophobic modification of HEC (hydroxyethylcellulose), 209*f*, 214, 250*f*
hydrophobic modification of HEC in DMAc, 245
infrared (IR) spectra of unmodified cationic and stearoyl-cationic guars, 248*f*, 249*f*
lipase-catalyzed synthesis of acetylated HEC, 205*f*
lipase-mediated TEMPO oxidation of primary alcohols, 259–260
lipases and proteases for enzymatic modifications of guar and HEC, 247, 250
materials, 212–213, 244
modifications, 16–17
modifying, 203–204
molecular weight reduction, 204–205
oligomerization of *N*-allylamine, 211*f*, 214
oxidation, 253–254
oxidation methods, 261–262
preparation of acetylated HEC, 213–214
preparation of maleated guar, 213
preparation of succinated HEC, 213
reaction categories, 204
reaction of cationic guar with vinyl stearate, 246
solvent effects on enzymatic reaction, 246–247
synthesis, 15–16
synthesis of glycidyl galactose using β-galactosidase, 210*f*
synthesis of maleated guar, 208*f*
synthesis of succinated HEC, 209*f*
uses, 203
See also Hydroxyethylcellulose (HEC); Reaction-diffusion of enzymes in biopolymer matrices
Poly(sucrose adipamide), chemoenzymatic synthesis, 144, 145*f*
Polytransesterification
ability of various lipases to catalyze pentadecalactone/trimethylene carbonate (PDL/TMC) copolymerization, 182
active pathway for catalyzing, 185
catalysis of, between polymer chains, 172–173
conversion of poly(PDL-*co*-TMC) from multiblock to random copolymers, 185
copolymerization of PDL and ε-caprolactone (CL) monomers, 173–176, 184–185
copolymerization of PDL and TMC by lipase, 183*t*
extent of transacylation, 178
^{1}H NMR spectrum of poly(PDL-*co*-TMC), 181*f*
lipase-catalyzed, of poly(PDL) with poly(CL), 177*t*

lipase-catalyzed PDL and CL copolymerization, 175*t*
lipase catalyzing transacylation for chains of mixed ester-carbonate linkages, 182
mechanism for transacylation of chains of mixed ester-carbonate linkages, 182, 184
mechanism for transacylation of PDL and TMC, 184
mechanism of, of polyester chains, 178–179
molecular weight and lipase catalyzed transacylation reactions, 178–179
PDL/TMC copolymerizations using lipases, 180
polyesters, 10
possible diad arrangements of PDL and CL, 173–174
potential transacylation reactions with polyester substrates of increased molecular weight, 176, 178
reactions between polymer chains, 172
reactions between preformed polyester chains, 176–179
reactivity ratios by Fineman–Ross method, 174
transesterification of polyester co-polycarbonate, 179–184
triads from poly(PDL-*co*-TMC), 180
Polytrimethylene terephthalate (PTT), biodegradable polymer, 189
Polyurethanes
characterization by laser light scattering, 286–287
effect of diffusion on rate, 292
enzymatic activity, 286
enzyme light scattering data, 291–292
experimental, 287–288
geometry of polyurethane substrate molecule, 291
light scattering measurements, 288, 291
light scattering of molecular weight of, substrate while under attack by polyurethanase, 293*f*
materials, 288, 291
Michaelis–Menton reaction, 292
molecular weight by multi-angle laser light scattering (MALLS), 292
polyurethanase enzymes, 288
Rayleigh ratio, 287
Zimm plot, 287–288
Zimm plot of light scattering data, 289*f*
Zimm plot of purified enzyme, 290*f*
Poly(vinyl alcohol). *See* Nucleoside branched poly(vinyl alcohol)
Porcine pancreatic lipase. *See* Enzymatic polymerizations
Proleather, regioselective monomer synthesis, 144
1,3-Propanediol, hydrolyzed cornstarch, 6, 20
Proteases
catalysis in polycondensation, 37, 38*f*, 39*f*
enzyme transfer, 224
esterification or transesterification reactions, 218–219
hydrophobic modification of guar and hydroxyethylcellulose (HEC), 247, 250
peptide bonds, 23
regioselective polymerization, 134, 136, 138
See also Vinyl sugar esters
Protein-based polymers, growth, 188
Protein engineering, tailoring enzymes, 51
Pseudomonas aeruginosa, polyhydroxyalkanoate synthase, 121
Pseudomonas oleovorans, polyhydroxyalkanoates, 144

Q

Quinones
reactions with chitosan, 234–235
tyrosinase-catalyzed chitosan crosslinking, 236

R

Ralstonia eutropha, controlling polyhydroxyalkanoate microstructures, 125
Rayleigh ratio, definition, 287
Reaction-diffusion model, enzymatic degradation process, 266–267
Reaction-diffusion of enzymes in biopolymer matrices
changes in enzyme diffusion coefficient and polymer mesh size, 276*f*
changes in enzyme diffusion time with polymer concentration, 276*f*
competition between diffusion and reaction, 276*f*, 277
degradation rate, 277
diffusion of enzyme molecules, 275
diffusion of enzymes, 275, 277
effect of substrate molecular weight, 278
effect of substrate structure, 278, 280
enzymatic degradation, 267, 269
enzymatic degradation kinetics, 270–273
enzymatic degradation rate at pH 7 vs. substrate concentration, 274*f*
enzymatic degradation rate vs. substrate molecular weight, 279*f*
experimental, 267, 269–270
gel permeation chromatography (GPC) measurement method, 269–270
inverse of number average molecular weight vs. enzymatic degradation time, 272*f*
materials, 267
normalized molecular weight changes with time during degradation of guar, 279*f*
normalized molecular weight changes with time vs. enzymatic degradation time, 281*f*
normalized viscosity of cationic guar vs. reaction time at different ionic strengths, 281*f*
polymer mesh size defining network, 275
reaction between enzyme and substrate, 270
reaction-diffusion model, 266–267, 273, 275–277
reaction time, 277
reaction time diffusion, 277
relation between molecular weight and degradation time, 271
side groups of derivatized guars, 268*f*
steady state viscometry, 269
structure of guar, 268*f*
triggered enzymatic degradation, 280, 282
viscosity of cationic guar vs. enzymatic degradation time, 280–280, 281*f*
viscosity of guar solution vs. shear rate during enzymatic degradation, 272*f*
Reactive oligomers
formation in polysaccharides, 206, 209, 212
lipase-catalyzed polymerization, 22–23
macromers, 22
Reactivity ratios, pentadecalactone and ε-caprolactone copolymerization, 174
Recursive ensemble mutagenesis (REM)
genetic algorithm technique, 55
library construction, 53–54
See also Screening enzymes

Regioselective polymerization
^{13}C NMR spectra of polymers, 137*f*
^{13}C NMR spectrum of sugar-containing polyester from sorbitol, 139*f*
crosslinkable polyesters, 138, 139*f*
divinyl esters as activated diacid substrate, 134
divinyl sebacate and sorbitol, 137*f*
divinyl sebacate and trials via lipase catalysis, 135*f*
lipase-catalyzed, of divinyl sebacate and triols, 135*t*
lipases and proteases as catalysts, 134
polymer microstructure by NMR techniques, 136, 137*f*
synthesis of polyesters, 134, 136
See also Enantioselective polymerization
Renewable chemicals. *See* Chitosan
Ring opening polymerization
β-butyrolactone by thermophilic lipases, 193–196
ε-caprolactone, 110–111, 146–147
combining with condensation, 12
copolymerization of 5-methyl-5-benzyloxycarbonyl-1,3-dioxan-2-one (MBC) and trimethylene carbonate (TMC), 164–165
lactones, 11
LiCl as catalyst for, of lactide, 197–200
mechanism for cyclic carbonates, 160
MBC, 162–163
percent MBC conversion vs. time, 163
polycaprolactone-modified hydroxyethylcellulose (HEC) by lipase-catalyzed, 196–197
polycarbonate synthesis, 11–12
TMC, 158–159
See also Macromers around sugar core; Polycarbonates
Rubus fruticosus
characterization of β-glucan by callose synthase from, 72*t*
polysaccharides, 16
See also Plant β-glucans

S

Screening enzymes
Abg *Agrobacterium faecalis* model enzyme system, 51
active site phylogeny, 53
combinatorial region of cassette to mutagenize Abg gene, 54*f*
enzymatic hydrolysis of indolyl derivative, 52*f*
feasibility of Kcat technology, 51
graphical user interface of Kcat instrument, 54–55
high-throughput solid-phase assay system, 51, 55
indolyl derivatives and colors of indigo products, 52*f*
kinetic analysis of Abg mutants, 54–55
liquid phase, 51
reaction type for Abg activity, 51, 52*f*
recursive ensemble mutagenesis (REM) library construction, 53–54
Self-condensation, polyesters, 9–10
Sequencing. *See* Cyclohexanol (chn) biodegradation genes
Silicon, bioscience, 24
Silicone acrylate, enzymatic reaction, 21–22
Sodium acrylate
polymerization in aqueous medium, 301
See also Free radical polymerization, peroxidase-mediated
Solid-phase assay
high-throughput, 51, 55
See also Screening enzymes

Solubilization, methods, 4–5
Solvents
evaluation, 149
organic, for enzyme-catalyzed reactions, 142–143
polytransesterification reaction, 149
solubilizing enzymes, 5
solvent-free transesterification, 147
transesterification in nonpolar organic, 144, 146*f*
Sonication, physical use, 151
Sorbitol
^{13}C NMR spectrum of sugar-containing polyester from, 139*f*
regioselective polycondensation, 136, 137*f*
Stars, research, 108
Static laser light scattering, polymer characterization, 286–287
Steady state viscometry, method, 269
Streptomyces sp. protease
conversion of nucleosides, 95
synthesis of vinyl sugar esters, 96*t*, 100, 101*t*
transesterification with divinyl adipate, 98
Substrate coupled approach, cofactors, 8
Subtilisin Carsberg
aqueous buffers to isooctane, 224*t*
enzyme transfer, 224*t*
Sugars. *See* Macromers around sugar core; Sorbitol; Vinyl sugar esters
Surfactants, solubilizing enzymes, 4–5
Synzyme, enzyme mimetic, 9

T

Temperature
horseradish peroxidase mediated free radical polymerization of methyl methacrylate, 303
polyesterification reaction, 150–151
TEMPO oxidation
experimental methods, 261–262
hydroxyethylcellulose (HEC), 255–259
poly(ethylene glycol) (PEG), 259*f*
primary alcohol groups, 259–260
See also Hydroxyethylcellulose (HEC)
Textiles, growth, 189
Thermal analysis, polycarbonate copolymer, 167–169
Thermophilic microorganisms
enzymes modifying hydroxy-containing compounds, 200
enzymes with unique characteristics, 190
Thiocapsa pfennigii, polyhydroxyalkanoate synthase, 122
Thymidine
enzymatic synthesis of vinyl, 97*f*
synthesis of vinyl sugar esters, 96*t*
See also Nucleoside branched poly(vinyl alcohol)
Transacylation reactions
between poly(caprolactone) and poly(pentadecalactone), 176–179
extent, 178
potential with polyester substrates of higher molecular weight, 176, 178
See also Polytransesterification
Transesterification
acylation of small molecules, 220
catalysis of, between polymer chains, 172–173
characterization methods, 223
description of enzyme-catalyzed, 218
enzymatic, 12
enzyme-catalyzed, of cellulosic solids, 221–222, 224–225
enzyme-catalyzed, on cellulose, 223, 228
enzymes transferring from aqueous buffers to isooctane, 224*t*
enzyme transfer, 224
enzyme transfer method, 221, 223

experimental, 221, 223
FTIR–KBr spectra of enzyme-catalyzed, on cellulose and regio-specifically substituted 2,3-O-methyl-cellulose and 6-O-trityl cellulose, 228*f*
FTIR spectra of cellulose, cotton fibers, and filter paper, 226*f*
lipase catalyzed, in nonpolar organic solvent, 144, 146*f*
materials, 221
mechanism of enzyme-catalyzed, 219
organic solvents, 219
regioselectivity of enzyme-catalyzed, 223, 227
solubilizing enzymes in hydrophobic organic solvents, 220
synthesis of macromolecules and small molecules, 220
vinyl acrylate and vinyl propionate, 225
vinyl esters as acylating agents, 219
water contact angles of cellulose and transesterified cellulose, 226*t*
water wetting contact angles of cellulose derivatives, 225
See also Direct polyesterification, enzyme-catalyzed
Transgalactosylation, modification of hydroxycellulose, 190, 192–193
1,2,3-Trichloropropane, degradation, 20
Trimethylene carbonate (TMC)
copolymerization with lactide, 159
copolymers of TMC and pentadecalactone, 179–184
polycarbonate synthesis, 11–12
polymerization, 157
See also 5-Methyl-5-benzyloxycarbonyl-1,3-dioxan-2-one (MBC); Polycarbonates; Polytransesterification
Tyrosinase
cell-free polymer processing, 223
chitosan crosslinking, 236
evidence for catalyzed oxidation of gallate esters, 239*f*

U

Umbelliferone stearate
hydrolysis on monolayers, 58
See also Enzyme kinetics on monolayers
Uridine
conversion by alkaline protease, 95
polymer containing, 94*f*
synthesis of vinyl sugar esters, 96*t*
See also Nucleoside branched poly(vinyl alcohol)

V

Valerate, approach to synthesizing controlled polyhydroxyalkanoate microstructures, 125, 126*f*
Vinyl acrylate
polymerization on sucrose in pyridine, 21
transesterification, 225
Vinyl esters, acylating agents, 219
Vinyl monomers
synthesis, 21–22
See also Free radical polymerization, peroxidase-mediated
Vinyl nucleoside derivatives
enzymatic synthesis, 96*f*
polymerization, 95, 97*f*
Vinyl nucleoside esters, synthesis, 95
Vinyl polymers, polymerization, 14–15
Vinyl propionate, transesterification, 225
Vinyl sugar esters
effect of dimethyl sulfoxide (DMSO) on enzymatic acylation, 102–103

enzymatic modification of sugars, 100
enzymatic synthesis, 100–102
polymerization, 103–104
regio-selective acylation of hexopyranoses in dimethyl formamide (DMF)/DMSO, 103*f*
sugar and divinyl adipate forming, 100*f*
synthesis by protease from *Streptomyces* sp., 101*t*
synthesis by proteases from *Streptomyces* sp. or *Bacillus subtilis*, 96*t*
synthesis of various, 99
Vinyl thymidine, enzymatic synthesis, 97*f*

W

Water, polytransesterification reaction, 149–150
Water contact angles, cellulose derivatives, 225, 226*t*
Water-resistant adhesive, conferring properties, 237
Water-soluble polymers, modifications, 17*t*
Wheat gluten, growth, 188
Whole-cell approaches
enrichment culture and screening, 5–6
genetic engineering, 6
plant cells and cell extracts, 6–7
using metabolic pathways, 5–7
Window of operation, graphical tool, 8

X

β-Xylobiosyl fluoride, xylan formation, 15

Z

Zimm plots
description, 287–288
light scattering data, 289*f*
purified enzyme, 290*f*
See Polyurethanes